S0-AHG-414

Advances in
ORGANOMETALLIC CHEMISTRY

VOLUME 34

Advances in Organometallic Chemistry

EDITED BY

F. G. A. STONE

DEPARTMENT OF CHEMISTRY
BAYLOR UNIVERSITY
WACO, TEXAS

ROBERT WEST

DEPARTMENT OF CHEMISTRY
UNIVERSITY OF WISCONSIN
MADISON, WISCONSIN

VOLUME 34

ACADEMIC PRESS, INC.
Harcourt Brace Jovanovich, Publishers
San Diego New York Boston
London Sydney Tokyo Toronto

This book is printed on acid-free paper. ♾

Academic Press, Inc.
1250 Sixth Avenue, San Diego, California 92101-4311

United Kingdom Edition published by
Academic Press Limited
24–28 Oval Road, London NW1 7DX

Library of Congress Catalog Number: 64-16030

International Standard Book Number: 0-12-031134-8

PRINTED IN THE UNITED STATES OF AMERICA
92 93 94 95 96 97 BB 9 8 7 6 5 4 3 2 1

Contents

Organometallic Chemistry of Palladium and Platinum with Poly(pyrazol-1-yl)alkanes and Poly(pyrazol-1-yl)borates

PETER K. BUYERS, ALLAN J. CANTY, and R. THOMAS HONEYMAN

Organometallic Chemistry of the N=N Group

HORST KISCH and PETER HOLZMEIER

Carbon–Oxygen Bond Activation by Transition Metal Complexes

AKIO YAMAMOTO

Charge-Transfer Complexes of Organosilicon Compounds

VALERY F. TRAVEN and SERGEI YU. SHAPAKIN

Structure, Color, and Chemistry of Pentaaryl Bismuth Compounds

KONRAD SEPPELT

Use of Water-Soluble Ligands in Homogeneous Catalysis

PHILIPPE KALCK and FANNY MONTEIL

Slowed Tripodal Rotation in Arene–Chromium Complexes: Steric and Electronic Barriers

MICHAEL J. McGLINCHEY

Contributors

Numbers in parentheses indicate the pages on which the author's contributions begin.

PETER K. BUYERS (1), Chemistry Department, University of Tasmania, Hobart, Tasmania, Australia 7001

ALLAN J. CANTY (1), Chemistry Department, University of Tasmania, Hobart, Tasmania, Australia 7001

PETER HOLZMEIER (67), Institut für Anorganische Chemie II, Universität Erlangen-Nürmberg, D-8520 Erlangen, Germany

R. THOMAS HONEYMAN (1), Chemistry Department, University of Tasmania, Hobart, Tasmania, Australia 7001

PHILIPPE KALCK (219), Laboratoire de Chimie des Procédés, Ecole Nationale Supérieure de Chimie, 31077 Toulouse Cédex, France

HORST KISCH (67), Institut für Anorganische Chemie II, Universität Erlangen-Nürmberg, D-8520 Erlangen, Germany

MICHAEL J. McGLINCHEY (285), Department of Chemistry, McMaster University, Hamilton, Ontario L8S 4M1, Canada

FANNY MONTEIL (219), Laboratoire de Chimie des Procédés, Ecole Nationale Supérieure de Chimie, 31077 Toulouse Cédex, France

KONRAD SEPPELT (207), Institut für Anorganische und Analytische Chemie, Freie Universität Berlin, 1000 Berlin 33, Germany

SERGEI YU. SHAPAKIN (149), Mendeleev Chemico-Technological Institute, Moscow, Russia

VALERY F. TRAVEN (149), Mendeleev Chemico-Technological Institute, Moscow, Russia

AKIO YAMAMOTO (111), Department of Applied Chemistry, School of Science and Engineering, Waseda University, Tokyo 169, Japan

ADVANCES IN ORGANOMETALLIC CHEMISTRY, VOL. 34

Organometallic Chemistry of Palladium and Platinum with Poly(pyrazol-1-yl)alkanes and Poly(pyrazol-1-yl)borates

PETER K. BYERS, ALLAN J. CANTY, AND R. THOMAS HONEYMAN

Chemistry Department
University of Tasmania
Hobart, Tasmania, Australia 7001

I
INTRODUCTION

The classical nitrogen donor ligands pyridine and 2,2′-bipyridyl have played a key role in the development of diverse areas of coordination and organometallic chemistry. Modification of these ligands has led to derivatives that are also now widely employed, e.g., introduction of substituents in positions adjacent to the nitrogen donor atom(s) to alter steric effects of the ligands. For bidentate ligands, an important early development was the synthesis of related ligands containing a bridging group between the pyridin-2-yl groups, $(py)_2X$, in particular, bis(pyridin-2-yl)amine ($X = NH$), bis(pyridin-2-yl)methanone ($X = CO$), and bis(pyridin-2-yl)-methane ($X = CH_2$).

The chemistry of ligands containing other nitrogen donor rings has also emerged concurrently with the pyridine-based ligands, but at a slower pace, and for pyrazole as the heterocycle a major impetus has been Trofimenko's early studies with poly(pyrazol-1-yl)alkanes (*1,2*) and poly(pyrazol-1-yl)borates (*2–6*). The poly(pyrazol-1-yl)alkane ligands have been important in the development of the organometallic chemistry of palladium and platinum, and recent new applications have included their role in stabilizing hydrocarbylpalladium(IV) compounds and in the synthesis of intramolecular coordination systems containing tripodal $[N{\sim}C{\sim}N]^-$ ligands coordinated to platinum(IV). This review includes these two new areas of poly(pyrazol-1-yl)alkane chemistry, and includes comparisons with earlier organopalladium and organoplatinum chemistry, and related chemistry where appropriate. In addition to poly(pyrazol-1-yl)alkanes and -borates, closely related polydentate ligands containing at least one pyrazol-1-yl (pz) group are included, in particular, ligands containing the pyridin-2-yl (py) and *N*-methylimidazol-2-yl (mim) groups.

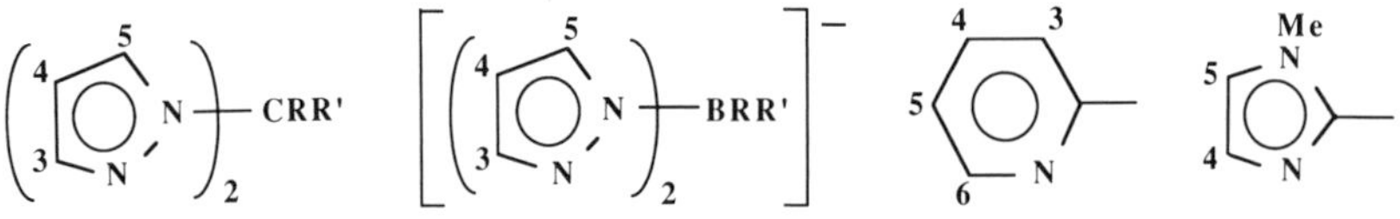

R, R′ = H, pz, alkyl, aryl

py

mim

The chemistry of complexes containing poly(pyrazol-1-yl)alkanes and -borates as nitrogen donor ligands is described in Section II, followed by a review of complexes in which these ligands or related ligands are present as intramolecular coordination systems.

II

POLY(PYRAZOL-1-YL)ALKANES AND -BORATES AS NITROGEN DONOR LIGANDS

A. (η^3-Allyl)palladium(II) Complexes

The first reported palladium(II) complexes of poly(pyrazol-1-yl) ligands were the η^3-allyl complexes of $(pz)_3BH^-$ and $(pz)_4B^-$, $Pd(C_3H_5)\{(pz)_3BR\}$ (**1** and **2**), obtained on reaction of the dimer $[PdCl(C_3H_5)]_2$ with $K[(pz)_3BR]$ in dichloromethane [Eq. (1)] (*7*). More recently, the

1 (R = pz, R' = H)

2 (R = R' = pz)

3 (R = R' = Et)

4 (R = R' = Pr^n)

5 (R = R' = Ph)

6 (R~R' =

$(pz)_2BR_2^-$ complexes **3–5** (*8*), **6** (*9*), **7**, and **8** (*10*) have been synthesized, and additional data for complex **2** have been reported (*9*). The bridging complexes **9–11** were formed from similar reactions in dimethylformamide and were isolated following addition of water and anion exchange (NH_4PF_6) Eqs. (2–4)] (*11*).

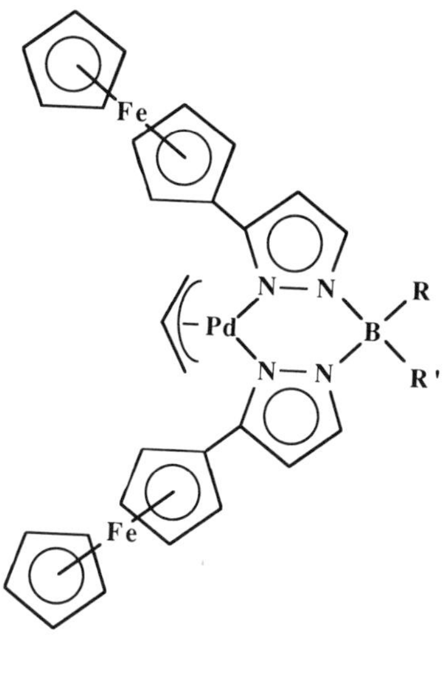

7 (R = R' = Et)

8 (R ~ R' =

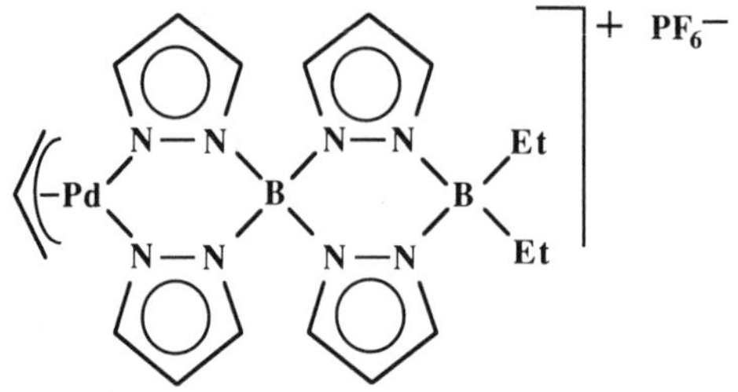

9

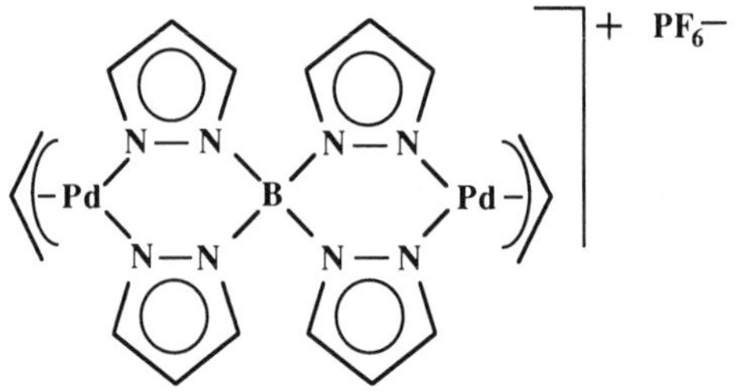

10

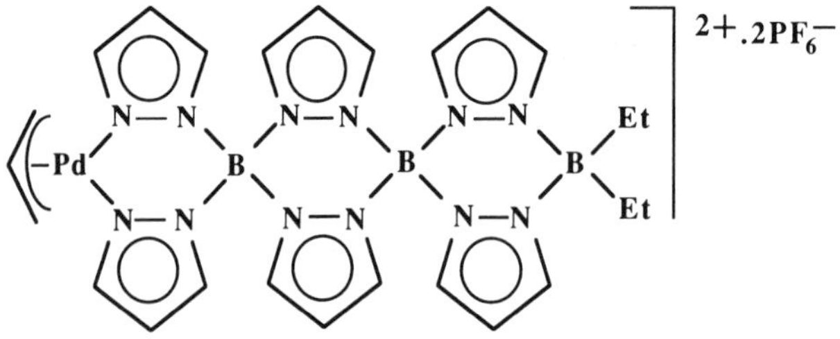

11

$$[PdCl(C_3H_5)]_2 + 2(pz)_3BR^- \rightarrow 2Pd(C_3H_5)\{(pz)_3BR\} + 2Cl^- \quad (1)$$

1,2

$$[PdCl(C_3H_5)]_2 + (pz)_4B^- \rightarrow [Pd(C_3H_5)\{(pz)_2B(pz)_2\}Pd(C_3H_5)]^+ + 2Cl^- \quad (2)$$

9

$$[PdCl(C_3H_5)]_2 + 2(pz)_2B(pz)_2BEt_2 \rightarrow 2[Pd(C_3H_5)\{(pz)_2BEt_2\}]^+ + 2Cl^- \quad (3)$$

10

$$[PdCl(C_3H_5)]_2 + 2[(pz)_2B(pz)_2B(pz)_2BEt_2]^+ \rightarrow$$
$$2[Pd(C_3H_5)\{(pz)_2B(pz)_2B(pz)_2BEt_2\}]^{2+} + 2Cl^- \quad (4)$$

11

Poly(pyrazol-1-yl)alkane complexes that are closely related to **1** and **2** have been prepared by an analogous method in methanol/water [Eq. (5)] (*1*), and they are more stable than their borate analogs. Thus, although

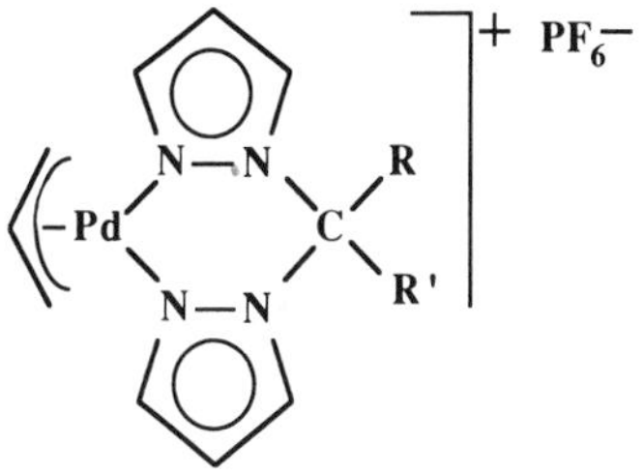

12 (R = R' = H)
13 (R = R' = pz)
14 (R = H, R' = pz)
15 (R = R' = Me)

$(pz)_2BH_2^-$ reduces Pd(II) to Pd(0) immediately and $(pz)_3BH^-$ does so upon moderate heating, the complexes $[Pd(C_3H_5)\{(pz)_2CRR'\}]PF_6$ (**12–15**) are stable beyond 200°C.

$$[PdCl(C_3H_5)]_2 + 2(pz)_2CRR' \rightarrow 2[Pd(C_3H_5)\{(pz)_2CRR'\}]^+ + 2Cl^- \qquad (5)$$

12–15

Variable-temperature ^{1}H NMR spectra for (η^3-allyl)palladium(II) complexes of poly(pyrazol-1-yl)alkanes and -borates are generally consistent with the presence of fluxional processes. For example, the cation $[Pd(C_3H_5)\{(pz)_2CMe_2\}]^+$ in the tetrafluoroborate salt (**16**) exhibits spectra at −70°C indicating the presence of two conformers in ~5:3 ratio (*12*),

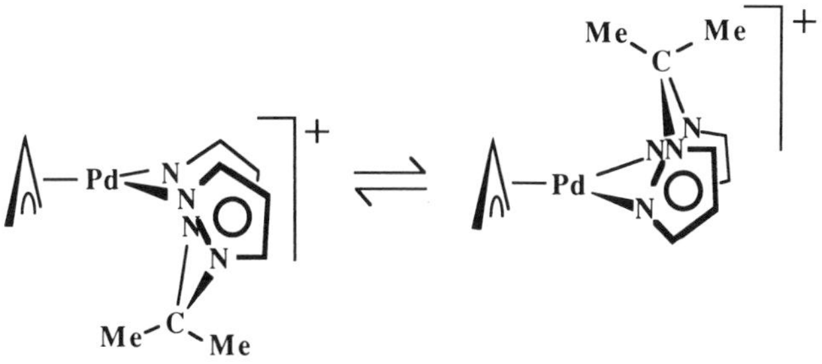

A **16** B

assigned as structures **A** and **B** by analogy with the facile inversion shown by related square planar complexes $PdX_2\{(pz)_2CMe_2\}$ [X = Cl (*13*), Me (*14*)] and $PdXMe\{(pz)_2CMe_2\}$ [X = Cl, Br, I (*14*)]. The major isomer is tentatively assigned structure **A**, because molecular models indicate that **B** is expected to have greater allyl ··· CMe interactions. In addition, ^{1}H NMR spectra of complexes containing one or two uncoordinated pz groups (**1, 2, 13,** and **14**) show one pz environment, indicative of rapid exchange of coordinated and uncoordinated groups (*1,7,15*), and this feature has been described as a "tumbling" process (*15*).

The binuclear complex **9** also exhibits stereochemical nonrigidity and well-defined limiting ^1H NMR spectra could be obtained, giving $\Delta G^{\ddagger} = 64.9$ kJ mol^{-1} and $E_a = 59.8$ kJ mol^{-1} (*11*). The spectra have been interpreted in terms of inversion of the "$B(pz)_2Pd$" bridge, or possibly a solvent-assisted dissociation of a Pd–N bond followed by displacement of one coordinated pz group by the now unattached one. The ^1H NMR spectra of **10** and **11** exhibit two types of pz groups, consistent with a static or dynamic structure for the cations (*11*).

For 2,2′-bipyridyl as the bidentate ligand, two series of complexes are obtained from syntheses depending on the mole ratio of $[PdX(allyl)]_2$ (X = Cl (*16*), Br, I (*17*)] and ligand [Eqs. (6) and (7)]. However, reaction of tetramethylethylenediamine (tmeda) with

$$[PdX(C_3H_5)]_2 + 2bpy \rightarrow 2[Pd(C_3H_5)(bpy)]X \tag{6}$$

$$[PdX(C_3H_5)]_2 + bpy \rightarrow [Pd(C_3H_5)(bpy)][PdX_2(C_3H_5)] \tag{7}$$

(η^3-allyl)palladium(II) halides in tetrahydrofuran results only in the formation of double salts, e.g., $[Pd(C_3H_5)(tmeda)][PdX_2(C_3H_5)]$ (*18*). The possible occurrence of similar phenomena appears to be unexplored for poly(pyrazol-1-yl)alkanes as ligands, although reaction of $[PdBr(C_3H_5)]_2$ with $(pz)_3CH$ in a 1:1 mole ratio in acetone does give the double salt $[Pd(C_3H_5)\{(pz)_3CH\}][PdBr_2(C_3H_5)]$ (**17**), indicating reactivity similar to that of 2,2′-bipyridyl (*12*).

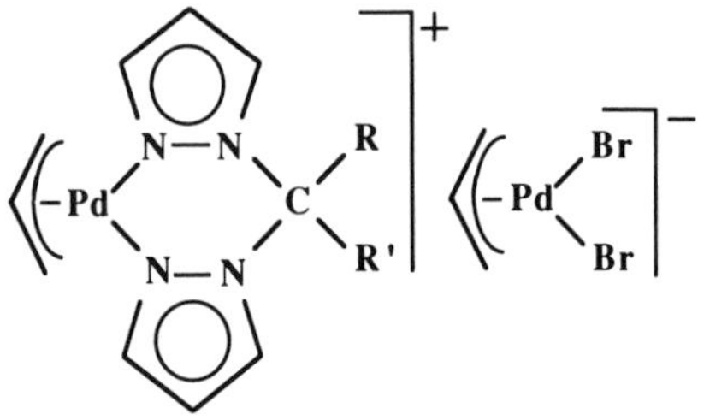

17 (R = H, R′ = pz)

18 (R = R′ = Me)

Double salts **17** and **18** are obtained on reaction of $PdMe_2\{(pz)_2CRR'\}$ (R = H; R′ = pz; R = R′ = Me) with 2-propenyl bromide in acetone (*12*), and for $(pz)_3CH$ the intermediate formation of a palladium(IV) complex **19** [Eq. (8)] is clearly seen when the reaction is monitored by ^1H NMR spectroscopy.

$$PdMe_2\{(pz)_3CH\} + C_3H_5Br \rightarrow \underset{\mathbf{19}}{[Pd(\eta^1\text{-}C_3H_5)Me_2\{(pz)_3CH\}]Br} \rightarrow$$

$$\underset{\mathbf{17}}{\tfrac{1}{2}[Pd(C_3H_5)\{(pz)_3CH\}][PdBr_2(C_3H_5)]} + \tfrac{1}{2}(pz)_3CH + Me\text{–}Me \qquad (8)$$

The complex $[Pd(C_3H_5)\{(pz)_2CMe_2\}][PdBr_2(C_3H_5)]$ (**18**) reacts with $AgBF_4$ to form the simple salt $[Pd(C_3H_5)\{(pz)_2CMe_2\}]BF_4$ (**16**).

The poly(pyrazol-1-yl)alkane double salts exhibit variable-temperature ^1H NMR spectral behavior similar to that of the simple salts, and in addition the spectra indicate exchange of allyl groups between the cation and anion (*12*). The exchange process is assumed to occur by an associative process, as demonstrated earlier for the tmeda complexes by kinetic and spectroscopic studies (*18*).

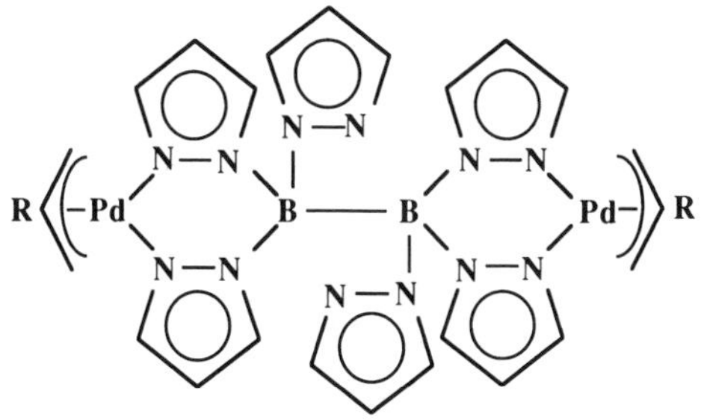

20 (R = H)

21 (R = Me)

Complexes **20** and **21** may be obtained on reaction of $K_2[(pz)_3B\text{–}B(pz)_3]$ with $[PdCl(C_3H_4R)]_2$, and the structural determination for **20** is shown in Fig. 1 (*19*). The complexes are fluxional, exhibiting exchange of coordinated and uncoordinated pz groups.

B. *Square Planar Palladium(II) and Platinum(II) Complexes*

The organometallic chemistry of palladium(II) and platinum(II) involving metal–carbon σ bonds has, in general, developed more slowly for nitrogen donor ligands than for phosphorus donor ligands. For palladium, there are only a few isolated reports of simple alkylpalladium(II) and arylpalladium(II) complexes prepared in the 20 years following the synthesis of $PdMe_2(bpy)$ by Calvin and Coates in 1960 (*20*). In contrast, progress with nitrogen donor groups as part of intramolecular coordination systems has been extensive. Compounds containing bidentate $[C{\sim}N]^-$ groups are

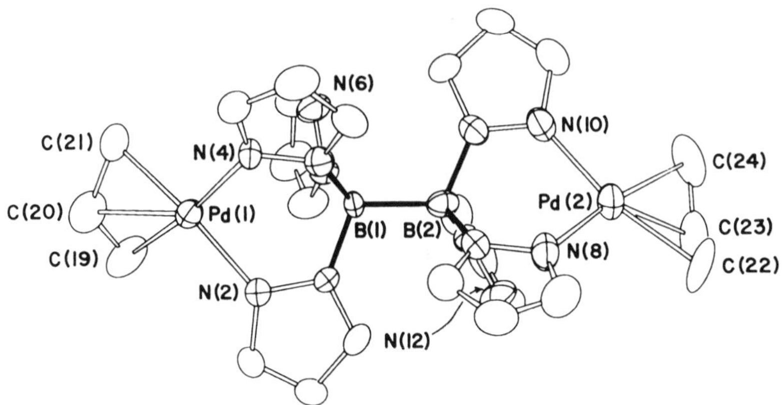

FIG. 1. Structure of $\{Pd(C_3H_5)\}_2\{(pz)_3B{-}B(pz)_3\}$ (**20**) (with permission from Ref. 19; copyright American Chemical Society).

often formed via cyclometallation reactions of "HC~N" by MCl_4^{2-} to give chloro-bridged complexes $[MCl(C{\sim}N)]_2$ (M = Pd, Pt). The chloro bridges may be subsequently replaced by $(pz)_3BH^-$ or $(pz)_4B^-$ [Eq. (9)] (*21–24*), by $(pz)_3CH$ or $(pz)_4C$ [Eq. (10)] (*22*), or by $(pz)_2CH_2$ [Eq. (11)] (*24*). The structure of **33** is shown in Fig. 2.

$$[MCl(C{\sim}N)]_2 + 2(pz)_3BR^- \rightarrow 2M(C{\sim}N)\{(pz)_3BR\} + 2Cl^- \qquad (9)$$

22–30, 33 (M = Pd)

31 (M = Pt)

$$[PdCl(C{\sim}N)]_2 + 2(pz)_3CR \rightarrow 2[Pd(C{\sim}N)\{(pz)_3CR\}]^+ + 2Cl^- \qquad (10)$$

32

$$[PdCl(C{\sim}N)]_2 \xrightarrow[\text{(ii) } Ag^+\ O_3SCF_3^-]{\text{(i) } 2\ (pz)_2CH_2} 2[Pd(C{\sim}N)\{(pz)_2CH_2\}]^+O_3SCF_3^- + 2AgCl \qquad (11)$$

34

The poly(pyrazol-1-yl)alkane complexes **32** are thermally more stable than the borate analogs **30** (*22*), as noted earlier for (η^3-allyl)palladium(II) species (*1*). Also, the complexes **22–30** and **32** and **33** exhibit ^1H NMR spectra indicating stereochemical nonrigidity (*21–24*), whereas the platinum complexes **31** are rigid on the NMR time scale (*22*), e.g., spectra of **30** (R = Me, R′ = H) show one set of pz resonances at 36°C and two at −34°C, but **31**(R = Me, R′ = H) displays three pz environments, even on heating to 85°C.

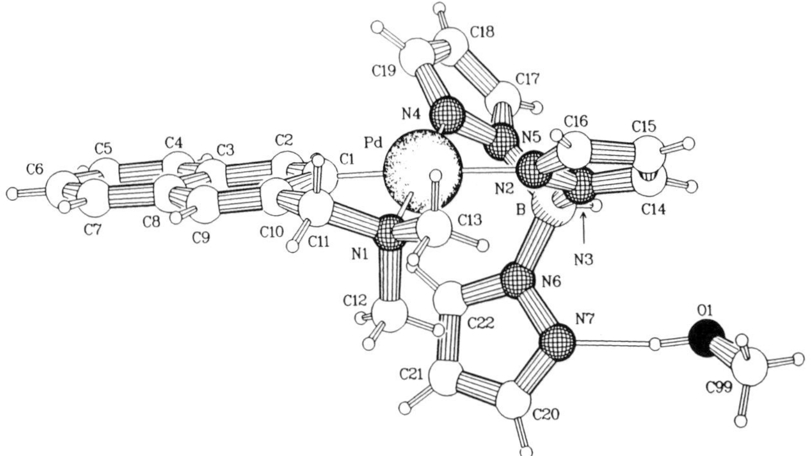

FIG. 2. Structure of $Pd(C_{13}H_{14}N)\{(pz)_3BH\} \cdot \frac{1}{2}MeOH$ (**33**) [with permission from G. van Koten *et al.* (*24*)].

Spectra of the fluxional complexes are readily interpretable in terms of three processes, with the occurrence or observation of the processes dependent on the ligands present and limitations of solvent and coalescence

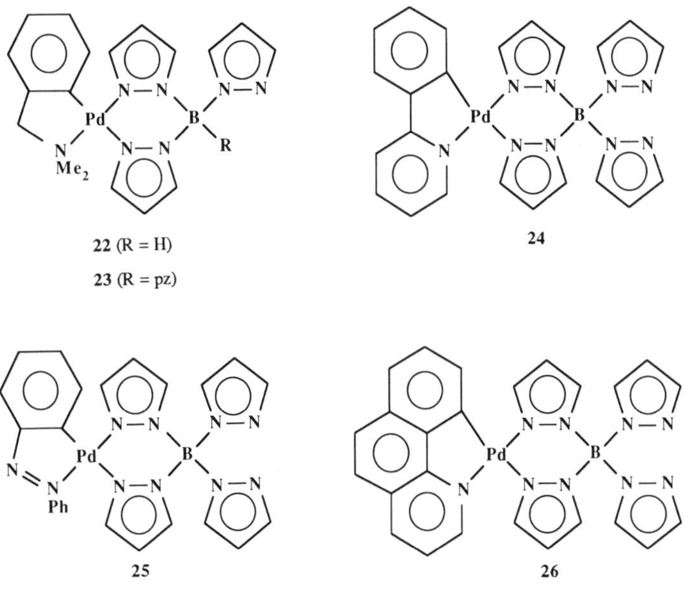

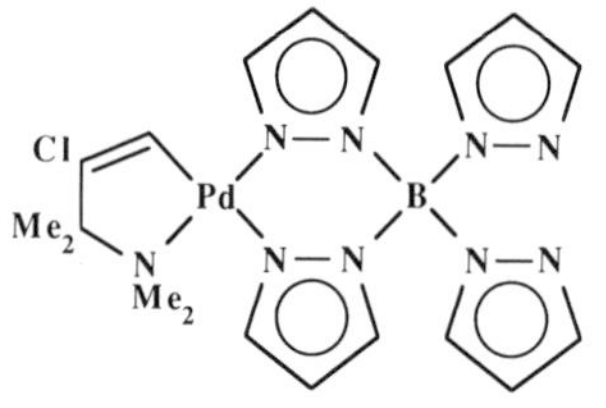

27

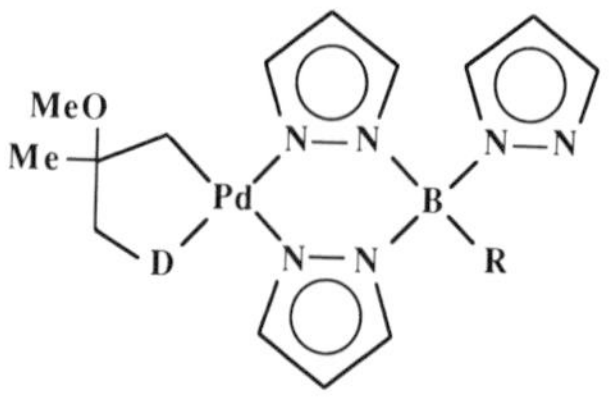

28 (D = NMe_2; R = H, pz)

29 (D = SBu^t; R = H, pz)

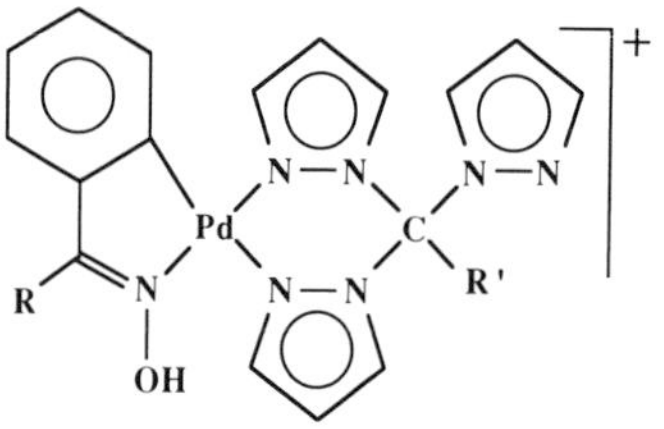

30 (M = Pd; R′ = H, pz)

31 (M = Pt; R′ = H, pz)

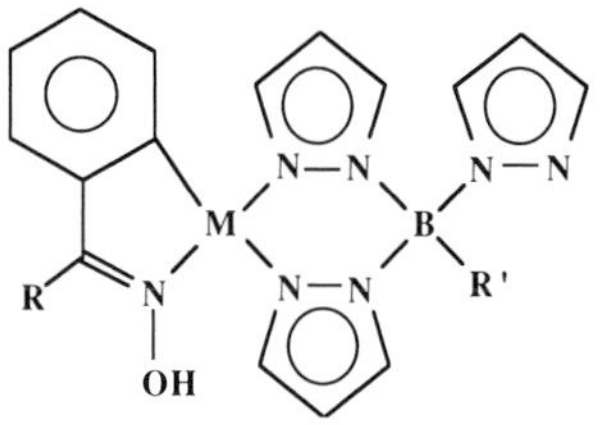

32 (R′ = H, pz)

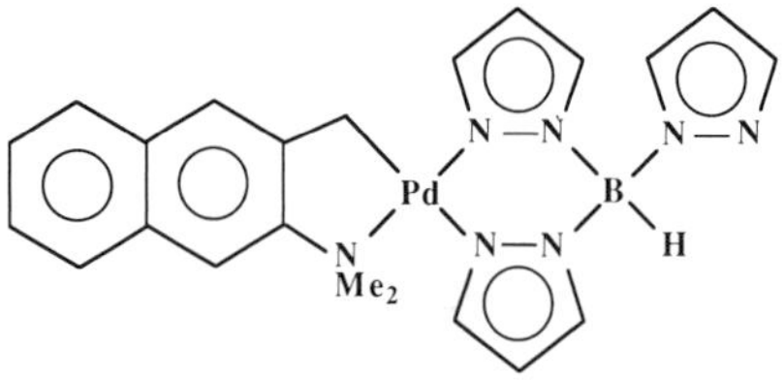

33

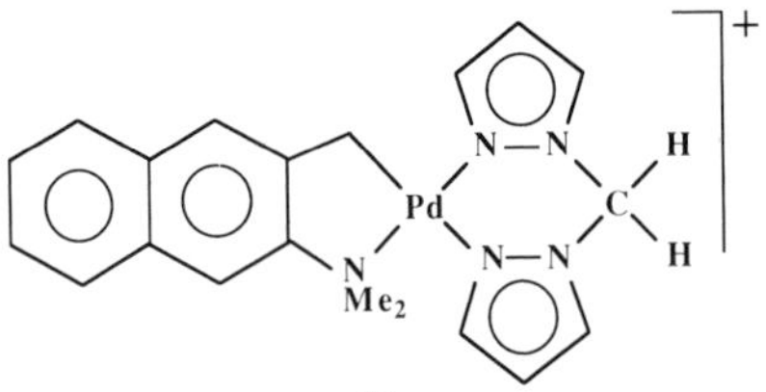

34

temperatures. A detailed line-shape analysis of the ^1H NMR spectra of **27** in $CDCl_3$ illustrates the fluxional processes available to these complexes (*23*) and to other square planar M(II) complexes mentioned in this review. At −30°C the spectrum of **27** shows four pz environments, corresponding to slow exchange on the NMR time scale. On warming, the proton resonances for two of the pz rings broaden, with resonances of a third pz ring broadening at higher temperature and involvement of the fourth pz ring occurring near 58°C. The most facile process corresponds to boat-to-boat inversion of the six-membered PdNNBNN ring, resulting in exchange of "axial" and "equatorial" uncoordinated pz rings. This process is illustrated for the related complex, $[Pd(C_3H_5)\{(pz)_2CMe_2\}]BF_4$ (**16**), which requires a temperature of approximately −70°C to observe slow exchange. The second process, occurring at a slower rate, involves incorporation of one of the coordinated pz rings in exchange with the uncoordinated pz rings. The coordinated ring undergoing exchange is assumed to be trans to the alkenyl group because the alkenyl group has a stronger trans influence than the nitrogen donor, and the exchange probably occurs via axial coordination as the Pd–N bond is broken. The slowest process, contributing to exchange of all four pz environments, requires exchange of the pz group trans to nitrogen, with the overall process envisaged as a tumbling motion. Another process, rotation about a metal–ligand C_{3v} axis, occurs in some five-coordinate complexes (Section II,C).

Fluxional motion of the $(pz)_3BH^-$ ligand in its complexes occurs more easily than that of its $(pz)_4B^-$ analog (*22,23*). A possible reason for this is illustrated by a structural study of the square planar dimethylgold(III) complex $AuMe_2\{(pz)_4B\}$ (**35**) (*25*). In this complex the "equatorial" uncoordinated pz ring plane forms an angle of 82.4° with the coordination plane. This orientation minimizes steric interaction between the ring and

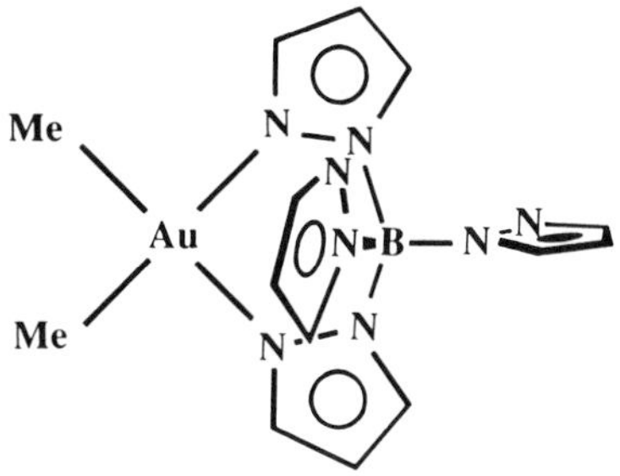

35

the H-3 protons of the coordinated rings, and also results in an orientation of the "axial" ring, as shown, to minimize steric interaction between the two uncoordinated rings. With a hydrogen atom as the equatorial group in $(pz)_3BH^-$ and $(pz)_3CH$ complexes, this effect is absent and a range of orientations for the axial group are more readily accessible, illustrated by $[AuMe_2\{(pz)_3CH\}]NO_3$ [axial group weakly coordinated (*26*)] and $AuMe_2\{(pz)_3BH\}$ [axial group intermediate between coordinated and as in **35** (*27*)]. These steric considerations suggest that complexes of $(pz)_3BH^-$ can achieve axial coordination, leading to exchange, with a lower activation energy than for complexes of $(pz)_4B^-$.

In a study related to intramolecular coordination complexes, it was found that *trans(P,N)*-$[PdCl(\mu\text{-pyCH}_2\text{-}N,C')(PPh_3)]_2$ does not react with

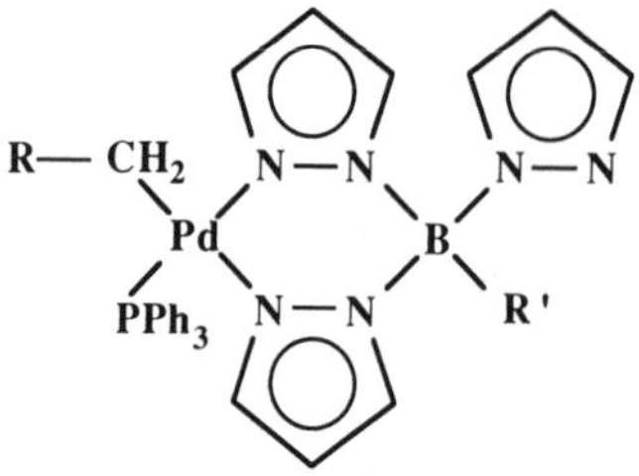

36 (R = 2-pyridyl, R'=H)

37 (R = 2-furfuryl, R'=H)

38 (R = 2-furfuryl, R'=pz)

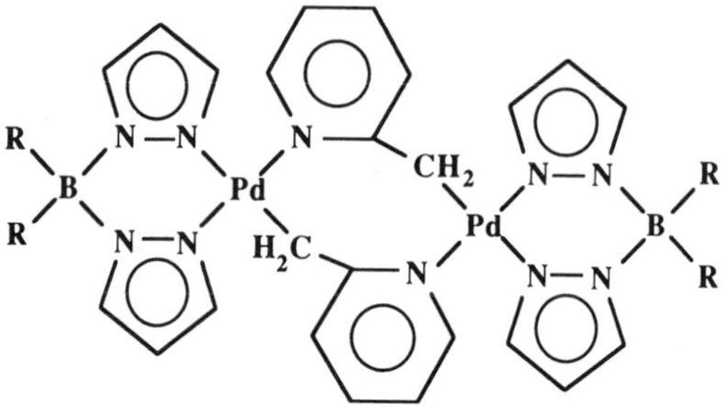

39 (R = H)

40 (R = pz)

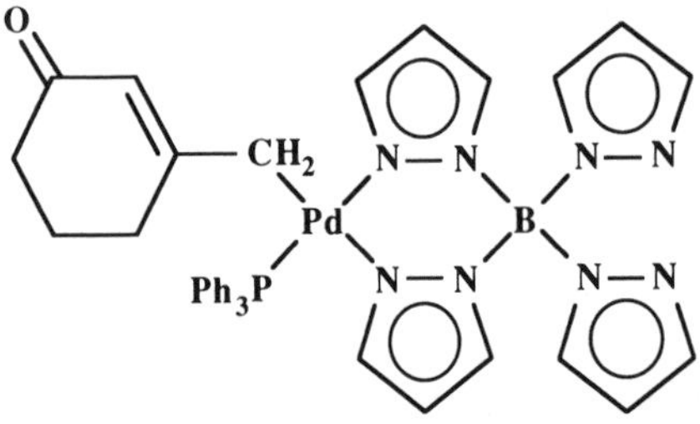

41

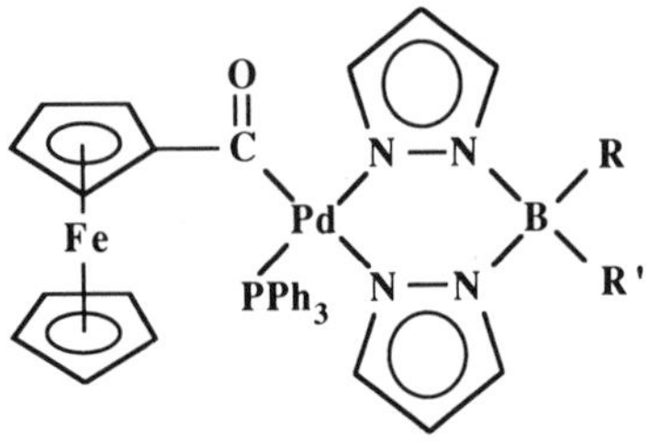

42 (R = R' = H)

43 (R = H, R' = pz)

44 (R = R' = pz)

$(pz)_3CH$ but does react with $(pz)_3BH^-$ and $(pz)_2BR_2^-$ to form mononuclear (**36**) and binuclear complexes (**39** and **40**) (*28*). Similarly, chloro and triphenylphosphine ligands are displaced from several mononuclear complexes $PdClR(PPh_3)_2$ by poly(pyrazol-1-yl)borate ions to form complexes **37, 38,** and **41–44** (*29–31*). Consistent with the discussion above, fluxional motion of the $(pz)_3BH^-$ ligand in **36, 37**, and **43** occurs more readily than that of the $(pz)_4B^-$ analogs, and the $(pz)_2BH_2^-$ complex (**42**) does not exhibit exchange of pyrazole environments. Complex **40** is rigid at 52°C in $CDCl_3$, and this effect may be partially attributed to steric effects associated with the 2-picolyl bridge (*28*). Measurement of redox potentials for the complexes **42–44** indicates that the ease of oxidation of the iron centers is **42** ⩾ **43** > **44** (*31*).

Square planar complexes with simple monoalkylmetal(II) groups, rather than intramolecular coordination (**22–34**) and related complexes (**36–44**), were first obtained for platinum bonded to poly(pyrazol-1-yl)borates. The complexes PtMe$\{(pz)_3BR\}$ [R = H (**45**), pz (**46**)] were obtained on addition

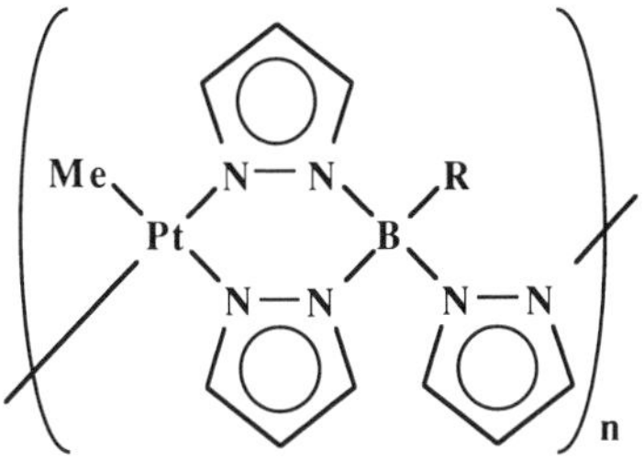

45 (R = H)

46 (R = pz)

of $AgPF_6$ to a solution of PtClMe(COD) (COD = 1,5-cyclooctadiene) in acetone, followed by removal of AgCl and addition of $(pz)_3BR^-$ [Eq. (12)] (*32,33*).

$$\text{PtClMe(COD)} \xrightarrow{Ag^+} [\text{PtMe(COD)(acetone)}]^+ \xrightarrow{(pz)_3BR^-} [\text{PtMe}\{(pz)_3BR\}]_n \quad (12)$$

45, 46

The complexes **45** and **46** exhibit low solubility and are almost certainly polymeric to permit square planar coordination (*33*). The tris(pyrazol-1-yl)methane analog [PtMe$\{(pz)_3CH\}$]PF_6 was isolated in a similar manner (*34*) and it may also be polymeric (**47**), but bidentate $(pz)_2CH_2$ gives

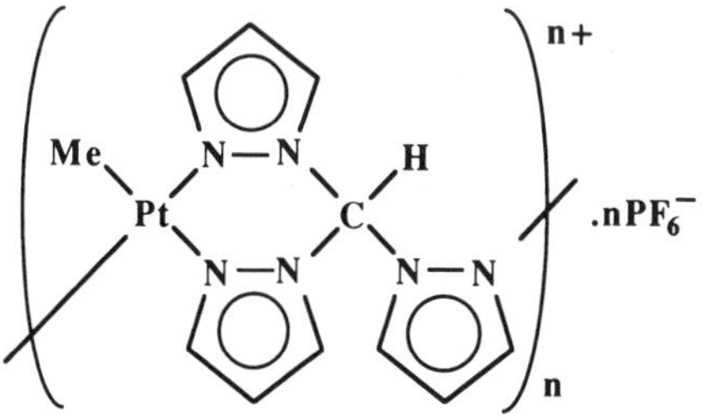

47

$[PtMe\{(pz)_2CH_2\}(acetone)^+(PF_6)^-]_2(COD)$ (*35*). Mononuclear and binuclear complexes are formed with bidentate bis(pyrazol-1-yl)borates (*36, 37*), e.g., as in Eq. (13) (*37*).

$$[PtMe(COD)(acetone)]^+ \xrightarrow{(pz)_2BR_2^-} \underset{\mathbf{48}}{PtMe\{(pz)_2BR_2\}(COD)} \xrightarrow{\Delta} \underset{\mathbf{49}}{\tfrac{1}{2}[PtMe\{(pz)_2BR_2\}]_2(COD)} \tag{13}$$

The polymer $[PtMe\{(pz)_3BH\}]_n$ (**45**) reacts with carbon monoxide or *tert*-butylisonitrile in dichloromethane to form $PtMe\{(pz)_3BH\}(CO)$ (**50**)

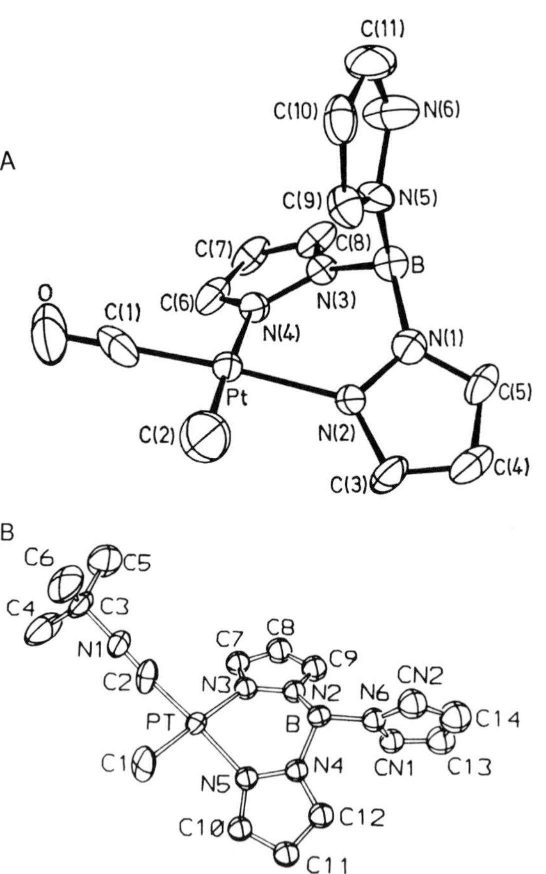

FIG. 3. Structures of (A) $PtMe\{(pz)_3BH\}(CO)$ (**50**) (with permission from Ref. 40; copyright Elsevier Sequoia S.A.), (B) $PtMe\{(pz)_3BH\}(CNBu^t)$ (**51**) (with permission from Ref. 42; copyright American Chemical Society), and (C) $PtMe\{(pz)_2BEt_2\}(PhC{\equiv}CMe)$ (**52**) (with permission from Ref. 43; copyright Elsevier Sequoia S.A.).

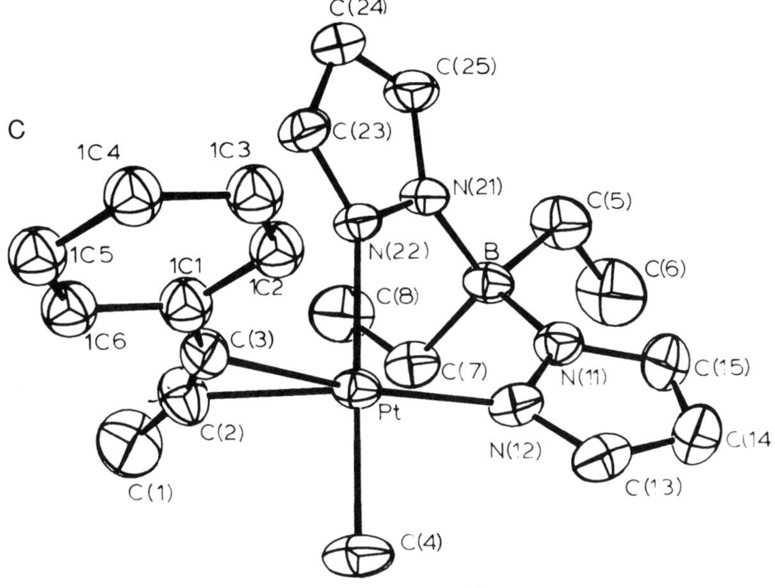

FIG. 3. *(continued)*

(*38*) or PtMe{$(pz)_3BH$}($CNBu^t$) (**51**) (*39*), which are square planar in the solid state (A and B, Fig. 3) (*40–43*) but five-coordinate in solution (Section II,C) (*39*). Similarly, complex **49** reacts with carbon monoxide, isocyanides, phosphines, and the alkynes PhC≡CR (R = Me, Ph) to form square planar complexes **52, 53,** and **55** (*36*), and the structure of one of

49 (R = Et, Ph)

48 (R = Et, Ph; L = COD)

52 (R = Et; L = CO, PPh_3, PMe_2Ph,

p-MeC_6H_4NC, PhC≡CPh, PhC≡CMe)

53 (R = Ph; L = PhC≡CMe)

54 (R = Me; L = PhC≡CPh)

the alkyne complexes is shown in Fig. 3C. The alkyne group is nonlinear, with the methyl and phenyl groups bent 17.7(10)° and 21.2(9)° away from platinum, respectively, and a Pt · · · H−C-7 interaction is 2.65 Å.

Although halogen abstraction from PtClMe(COD) is required for the synthesis of **45–53** and **55**, the complex PtMe{$(pz)_2BMe_2$}(PhC≡CPh) (**54**) has been obtained directly by addition of Na[$(pz)_2BMe_2$] to PtClMe(COD) in tetrahydrofuran followed by addition of PhC≡CPh (*44*). A series of gallate derivatives (**56**) were obtained by an analogous procedure, and ^{1}H NMR spectra of **50–56**, including variable-temperature studies for PtMe{$(pz)_2GaMe_2$}(PPh_3) (*44*), indicate the presence of rapid inversion of the chelate rings.

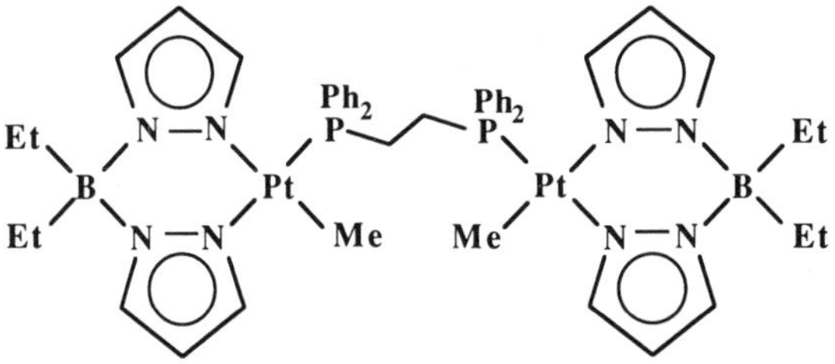

55

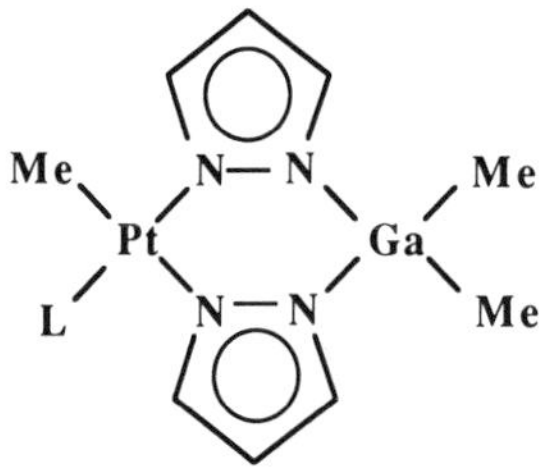

56

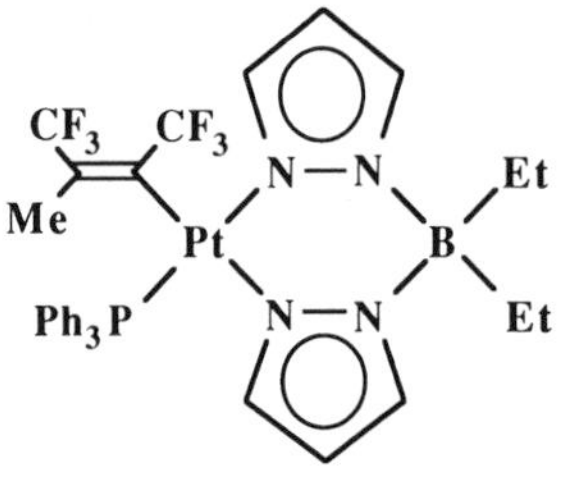

57

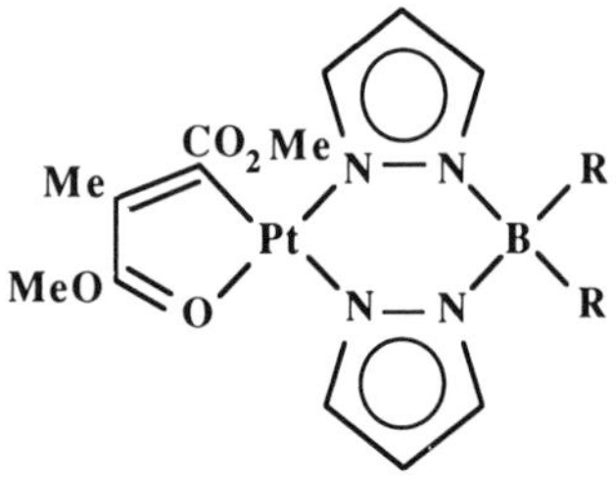

58

The structural studies of complexes **50, 51**, and **52** (L = PhC≡CMe) (Fig. 3), together with other structures described in this review and the structures of simple coordination complexes such as $PdCl_2${$(pz)_2CMe_2$} (*13*), [Pd{$(pz)_2CH_2$}$_2$][BF_4]$_2$ (*13*), [Pd{$(pz)_3CH$}$_2$][BF_4]$_2$ (*45*), and Pd{$(pz)_3$-BH}$_2$ (*45*), illustrate the typical boat conformation usually adopted by six-membered chelate rings in poly(pyrazol-1-yl)alkane and borate complexes.

Reaction of **49** with alkynes containing electron-withdrawing substituents results in insertion reactions: hexafluorobut-2-yne gives a product isolated as a triphenylphosphine derivative (**57**), and reaction of dimethylacetylene dicarboxylate results in formation of two products, one of which is assigned structure **58** (*37*).

The mono(organo)palladium(II) phosphine complexes **36–38** and **41–44** may be regarded as palladium(II) analogs of the monomethylplatinum(II) complexes **52** (L = PPh_3 or PMe_2Ph), and a number of simple monomethylpalladium(II) complexes PdXMe(L) [X = halide, L = poly-(pyrazol-1-yl)alkane] (**59–69**) (*14, 46–50*) have been isolated. Several

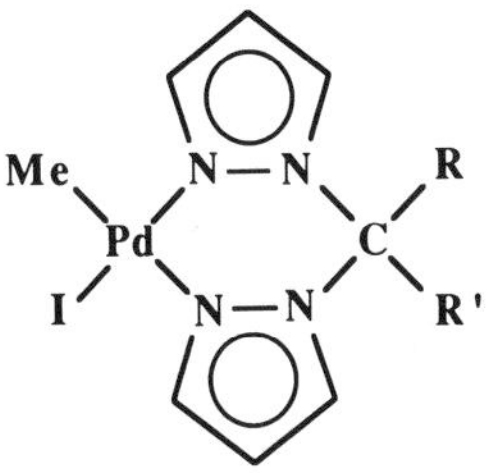

59 (R = R' = H)

60 (R = R' = Me)

61 (R = H, R' = pz)

62 (R = R' = pz)

routes were developed for synthesis of the iodo complexes (*14*), with the simplest involving reaction of the iodo-bridged dimer $[PdIMe(SMe_2)]_2$ with the ligand at ambient temperature [Eq. (14)].

$$[PdIMe(SMe_2)]_2 + 2L \rightarrow \underset{\textbf{59–61, 63–67}}{PdIMe(L)} + 2SMe_2 \tag{14}$$

Some of the complexes have also been isolated by addition of the ligand at −40°C to the solution obtained on reaction of MgIMe (**59–61** and **63**) or LiMe (**59–63**) with *trans*-$PdCl_2(SMe_2)_2$ at −60°C [Eq. (15)]. For the

$$trans\text{-}PdCl_2(SMe_2) \xrightarrow[\text{or (i) LiMe/LiI/MeI, (ii) L}]{\text{(i) 2MgIMe, (ii) L}} PdIMe(L) \tag{15}$$

syntheses of PdIMe(L) involving LiMe, the LiMe must be prepared from lithium and iodomethane or, if made from lithium and chloromethane,

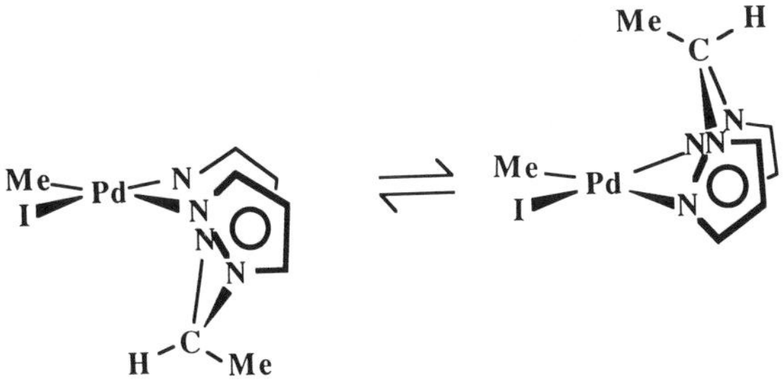

63

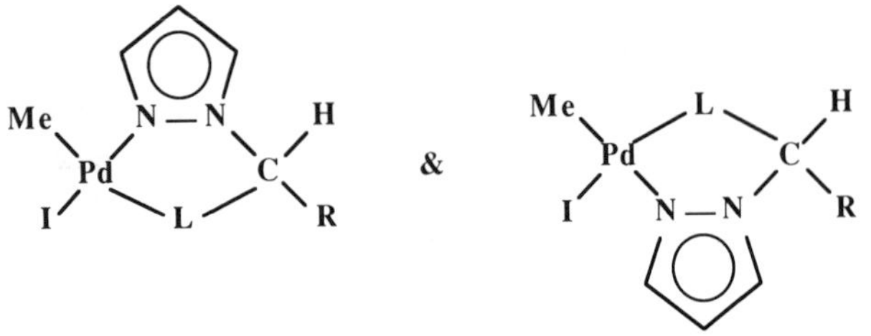

64 (L = py, R = H; 4:1 ratio)

65 (L = mim, R = H; 4:1 ratio)

some iodomethane must be added during the synthesis. It is assumed that the presence of residual MeI in the LiMe reagent results in an oxidative addition–reductive elimination (of ethane) sequence to generate a Pd^{II}IMe species prior to or after addition of the ligand. The Grignard and organolithium routes, and the procedure of Eq. (14), are assumed to be applicable for all of the complexes **59–67**, although the Grignard route gives lower yields and the products usually required recrystallization.

The complex PdIMe{$(pz)_2CH_2$} has also been obtained by oxidative addition of iodomethane to $Pd_2(dba)_3 \cdot CHCl_3$ in the presence of

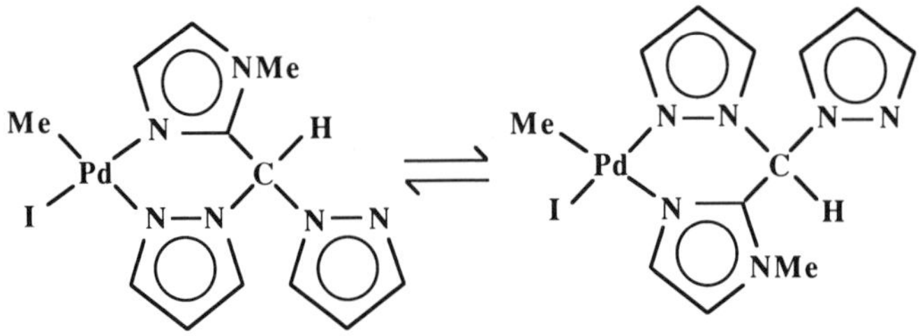

66 (6:1 ratio)

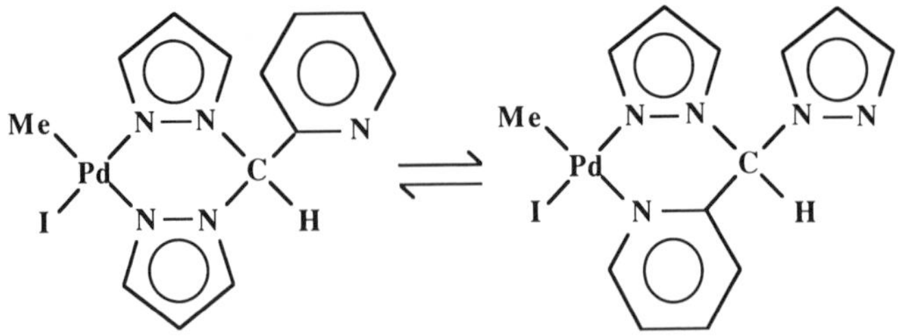

67 (6:1 ratio)

$(pz)_2CH_2$, but in view of the low yield obtained (11%) this route was not attempted for the other complexes. The chloro and bromo complexes $PdXMe\{(pz)_2CMe_2\}$ (**68** and **69**) were obtained on addition of $(pz)_2CMe_2$

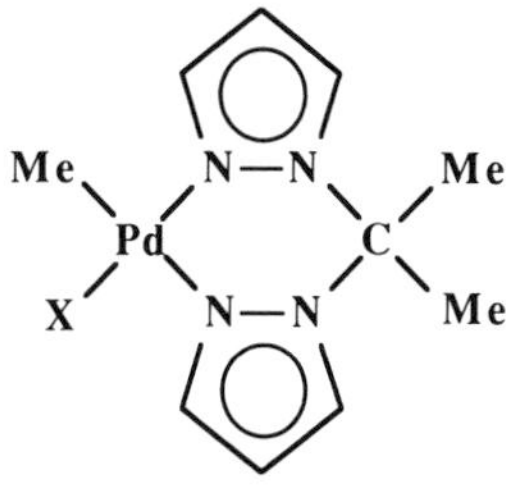

68 (X = Cl)

69 (X = Br)

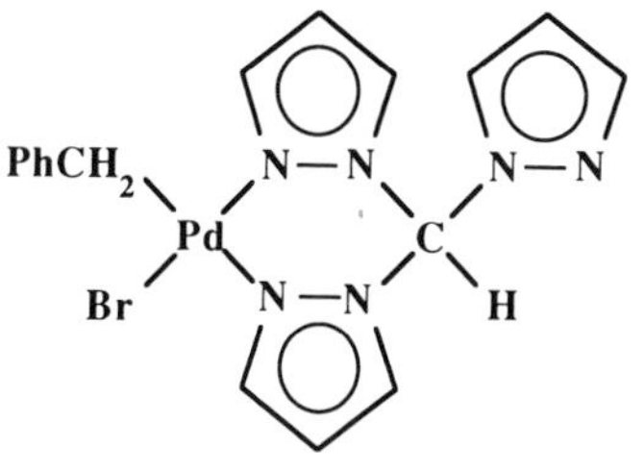

70

to solutions of $[PdXMe(SMe_2)]_2$ in aqueous acetonitrile, the latter being generated by removal of iodide from $[PdIMe(SMe_2)]_2$, followed by addition of KX (*14*). Complexes of unsymmetrical ligands occur as structural isomers (**64–67**), with the isomer ratios estimated from ^1H NMR spectra; rapid exchange between isomers occurs for the complexes of ligands containing three donor groups (**66** and **67**). The benzylpalladium(II) complex **70** was obtained on reductive elimination of ethane from the palladium(IV) complex $PdBrMe_2(CH_2Ph)\{(pz)_3CH\}$ (*12*).

Dimethylpalladium(II) analogs of complexes **59–67** have been prepared (**71–74** and **77–81**), and are generally best obtained using a modification

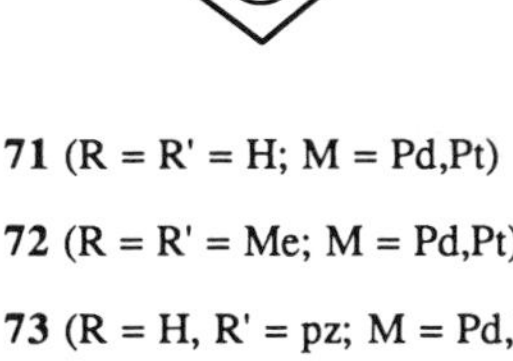

M = Pd, Pt

71 (R = R' = H; M = Pd,Pt)

72 (R = R' = Me; M = Pd,Pt)

73 (R = H, R' = pz; M = Pd,Pt)

74 (R = R' = pz; M = Pd,Pt)

75 (R = H, R' = Ph; M = Pt)

76 (R = H, R' = thien-2-yl, M = Pt)

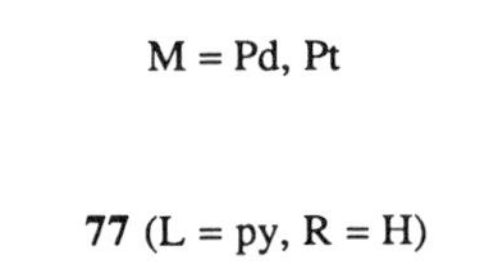

77 (L = py, R = H)

78 (L = mim, R = H)

80 (L = mim, R = pz)

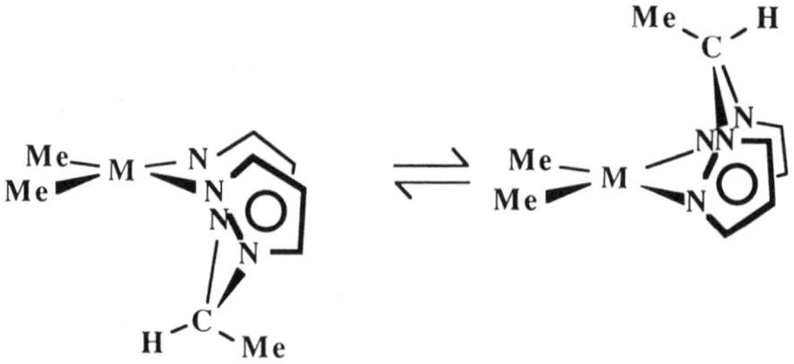

Me Me M N N N N C H Me ⇌ Me Me M N N N N C Me H

79 (M = Pd, Pt)

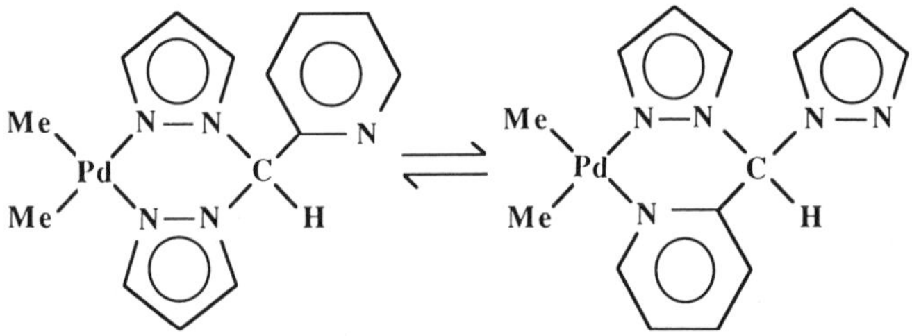

Me Me Pd N—N N—N C N H ⇌ Me Me Pd N—N N C N—N H

81 (3:1 ratio)

of the MeLi procedure outlined for the $Pd^{II}IMe$ complexes [Eq. (15)] (*14, 46, 47, 49*). In this case, "methyl halide-free" LiMe is used so that $Pd^{II}Me_2$ species are formed in solution without the subsequent oxidative addition–reductive elimination that appears to occur for the procedure in Eq. (15).

The $Pd^{II}Me_2$ complexes of (pz)(py)CH_2, (pz)(mim)CH_2, $(pz)_2$(mim)-CH, and $(pz)_2$(py)CH were obtained on addition of a solution of the ligand in acetone or benzene to solid $[PdMe_2(pyridazine)]_n$ (*14, 49, 50*), but this procedure is not as satisfactory for the other poly(pyrazol-1-yl)-alkane ligands because they do not contain the more basic pyridine or *N*-methylimidazole donors that ensure efficient displacement of pyridazine.

The low basicity of poly(pyrazol-1-yl)alkane ligands has also presented difficulties for the synthesis of dimethylplatinum(II) complexes. The complex $PtMe_2$(COD) is an excellent precursor for the synthesis of phosphine complexes (*51*), but it is less satisfactory for nitrogen donors (*52*). The norbornadiene (NBD) complex $PtMe_2$(NBD) is more suitable for the synthesis of complexes containing pyridine and related ligands, possibly owing to the smaller "bite" of NBD (*52*). However, dialkylsulfide-bridged complexes $[PtMe_2(SR_2)]_2$ appear to be ideal substrates (*53–55*), with the application of this type of reagent pioneered by Vrieze and co-workers (*53*). The $Pt^{II}Me_2$ complexes **71–80** were readily synthesized from

$[PtMe_2(SEt_2)]_2$ (*49,50,56,57*), and as an illustration of the utility of this complex it is reported that reaction of $PtMe_2(COD)$ with $(pz)_3CH$ in refluxing benzene for 30 hours gave $PtMe_2\{(pz)_3CH\}$ in ~69% yield (*57–59*) but reaction of $[PtMe_2(SEt_2)]_2$ with $(pz)_3CH$ under the same conditions for 10 minutes gave the complex in ~90% yield (*57*). Several diphenylplatinum(II) complexes **82–89**, including $PtPh_2\{(pz)_3CH\}$ (*57*), were obtained by a similar procedure (*60*).

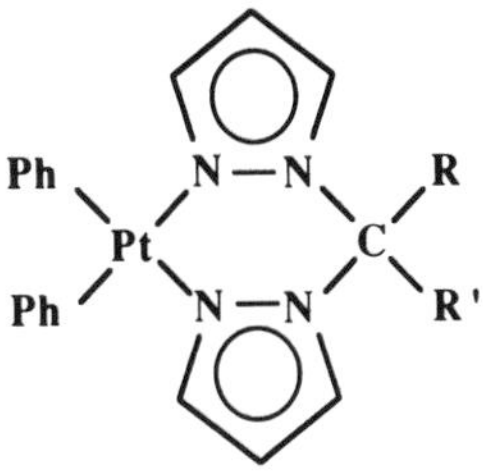

82 (R = R' = H)

83 (R = R' = Me)

84 (R = H, R' = Me)

85 (R = H, R' = pz)

86 (R = H, R' = thien-2-yl)

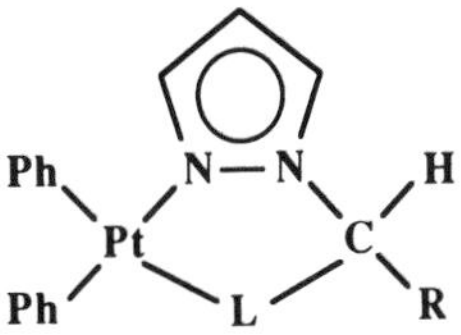

87 (L = py, R = H)

88 (L = mim, R = pz)

89 (L = mim, R = pz)

The dimethylmetal(II) complexes undergo fluxional processes similar to those of PdXMe(L), in particular, ring inversion for the complexes of bidentate ligands [$(pz)_2CH_2$, $(pz)_2CMe_2$, $(pz)_2CHMe$, $(pz)(py)CH_2$, and $(pz)(mim)CH_2$], and exchange between conformers (**63** and **79**) and isomers (**67** and **81**) (Fig. 4), although insolubility prevented NMR studies of $PtMe_2(L)$ [L = $(pz)_2CH_2$, $(pz)_3CH$, $(pz)_2CHPh$]. Rapid exchange of all pz groups occurs for $PdIMe\{(pz)_3CH\}$ and $PdIMe\{(pz)_4C\}$ at ambient temperature. For $PdIMe\{(pz)_3CH\}$, spectra at −90°C show sharp resonances for only one pz ring, with the other two environments partially resolved into six broad resonances, indicating that one ring exchanges more slowly. The complex $PdIMe\{(pz)_4C\}$ gives spectra at −70°C showing two PdMe (in 2:1 ratio) and eight pz ring environments (*49*), consistent with two "frozen" structures similar to that of $AuMe_2\{(pz)_4B\}$ in the solid state (**35**). The complex $PdMe_2\{(pz)_4C\}$, with higher symmetry, shows similar NMR behavior and exhibits four ring and two PdMe environments (in 1:1 ratio) at −70°C in $(CD_3)_2CO$.

The new ligands $(pz)_2(py)CH$ and $(pz)_2(mim)CH$ are obtained by the facile condensation of $(pz)_2CO$ with (py)CHO (commercially available)

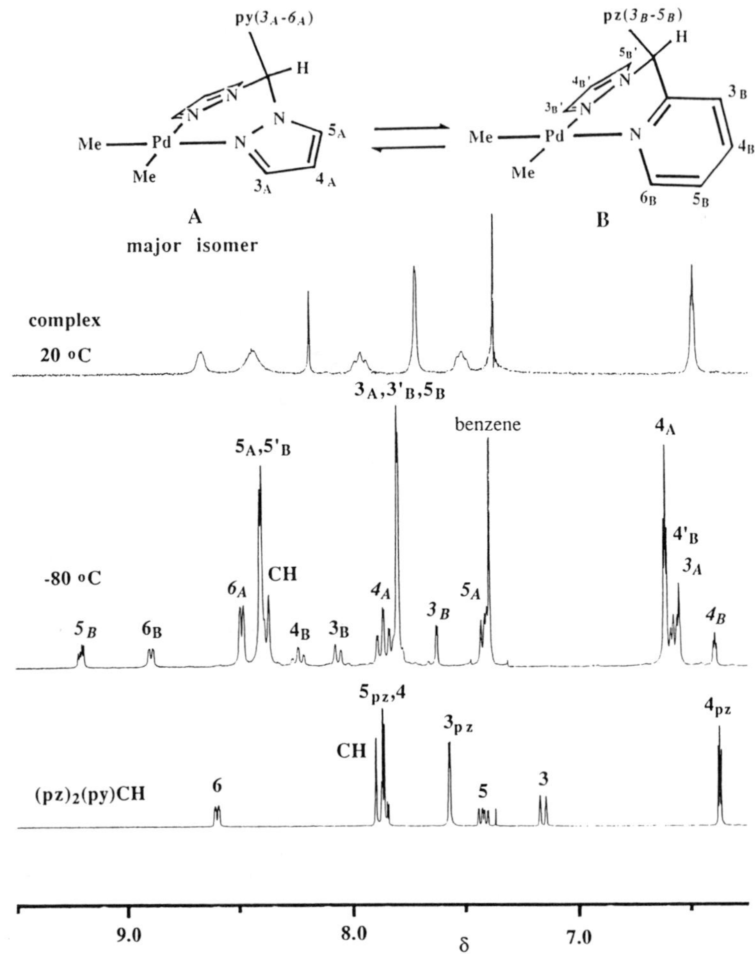

FIG. 4. ^{1}H NMR spectra of $(pz)_2(py)CH$ and its $Pd^{II}Me_2$ complex in $(CD_3)_2CO$, illustrating the variable-temperature ^{1}H NMR behavior exhibited by complexes of tripod ligands and the formation of isomers (with permission from Ref. 49; copyright Elsevier Sequoia S.A.).

or mimCHO (*49*), and structural studies of $[M\{(pz)_2(py)CH\}_2](NO_3)_2$ (M = Fe, Co, Ni, Cu, Zn) have recently been reported (*61*). The absence of C_{3v} symmetry for these ligands results in interesting NMR effects and isomerism, as illustrated for $PdMe_2\{(pz)_2(py)CH\}$ (**81**) in Fig. 4. The complex $PdIMe\{(pz)_2(py)CH\}$ (**67**) exhibits similar isomerism, with the isomers in ~6:1 ratio. The complex $PdMe_2\{(pz)_2(mim)CH\}$ (**80**) forms only one isomer, with spectra showing sharp mim resonances ~0.2–0.7 ppm downfield from the free ligand values at ambient temperature, and four broad (barely visible) resonances for the pz rings, suggesting rapid exchange of coordinated and uncoordinated pz rings. On cooling, resonances for the two pz rings are resolved, with one of them showing the expected downfield shifts of the H-3 and H-4 protons on the coordinated ring. The analogous platinum(II) complex and $PtPh_2\{(pz)_3CH\}$ are rigid on the NMR time scale. The complex $PdIMe\{(pz)_2(mim)CH\}$ exhibits spectra similar to those of $PdMe_2\{(pz)_2(mim)CH\}$, but of greater complexity owing to the presence of two isomers in ~6:1 ratio (**66**) (*49*). The $Pt^{II}Ph_2$ complexes of $(pz)_2CH_2$, $(pz)_2CMe_2$, $(pz)(py)CH_2$, and $(pz)(mim)CH_2$ give variable-temperature ^1H NMR spectra, indicating the occurrence of ring inversion, but, unlike the $M^{II}Me_2$ and $Pd^{II}IMe$ analogs, $PtPh_2\{(pz)_2$-$CHMe\}$ occurs as only one conformer.

The complex $PdBr(CH_2Ph)\{(pz)_3CH\}$ (**70**) is also fluxional, with three pz resonances at ambient temperature resolved into nine at −60°C. A CH_2 singlet at ambient temperature is resolved into two doublets at −60°C, indicating that the benzyl group has a preferred conformation with inequivalent benzylic protons (*12*).

C. *Five-Coordinate Platinum(II) Complexes*

The complex $PtMe\{(pz)_3BH\}(CO)$ (**50**) is square planar in the solid state (Fig. 3), but detailed NMR studies (*32,38,39*), including line-shape analyses for variable-temperature $^{13}C\{^1H\}$ spectra (*39*), indicate that this complex and $PtMe\{(pz)_4B\}(CO)$ (**91**) are five-coordinate in solution (Fig. 5). The complex $PtMe\{(pz)_3BH\}(CO)$ (**50**) gives a single set of pz resonances in the ^1H NMR spectrum, and $PtMe\{(pz)_4B\}(CO)$ (**91**) gives pz resonances in 3:1 ratio, indicating that the uncoordinated ring in **91** does not exchange with the coordinated rings. Coupling between ^{195}Pt and the H-3 and H-4 protons of the coordinated pz rings is observed, indicating that the fluxional process does not involve Pt–N bond breaking, and thus does not involve the tumbling or associative–dissociative exchange mentioned earlier. Rotation about the "Pt–$(pz)_3BR$" C_{3v} axis is proposed, as established earlier for $Mo(\eta^3\text{-allyl})\{(pz)_3BR\}(CO)_2$ (R = H, pz) by line-shape analyses (*62*). Similar results were obtained for the phosphite (**92**) and isonitrile complexes $PtMe\{(pz)_3BH\}(L)$ (**51** and **90**) (*39*) and, like the carbonyl (**50**), the $CNBu^t$ complex has been shown to be square planar in the solid state (Fig. 3B) (*42*).

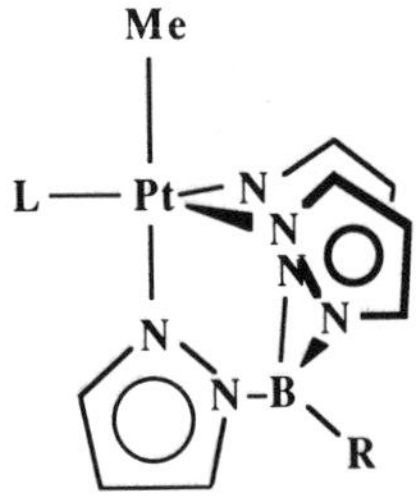

A 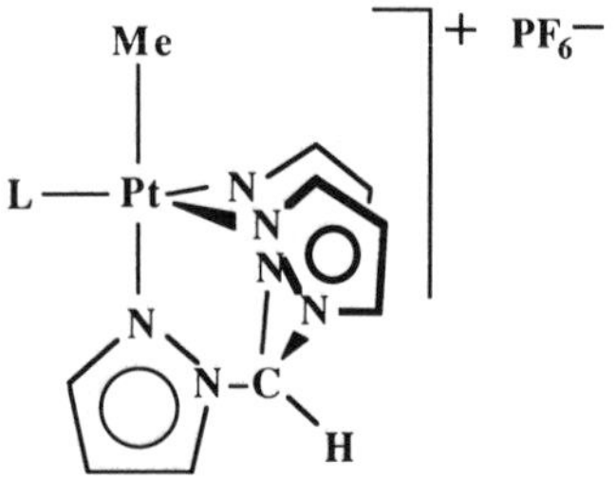 **B**

50 (L = CO, R = H) **92** [L = $P(OMe)_3$, R = H] **96** (L = CO)

51 (L = $CNCMe_3$, R = H) **93** (L = alkyne) **97** (L = $CH_2{=}CH_2$)

90 (L = CNC_6H_{11}, R = H) **94** (L = alkene) **98** (L = alkyne)

91 (L = CO, R = pz) **95** (L = allene)

FIG. 5. Complexes for which NMR spectra indicate five-coordination in solution. Representative examples of **A** are square planar in the solid state (Fig. 3), and examples of **B** are five-coordinate in the solid state (Fig. 6).

Closely related alkyne, alkene, and allene complexes (**93–95**) have been obtained on reaction of PtMe{$(pz)_3BH$} with the ligands in dichloromethane (*32, 33, 38, 63, 64*), and representative alkyne and alkene complexes have been characterized by X-ray crystallography (Fig. 6) (*65*). The complexes have similar geometries and may be regarded as trigonal bipyramidal in the solid state, with the ligands $CF_3C{\equiv}CCF_3$ or $CF_2{=}CF_2$ symmetrically bonded and occupying one equatorial site. The Pt(η^2-$CF_3C{\equiv}CCF_3$) geometry in the alkyne complex is similar to that in the related platinum(0) complex $Pt(PPh_3)_2(CF_3C{\equiv}CCF_3)$ (*66*). In contrast to the carbonyl (**50** and **91**) and other complexes (**51, 90,** and **92**), these complexes are rigid in solution, e.g., the ^1H NMR spectra for the symmetrical alkene and alkyne complexes show pz environments in 2:1 ratio with presence of ^1H–^{195}Pt coupling. Unsymmetrical alkenes form geometrical isomers, with the ratio of isomers dependent on the steric requirements of the alkene substituents. The ^1H NMR spectrum of PtMe{$(pz)_3BH$}($CH_2{=}CHCO_2Me$) was studied up to 150°C (*38*). The two PtMe resonances in 2:1 ratio at ambient temperature broaden and coalesce (120°C) to form a single broad peak that sharpens at 150°C. The fluxionality is probably due to equilibration of the two isomers rather than rotation of the $(pz)_3BH^-$ ligand, because similar variable-temperature studies for complexes of simple symmetrical alkenes do not result in coalescence of the resonances attributed to the two different coordinated pz rings.

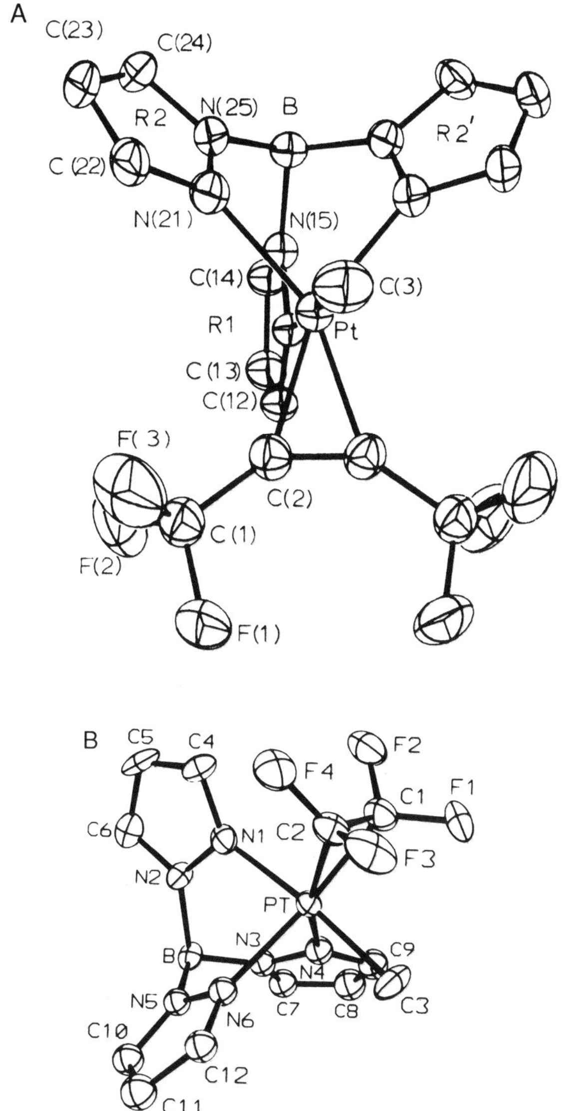

FIG. 6. Structures of (A) PtMe{$(pz)_3BH$} ($CF_3C{\equiv}CCF_3$) (with permission from Ref. 65; copyright American Chemical Society) and (B) PtMe{$(pz)_3BH$}($CF_2 = CF_2$) (with permission from Ref. 66; copyright International Union of Crystallography).

Tris(pyrazol-1-yl)methane analogs of **50, 93**, and **94** are readily obtained on addition of CO, ethylene, or alkynes to a suspension of [PtMe-{(pz)$_3$CH}]PF$_6$ (**47**) in dichloromethane to form **96–98** (*34*). The carbon monoxide complex exhibits fluxionality in solution similar to that of the borate analog, but, unlike rigid PtMe{(pz)$_3$BH}(L) (L = CH$_2$=CH$_2$ (*38*), MeC≡CPh (*33*)], the analogous cations [PtMe{(pz)$_3$CH}(L)]$^+$ have stereochemically nonrigid (pz)$_3$CH at ambient temperature. The CH$_2$=CH$_2$ and MeC≡CPh groups also appear to be rotating freely. However, the cations [PtMe{(pz)$_3$CH}(L)]$^+$ (L = CF$_3$C≡CCF$_3$, MeO$_2$CC≡CCO$_2$Me) are static, and thus for the (pz)$_3$CH complexes, the NMR spectra qualitatively indicate that the barrier to stereochemical nonrigidity decreases in the order L = MeO$_2$CC≡CCO$_2$Me ~ CF$_3$C≡CCF$_3$ > MeC≡CPh > CO > CH$_2$=CH$_2$.

D. *Palladium(IV) and Platinum(IV) Complexes*

Poly(pyrazol-1-yl)borate complexes containing organoplatinum(IV) groups, first reported in 1972, were obtained as part of an early investigation of nitrogen donor dimethylplatinum(IV) chemistry (*67*) following the first report of complexes of this type (*68*). Reaction of AgPF$_6$ with

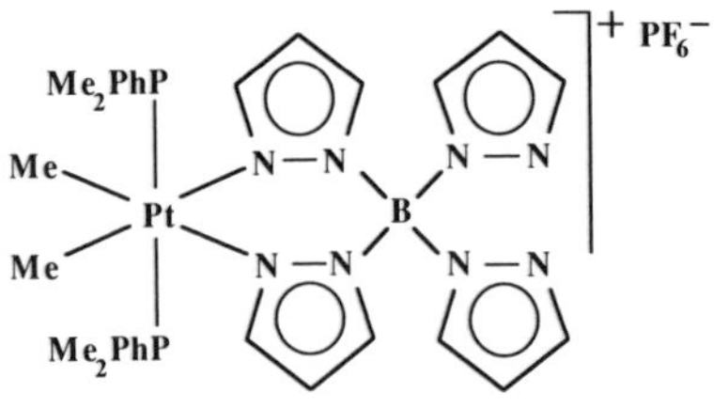

99

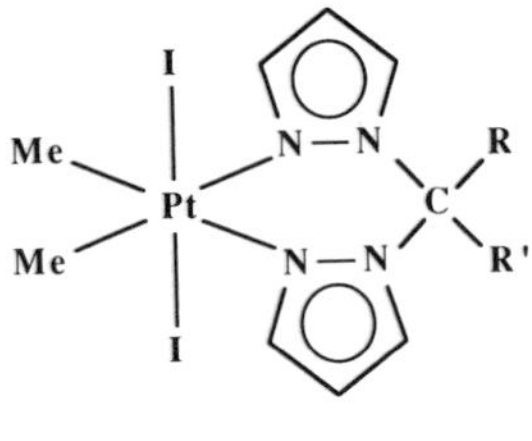

100 (R = R' = H)

101 (R = R' = Me)

102 (R = H, R' = pz)

103 (R = H, R' = thien-2-yl)

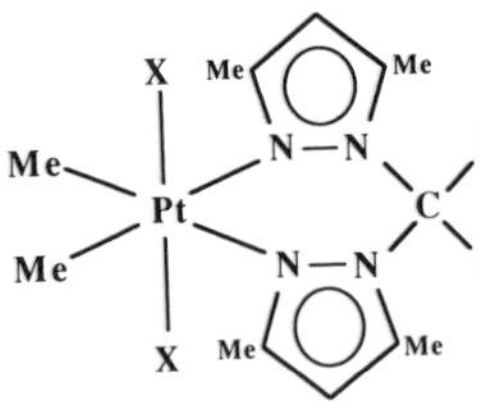

104 (X = I, NO$_3$)

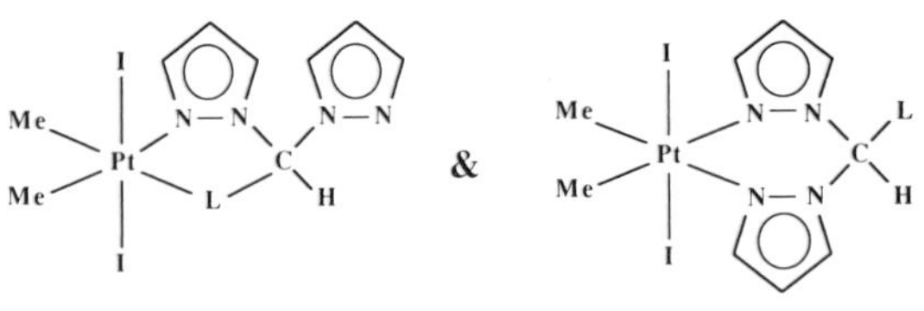

105 (L = mim; 2:1 ratio)

106 (L = py; 2:1 ratio)

$PtI_2Me_2(PMe_2Ph)_2$ in 2:1 mole ratio in acetone followed by removal of AgI and addition of $(pz)_4B^-$ gave the cation $[PtMe_2(PMe_2Ph)_2\{(pz)_4B\}]^+$ (**99**) as the hexafluorophosphate salt. The bis(pyrazol-1-yl)alkane complexes **100, 101**, and **104** (X = I) have been obtained by oxidative addition of iodine to $PtMe_2(COD)$ in the presence of the ligands [Eq. (16)] (*35*),

$$PtMe_2(COD) + L + I_2 \xrightarrow{CH_2Cl_2} PtI_2Me_2(L) \quad (16)$$

100, 101, 104 (X = I)

and this reaction system involves oxidation of platinum(II) prior to reaction with L. Structural studies of two complexes have been reported, and they are shown in Fig. 7 (A and B). Addition of silver nitrate to the iodo complex (**104**) gave the nitrato derivative, and in a study of the decomposition of a representative complex, $PtI_2Me_2\{(pz)_2CH_2\}$, ethane was formed at ~160°C (*35*).

An extension of the approach of Eq. (16) to include tridentate ligands, and using $[PtMe_2(SEt_2)]_2$ as the platinum(II) substrate, gave the neutral complexes $PtI_2Me_2\{(pz)_2(L)CH\}$ [**102, 103** (Fig. 7C), **105, 106**]. Three of these complexes (**102, 105**, and **106**) could be converted to cations (**107, 109**, and **110**) on heating in the solid state or in $CDCl_3$ (*69*). Further, complexes **105, 106, 109**, and **110** occur in solution as a mixture of two structural isomers, with **105, 106**, and **110** displaying the statistical 2:1

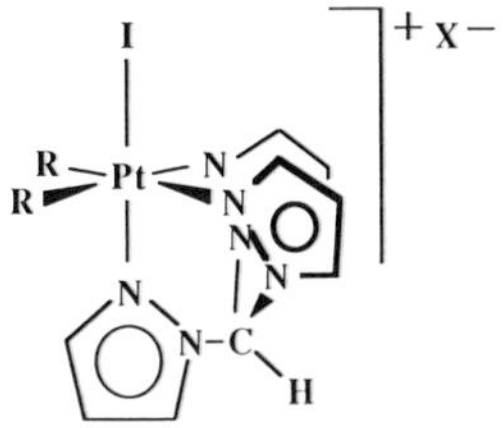

107 (R=Me, $X^- = I^-$)

108 (R=Ph, $X^- = 1/2(I^-.I_3^-)$)

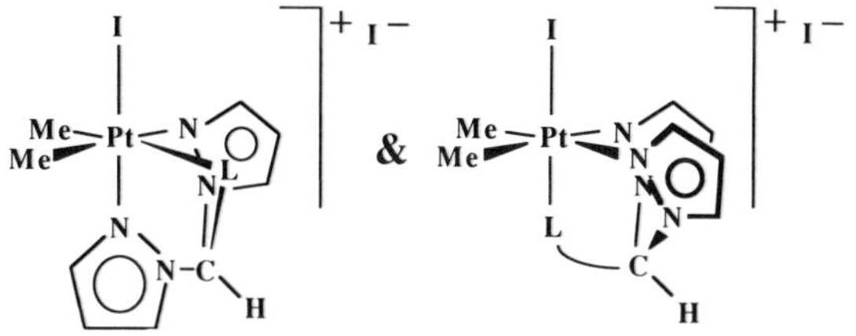

109 (L = mim; 1:1 ratio)

110 (L = py; 2:1 ratio)

ratio, whereas **109** exhibits a ratio of 1:1. The reaction of $[PtPh_2(SEt_2)]_2$ with $(pz)_3CH$ and I_2 at ambient temperature gives the cation $[PtIPh_2\{(pz)_3CH\}]^+$ (**108**) directly (*60*).

Trimethylplatinum(IV) also forms stable complexes with poly(pyrazol-1-yl)borate and -alkane ligands. The first reported complexes $PtMe_3\{(pz)_3BR\}$ (R = H, pz) (**111**) were obtained from the direct reaction of $[PtIMe_3]_4$ with $(pz)_3BR^-$ (*70*), and in a recent report $PtMe_3\{(3,5\text{-}Me_2pz)_3BH\}$ (**112**) was obtained in a similar way (*71*). The high tendency for

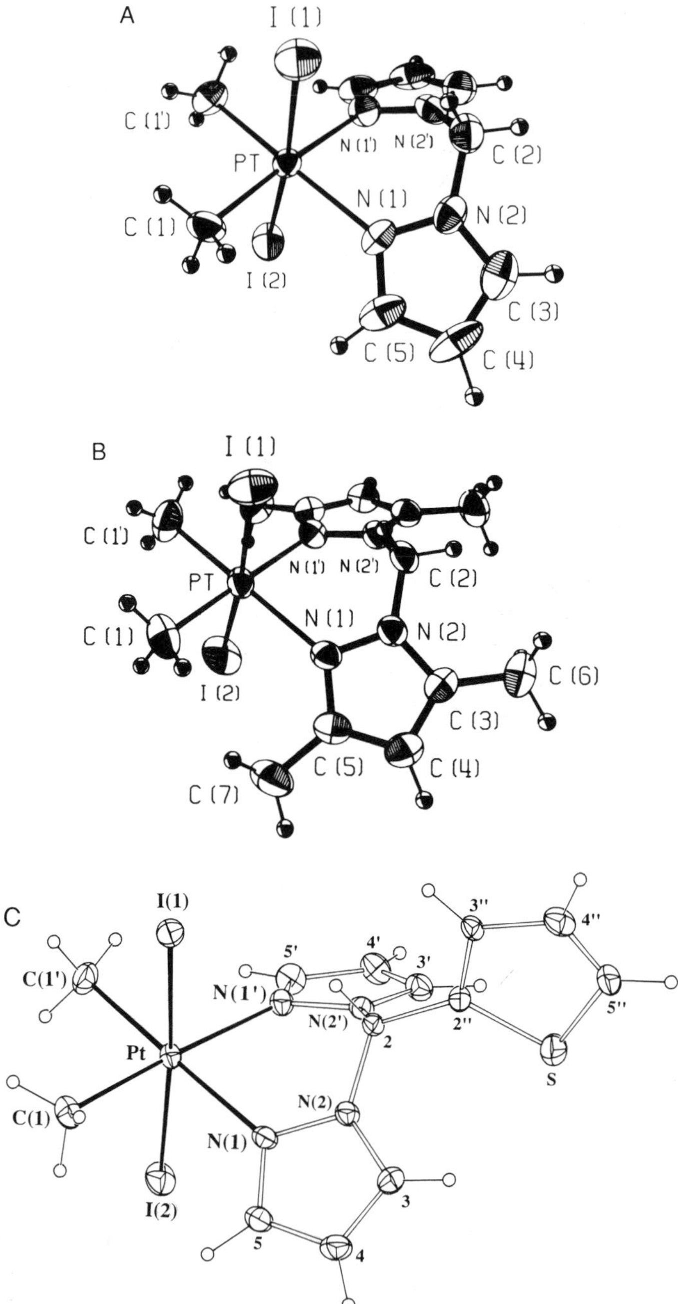

FIG. 7. Structures of (A) $PtI_2Me_2\{(pz)_2CH_2\}$ (**100**) and (B) $PtI_2Me_2\{(3,5\text{-}Me_2pz)_2CH_2\}$ (**104**) (with permission from Ref. 35; copyright American Chemical Society) and (C) $PtI_2Me_2\{(pz)_2(thi)CH)\}$ (**103**) (thi = thien-2-yl) (with permission from Ref. 69; copyright Elsevier Sequoia S.A.).

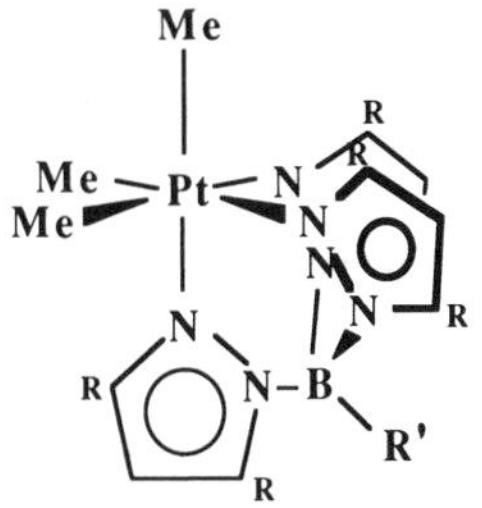

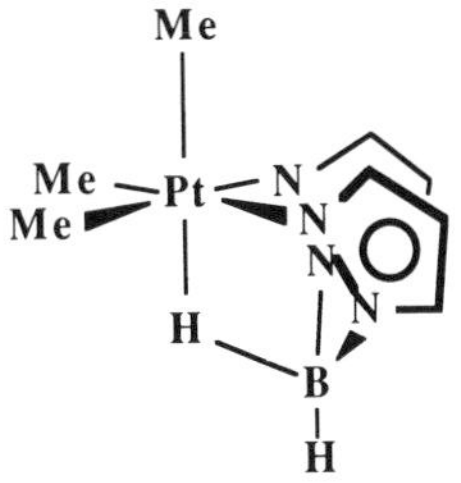

111 (R = H; R' = H, pz)

112 (R = Me; R' = H)

113

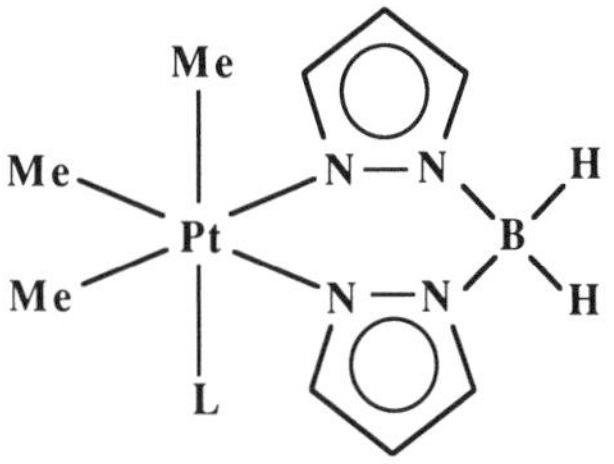

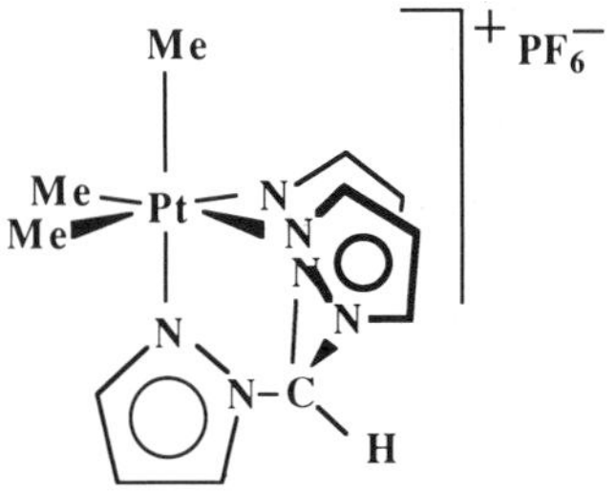

114 [L = $P(OMe)_3$]

115 (L = pzH, 3,5-Me_2pz)

116 (L = CO)

117

formation of the complexes was illustrated by displacement of the η^5-$C_5H_5^-$ group from $PtMe_3Cp$ [Eq. (17)]. In a surprising result from this study, it was found that $PtMe_3\{(pz)_4B\}$ is fluxional with exchange of all four pz groups (*70*). Reaction of $PtMe_3$ $\{(3,5\text{-}Me_2pz)_3BH\}$ (**112**) with bromine resulted in bromination of the pyrazole rings at the 4-position (*70*).

$$PtMe_3Cp + (pz)_3BH^- \rightarrow \underset{\textbf{111}\ (R = H)}{PtMe_3\{(pz)_3BH\}} + Cp^- \tag{17}$$

The complex $PtMe_3\{(pz)_2BH_2\}$ (**113**) was obtained on addition of $K[(pz)_2BH_2]$ to the solution resulting from reaction of $[PtIMe_3]_4$ with $AgPF_6$ (*70*). Complex **113** exhibits two PtMe resonances in its ^1H NMR spectrum, in the ratio 2 : 1; it is assigned the structure shown, with a strong infrared absorption at 2039 cm^{-1} ascribed to the "PtH_2B" moiety. The complex reacts with trimethyl phosphite to form **114**, and this complex could also be prepared from $[PtIMe_3]_4$ and $K[(pz)_2BH_2]$ in the presence of

$P(OMe)_3$. The latter procedure was used to prepare **115**, and a white solid formulated as $PtMe_3\{(pz)_2BH_2\}(CO)$ (**116**) (ν_{CO} 2125 cm^{-1}) was obtained with carbon monoxide but it readily loses CO to form **113**.

Poly(pyrazol-1-yl)alkane complexes of $Pt^{IV}Me_3$ have also been reported, with the initial work by Clark *et al.* (*72*) leading to the synthesis of complexes **117** and **122–125**, followed by Canty and co-workers (*47,56–60*)

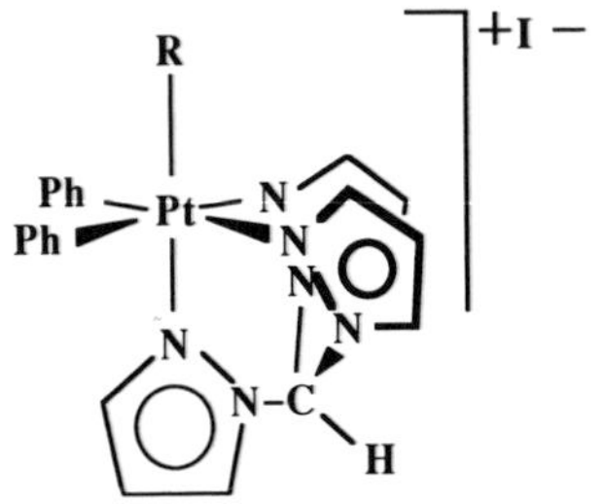

118 (R = Me)

119 (R = Et)

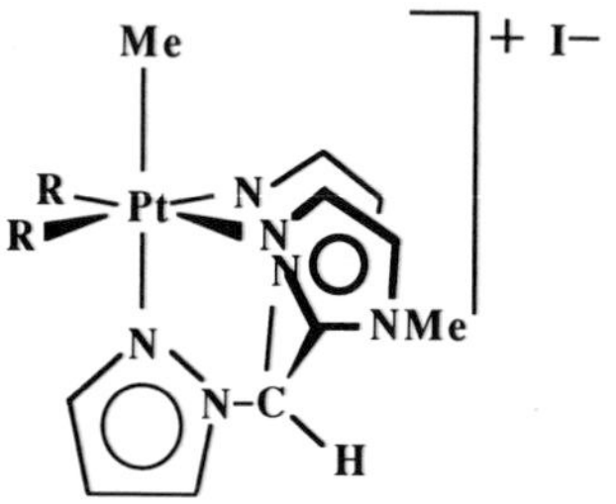

120 R = Me)

121 (R = Ph)

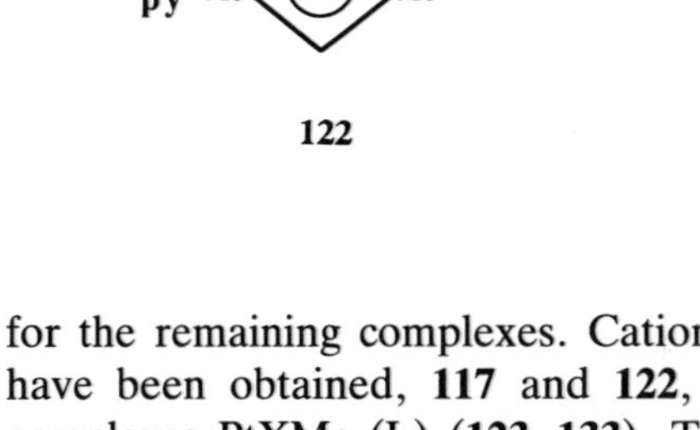

122

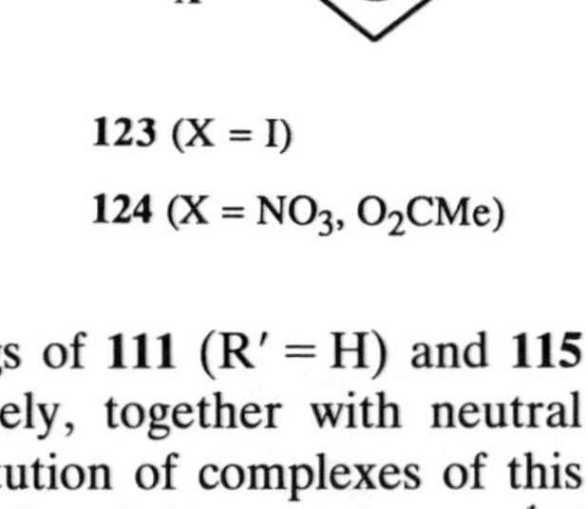

123 (X = I)

124 (X = NO_3, O_2CMe)

for the remaining complexes. Cationic analogs of **111** (R′ = H) and **115** have been obtained, **117** and **122**, respectively, together with neutral complexes $PtXMe_3(L)$ (**123–133**). The constitution of complexes of this type is readily determined by a combination of variable-temperature ^1H NMR spectroscopy and osmometric molecular weight and conductivity measurements. Crystal structures of **123** and **127** are shown in Fig. 8 (*73*).

The complex $[PtMe_3\{(pz)_3CH\}]PF_6$ (**117**) was prepared by addition of $(pz)_3CH$ to a solution containing $[PtMe_3(acetone)_3]^+$, generated from the reaction of $[PtIMe_3]_4$ with $AgPF_6$ in acetone (*34*). The chloro (**125**) and

125 (X = Cl, I; R = R' = H)

126 (X = I; R = R' = pz)

127 (X = I; R = Me, R' = H)

128 (X = I; R = Ph, R' = H)

129 (X = I, L = py)

130 (X = I, L = mim)

131 (X = O_2CMe, L = pz)

132 (X = O_2CMe, L = py)

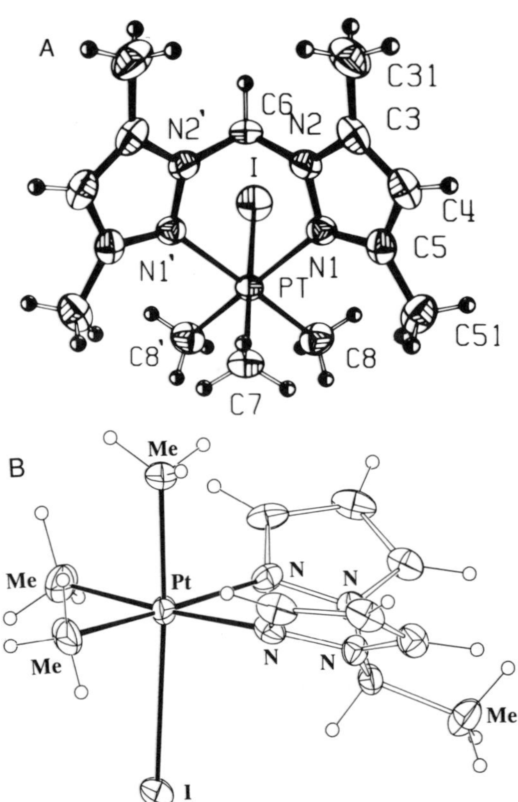

FIG. 8. Structures of (A) $PtIMe_3\{(3,5\text{-}Me_2pz)_2CH_2\}$ (**123**) (with permission from Ref. 72; copyright Elsevier Sequoia S.A.) and (B) $PtIMe_3\{(pz)_2CHMe\}$ (**127**) [with permission from the authors (*73*)].

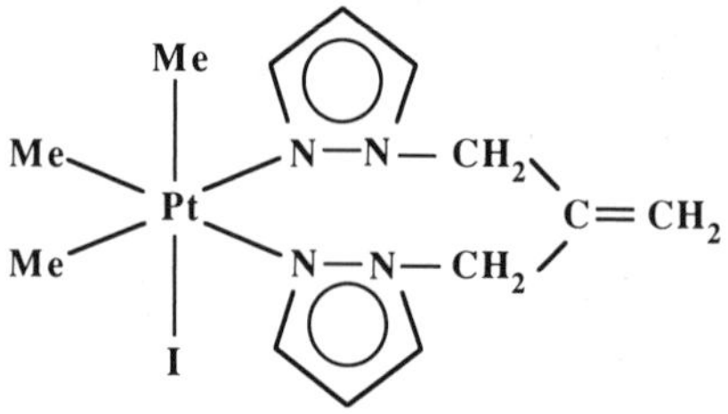

133

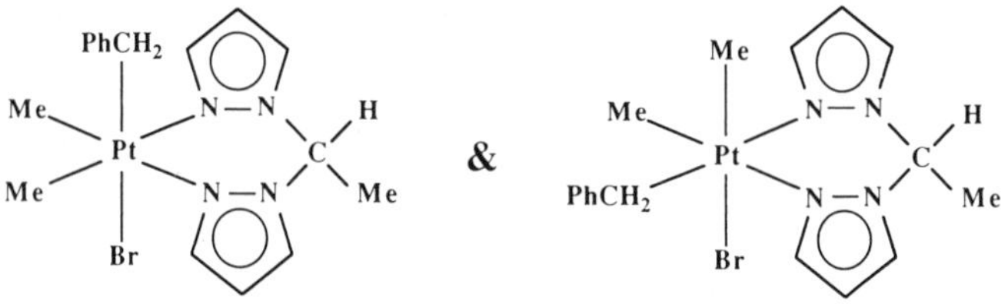

134

A **B**

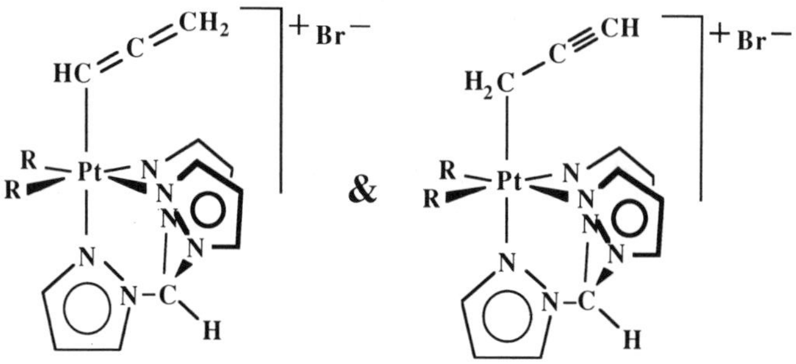

135 (R = Me; 2:1 ratio)

136 (R = Ph; 3:1 ratio)

iodo complexes (**123** and **125**) were obtained on reaction of $[PtXMe_3]_4$ with the ligands in benzene.

Most of the iodo complexes (**119, 125, 126, 127, 129, 130,** and **133**) were obtained by oxidative addition of organohalides to $PtR_2(L)$, except for $PtIMe_3\{(pz)_2CHPh\}$ (**128**) and **135**, which were obtained by an oxidative addition procedure similar to that of Eq. (16), and using $[PtMe_2(SEt_2)]_2$ as

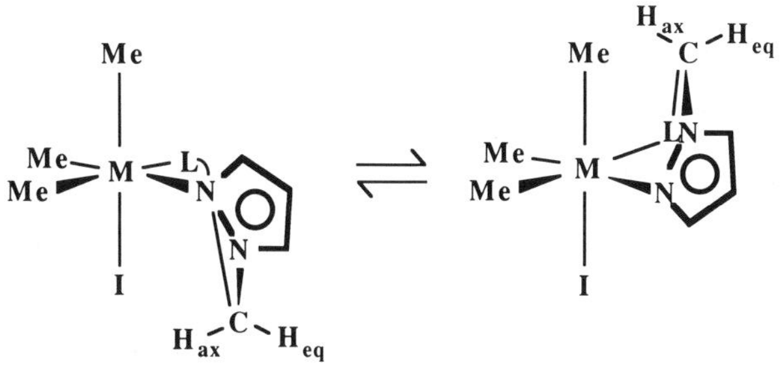

A B

125 (L = pz, M = Pt) **137** (L = pz, M = Pd)

129 (L = py, M = Pt) **138** (L = py, M = Pd)

130 (L = mim, M = Pt) **139** (L = mim, M = Pd)

the platinum(II) substrate. Complex **134** forms as a mixture of isomers in 1:1 ratio, with characteristic ^{1}H NMR resonances for the $PtCH_2Ph$ protons of the trans (**A**) and cis (**B**) isomers similar to those reported for the isomerization of the tetramethylethylenediamine complex *trans*-$PtBrMe_2$-(CH_2Ph)(tmeda) to *cis*-$PtBrMe_2(CH_2Ph)$(tmeda) (*74*).

Reaction of $PtIMe_3\{3,5\text{-}Me_2pz)_2CH_2\}$ (**123**) with $AgPF_6$ in acetone, followed by filtration and addition of pyridine, gave the cationic complex $[PtMe_3\{(3,5\text{-}Me_2pz)CH_2\}(py)]PF_6$ (**122**). The nitrato and acetato complexes (**124, 131,** and **132**) were obtained on reaction of the iodo complexes with an excess of the appropriate silver salt in chloroform, followed by filtration and evaporation of the filtrate.

The complexes formed by oxidative addition of $HC{\equiv}C{-}CH_2Br$ (**135** and **136**) occur as allenyl and propynyl complexes in 2:1 (**135**) and 3:1 (**136**) ratio in $CDCl_3$, with the isomers clearly identified by ^{1}H NMR spectroscopy (Fig. 9) (*60*). Similar isomerizations on oxidative addition of propynyl halides to $Pt(PPh_3)_4$ (*75, 76*) and $Pd(PPh_3)_4$ (*77*) to form metal-(II) complexes have been reported.

The Pt–I distance for $PtIMe_3\{(3,5\text{-}Me_2pz)_2CH_2\}$ (Fig. 8A), 2.843(1) Å, is one of the longest Pt^{IV}–I bond distances reported, and provides an excellent illustration of the strong trans influence of the methyl group. A similar Pd^{IV}–I distance, 2.834(1) Å, occurs in $PdIMe_3$(bpy) (*78, 79*). The Pt–N bonds trans to a methyl group in $PtIMe_3\{(3,5\text{-}Me_2pz)_2CH_2\}$, 2.364(4) Å, are exactly the same length as in $PtI_2Me_2\{(3,5\text{-}Me_2pz)_2CH_2\}$ and are longer than in the less crowded $PtI_2Me_2\{(pz)_2CH_2\}$ [2.183(3) Å); Fig. 5A], $PtI_2Me_2\{(pz)_2(thi)CH\}$ [2.181(5) and 2.192(6) Å; Fig. 5B], and $PtIMe_3$ $\{(pz)_2CHMe\}$ [2.15(1) and 2.16(1) Å; Fig. 8B]. The Pt–C distance

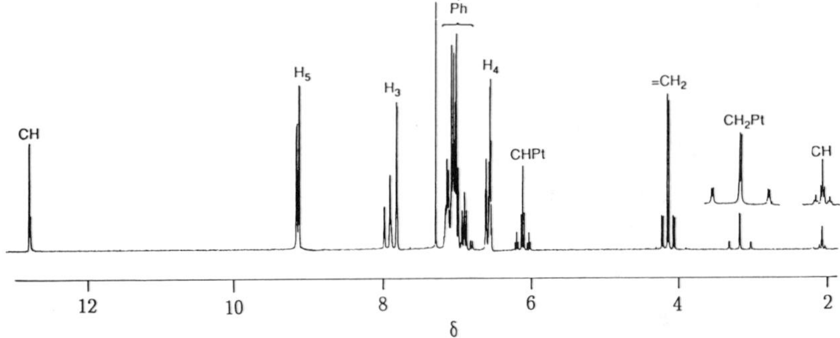

FIG. 9. ^{1}H NMR spectrum of the product of oxidative addition of HC≡C–CH_2Br to $PtPh_2\{(pz)_3CH\}$, illustrating the occurrence of allenylplatinum(IV) and alkynylplatinum(IV) cations (**136**) in ~3:1 ratio, $[PtPh_2(CH{=}C{=}CH_2)\{(pz)_3CH\}]^+$ [^2J(HPt) 49.5, ^4J(HPt) 47, ^4J(HH) 6.2 Hz] and $[PtPh_2(CH_2{-}C{\equiv}CH)\{(pz)_3CH\}]^+$ [^2J(HPt) 88.6, ^4J(HPt) 21.8, ^4J(HH) 2.8 Hz] (with permission from Ref. 60). Similar spectra are obtained for the dimethylplatinum(IV) complexes (**135**) (57).

trans to iodine in $PtIMe_3\{(3,5\text{-}Me_2pz)_2CH_2\}$ [2.077(6) Å; Fig. 8A] is significantly longer than the Pt–C distance trans to nitrogen [2.032(5) Å], but a similar effect does not occur in $PtIMe_3\{(pz)_2CHMe\}$ (Fig. 8B).

The development of organopalladium(IV) chemistry has been slow compared with organoplatinum(IV) chemistry. Several pentafluorophenylpalladium(IV) complexes were reported in 1975 (*80*), and the first hydrocarbylpalladium(IV) complex, $PdIMe_3$(bpy), was characterized in 1986 (*78*). The palladium(IV) complexes are usually obtained by oxidative addition of organohalides to palladium(II) substrates, and the most stable complexes isolated to date involve planar bpy or phen in $PdBrMe_2(CH_2Ar)(L)$ (*81*), tripodal $(mim)_2(py)CH$ in $[PdMe_2R\{(mim)_2(py)CH)]X$ (RX = MeI, EtI, Pr^nI, $PhCH_2Br$, $CH_2{=}CHCH_2I$, $CH_2{=}CHCH_2Br$) (*12,82*), or tripodal $(pz)_3CH$ in $[PdMe_3\{(pz)_3CH\}]BF_4$ (*79*).

Organoplatinum(IV) complexes of nitrogen donor ligands are very stable, and concurrent studies of palladium and platinum chemistry have been valuable in investigating the chemistry of unstable organopalladium(IV) compounds (*50,56,74,83*). Oxidative addition of iodomethane to representative $Pd^{II}Me_2$ complexes of bidentate poly(pyrazol-1-yl)alkanes has been studied by ^{1}H NMR spectroscopy at temperatures <−30°C, and, because the palladium(IV) complexes formed, $PdIMe_3(L)$ (**137–139**), were unstable toward reductive elimination of ethane, they could not be isolated [Eq. (18)] (*50,56*).

$$PdMe_2(L) + MeI \rightarrow PdIMe_3(L) \rightarrow PdIMe(L) + C_2H_6 \quad (18)$$

71, 77, 78 **137–139** **59, 64, 65**

Analogous platinum(IV) complexes have been isolated as stable solids [**125** (X = I), **129,** and **130**], and for these complexes spectra may be obtained at ambient temperature, except for insoluble $PtIMe_3\{(pz)_2CH_2\}$. Variable-temperature spectra of the palladium(IV) and platinum(IV) analogs in $(CD_3)_2CO$ are almost identical, except for $^1H-^{195}Pt$ coupling, and for the (pz)(py)CH_2 complexes the spectra may be readily interpreted in terms of conformational inversion of the chelate ring. Spectra are well resolved at −70°C, with conformer **A** favored over **B** for both palladium **138** (~7:1) and platinum **129** (~5:1). The complexes $PdIMe_3\{(pz)_2CH_2\}$ and $MIMe_3\{(pz)(mim)CH_2\}$, however, undergo rapid inversion even at low temperature (*56*).

Variable-temperature ^1H NMR studies have revealed isomerism for other platinum(IV) complexes, e.g., spectra of $PtIMe_3\{(pz)_2CHPh\}$ (**128**) at low temperature show the presence of two isomers in 3:1 ratio, with a marked upfield shift of 0.9 ppm for the axial PtMe resonance of the minor isomer compared with the major isomer suggesting that the minor isomer has the phenyl group near the axial methyl group (*60*). Similar nonrigidity and/or isomerization may well occur for the other related platinum(IV) complexes, e.g., **100, 114–116, 125** (X = Cl), **131,** and **132**, although this aspect of their chemistry does not appear to have been fully explored.

Tris(pyrazol-1-yl)methane and the closely related ligands $(pz)_2$(mim)CH and $(pz)_2$(py)CH have played a key role in the initial development of organopalladium(IV) chemistry. The complexes $[PdMe_3(L)]I$ (**140** and **141**) are among the first isolated cations for palladium(IV), and were readily obtained on addition of L to a suspension of $[PdMe_2(pyridazine)]_n$ in acetone followed by iodomethane at 0°C, or on direct oxidative addition of iodomethane to $PdMe_2(L)$ (*47, 79*). The platinum(IV) complex $[PtMe_3\{(pz)_3CH\}]I$ cannot be synthesized by oxidative addition, owing to the occurrence of a cyclometallation reaction (Section III), but may be readily obtained by the procedure of Eq. (16) using $[PtMe_2(SEt_2)]_2$ as the platinum(II) substrate (*79*).

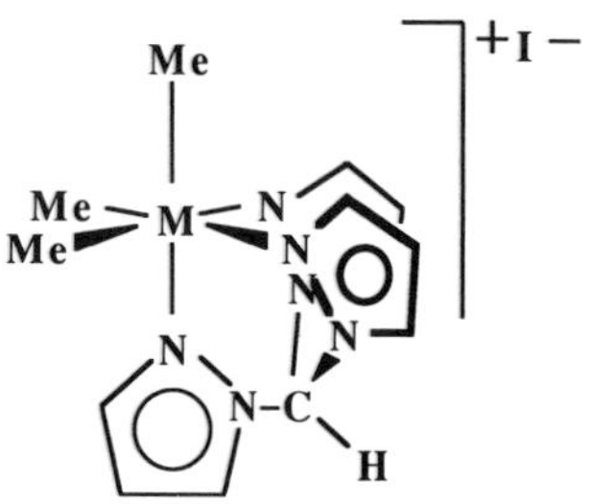

140 (M = Pd, Pt)

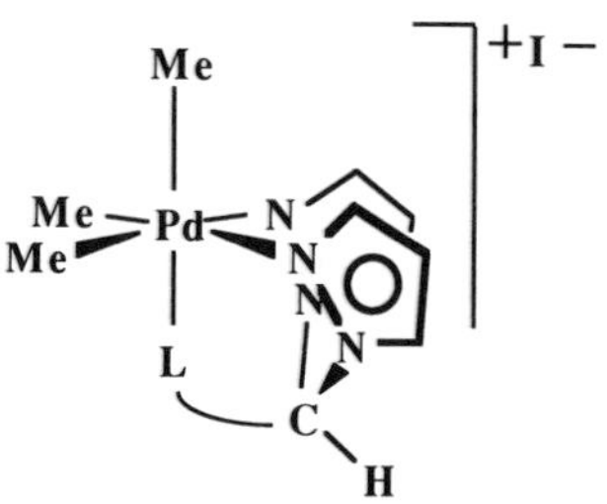

141 (L = mim, py)

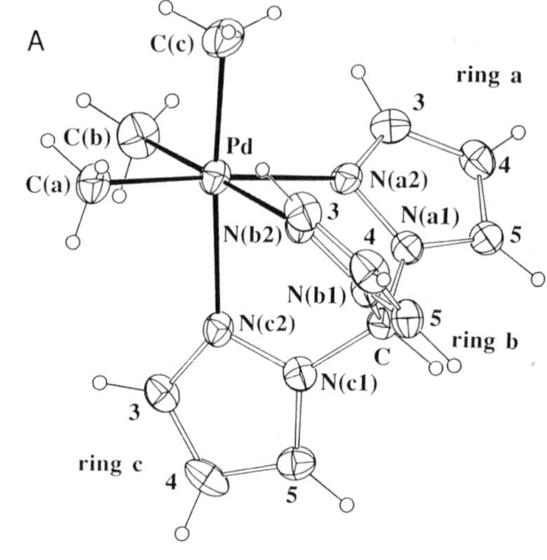

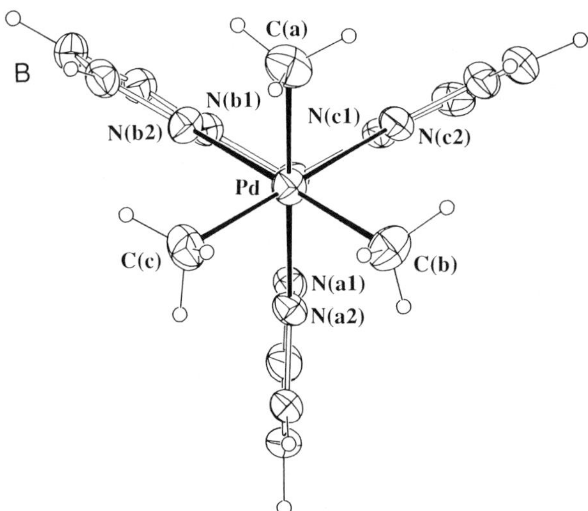

FIG. 10. Structure of $[MMe_3\{(pz)_3CH\}]^+$ (M = Pd, Pt) in the iodide salts (**140**). Diagram for M = Pd. Projection B emphasizes the noncrystallographic C_{3v} symmetry (with permission from Ref. 79; copyright American Chemical Society).

Crystals of the complexes **140** are isostructural (Fig. 10), allowing a direct comparison of geometries when the effects of crystal packing are constant (*79*). However, the absence of crystallographic C_{3v} symmetry for the cations implies that packing effects at individual Me groups and pz rings may be different, perhaps partly accounting for the minor but significant variation in M–C and M–N distances in the complexes. The M^{IV}–C bond lengths are essentially identical, ~2.04–2.05 Å, but the Pd^{IV}–N distance is ~0.04 Å longer than that of Pt^{IV}–N. A similar difference for M^0–P and M^{II}–P bond lengths has been reported for the complexes $M(PBu^t_2Ph)_2$ (*84*) and *cis*-$MMe_2(PMePh_2)_2$ (*85*), respectively. Variations in M–C and M–N distances in the $(pz)_3CH$ complexes suggest that a detailed comparison of bond lengths for Pd(IV) and Pt(IV), as given for M(0) and (MII) (*86*), is not warranted. However, the structural determinations of the M(IV) complexes complete a remarkable series—isostructural Pd/Pt pairs of complexes for three oxidation states—and in the complexes the Pd–P or Pd–N distances are ~0.03–0.04 Å longer than those of Pt–P or Pt–N.

^1H NMR spectra of the complexes $[PdMe_3\{(pz)_3CH\}]I$ and $[PdMe_3\{(pz)_2(L)CH\}]I$ (L = mim, py) in $CDCl_3$ indicate that the ligand bridgehead methine proton exchanges with the deuterium of $CDCl_3$ over several hours, and the exchange may be reversed on addition of $CHCl_3$ to the solid obtained on removal of $CDCl_3$–$CHCl_3$. In another NMR effect, the complex $[PdMe_3\{(pz)_3CH\}]BF_4$ (**142**), obtained from the iodide and $AgBF_4$ (*79*), has CH and H-5 resonances ~3.1 and ~0.45 ppm more shielded than in the iodide analog, suggesting that cation···anion interactions occur predominantly in this region of the cations.

Addition of benzyl bromide to $PdMe_2\{(pz)_3CH\}$ at 0°C in acetone gave the isolable complex $[PdMe_2(CH_2Ph)\{(pz)_3CH\}]Br$ (**143**); an upfield shift of ~0.7 ppm is observed for the H-3 protons of the rings trans to PdMe,

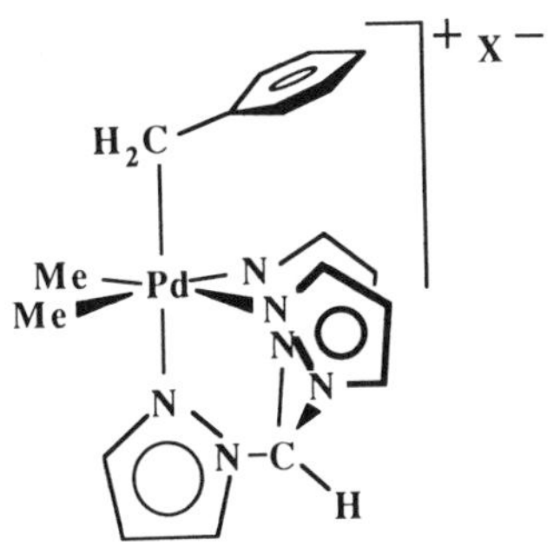

143 (X = Br, BF_4)

when compared with H-3 for the ring trans to $PdCH_2Ph$, consistent with orientation of the adjacent benzyl ring near these protons (*12*). A similar orientation away from the $Pd^{IV}Me_2$ group is shown in the X-ray structure of $PdBrMe_2(CH_2\text{-}p\text{-}C_6H_4Br)(phen)$ (*81*). The spectrum of the BF_4^- salt is very similar to that of $[PdMe_2(CH_2Ph)\{(pz)_3CH\}]Br$, except that the CH and H-5 resonances show the same feature exhibited by $[PdMMe_3\text{-}\{(pz)_3CH\}]X$ ($X = I, BF_4$) (see above), interpreted in terms of a cation $\cdots$ anion interaction near the methine proton.

The complex $[PdMe_3\{(pz)_3CH\}]BF_4$ is far more stable than the iodide toward reductive elimination of ethane, and a detailed kinetic study of reductive elimination from $PdIMe_3(bpy)$ (*87*) indicates why this might be so and provides an insight into the role of $(pz)_3CH$ and $(pz)_3BH^-$ in the stabilization of high oxidation states. Kinetic studies indicate that the reductive elimination of ethane from $PdIMe_3(bpy)$ is concerted, occurring mainly after ionization of the iodo group, but with a minor proportion of the complex undergoing elimination directly from $PdIMe_3(bpy)$. Kinetic parameters obtained under conditions of a large excess of iodide indicate a polar intermediate or transition state for the latter path, probably involving at least partial ionization of iodide, $[Me_3(bpy)Pd^{\delta+} \cdots I^{\delta-}]$, before reductive elimination of ethane. The higher stability of the BF_4^- salt indicates that reductive elimination may proceed by a similar path for the iodide salt, with formation of an intermediate with bidentate $(pz)_3CH$, $[Me_3\{(pz)_3CH\text{-}N,N'\}Pd^{\delta+} \cdots I^{\delta-}]$. A "model" for this intermediate is the stable platinum(IV) complex $PtIMe_3\{(pz)_4C\text{-}N,N'\}$ (**126**). These results indicate that the enhanced stability of $[PdMe_3(tripod)]I$ compared to that of $PdIMe_3$-(bidentate) at least partly results from the requirement for dissociation of one donor group of the tripod ligand, because the complexes of bidentate ligands may reductively eliminate directly via either ionization or partial ionization of the iodo group.

Addition of iodoethane to $PdMe_2\{(pz)_3CH\}$ gave complex spectra, including resonances attributable to $[PdEtMe_2\{(pz)_3CH\}]I$, and decomposition to palladium metal also occurred (*12*). Oxidative addition of 2-propenyl bromide gave spectra readily assigned as $[PdMe_2(\eta^1\text{-}C_3H_5)\text{-}\{(pz)_3CH\}]Br$ (**19**), prior to clean reductive elimination of ethane to form $[Pd(\eta^3\text{-}C_3H_5)\{(pz)_3CH\}][PdBr_2(\eta^3\text{-}C_3H_5)]$ (**17**) [Eq. (8)], but an analogous reaction with 2-propenyl iodide gave a different pathway for reductive elimination [Eq. (19)].

$$PdMe_2\{(pz)_3CH\} + C_3H_5I \rightarrow [Pd(\eta^1\text{-}C_3H_5)Me_2\{(pz)_3CH\}]I \rightarrow \tfrac{1}{2}[PdI(\eta^3\text{-}C_3H_5)]_2 + (pz)_3CH + Me\text{-}Me \quad (19)$$

III

INTRAMOLECULAR COORDINATION SYSTEMS INVOLVING POLY(PYRAZOL-1-YL)ALKANES AND RELATED LIGANDS

A. *Palladium(II) and Platinum(II)*

The stabilization of metal–carbon σ bonds by incorporation of a donor (D) atom into the organic group, M–C~D, is a key feature in the development of organometallic chemistry, and for palladium and platinum the synthesis of complexes of this type via cyclometallation reactions has been an active area of investigation since the early studies of Cope and Siekman (*88*). Thus, reagents containing rings connected directly to a donor ring, or via a bridging group, e.g., **144** (*89,90*) and **145** (*91,92*), respectively, may be palladated at the ortho position of the aryl ring to form **146** [Eq. (20)].

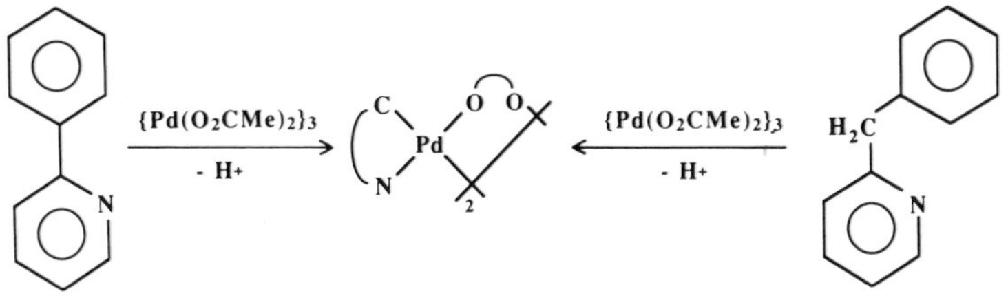

$$\{Pd(O_2CMe)_2\}_3,\ -H^+ \qquad\qquad \{Pd(O_2CMe)_2\}_3,\ -H^+ \qquad (20)$$

144 **146** **145** (R = H, Ph)

Several studies of the cyclometallation of pyrazole and closely related reagents, similar to that of 2-phenylpyridine (**144**) and 2-benzylpyridine (**145**), have been reported (*93–98*) (Fig. 11), commencing with Trofimenko's account of the reaction of *N*-phenylpyrazole with Na_2PdCl_4 to form **147** (*93*). Complex **148** was formed by cyclopalladation of *N*-benzylpyrazole by palladium(II) acetate followed by acetate–chloride exchange (*98*). Cyclometallation of *N*-(3-thienyl)pyrazole occurs to form the complex **149**, with the isomers **A** and **B** in 3:1 ratio (*96*). Complexes of this type may be characterized by bridge splitting reactions to form, for example, acetylacetonato complexes, Pd(C~N)(acac) (*93,95,98,99*; Fig. 11), $[Pd(C{\sim}N)(py)_2]PF_6$(*93*), $[Pd(C{\sim}N)(NH_2CH_2CH_2NH_2)]Cl$ (*95*), and Pd(C~N)(L)Cl [L = py (*95*), PR_3 (*95,96*)]. Poly(pyrazol-1-yl)alkanes and -borates may also be used to cleave chloro-bridged complexes, as discussed earlier [Eqs. (9) and (10), complexes **22–33**]. Reaction of LiC_6H_4-CH_2NMe_2 with **147** gave the complex **150** (*93*), and similar platinum(II)

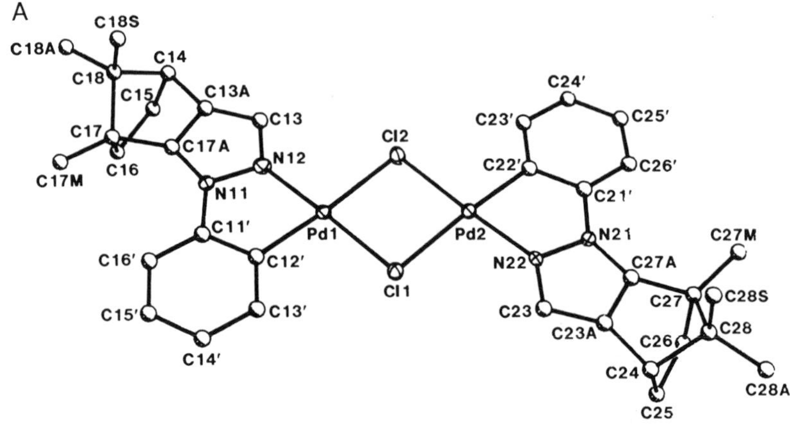

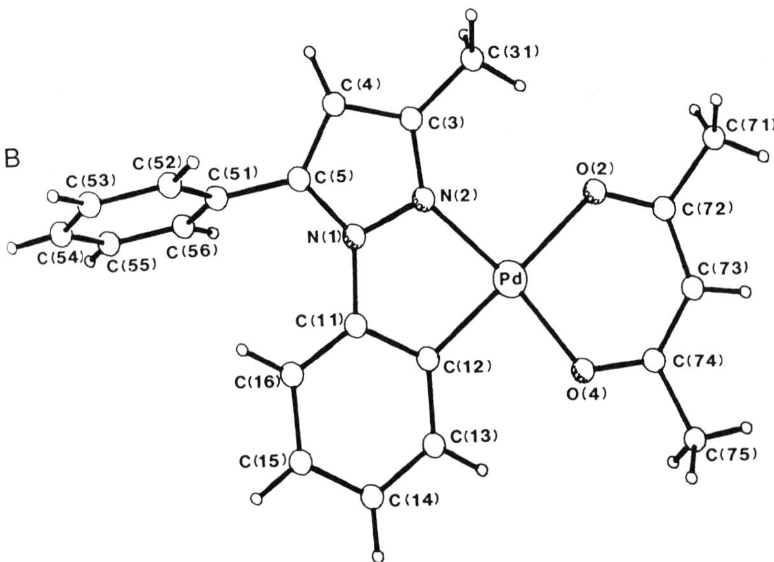

FIG. 11. Structure of (A) the product of palladation of (4*S*, 7R)-7,8,8-trimethyl-1-phenyl-4,5,6,7-tetrahydro-4,7-methano-1*H*-indazole (with permission from Ref. 97; copyright Elsevier Sequoia S.A.) and (B) acetylacetonato[2-(3-methyl-5-phenylpyrazol-1-yl)-phenyl-*C*,*N*]palladium(II), an example of Pd(C~N)(acac) (with permission from Ref. 98; copyright Elsevier Sequoia S.A.).

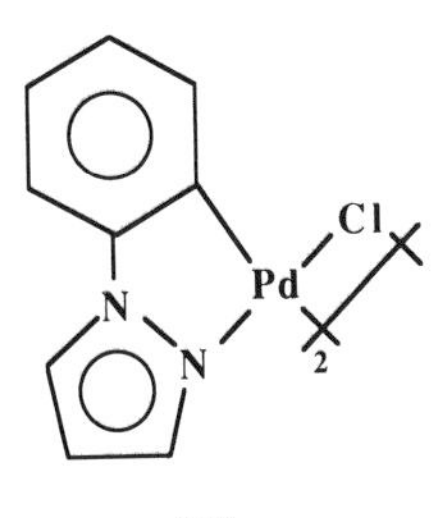

147

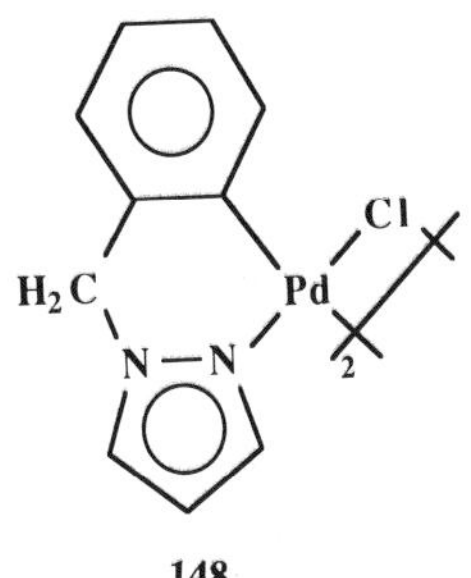

148

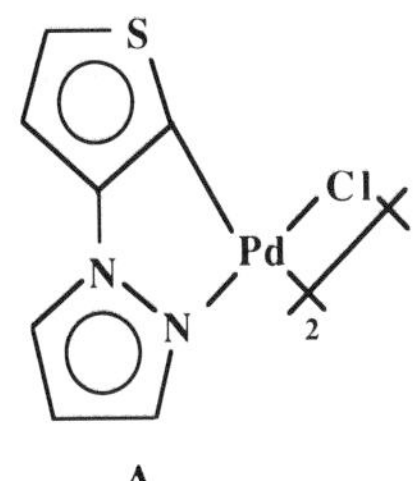

A

&

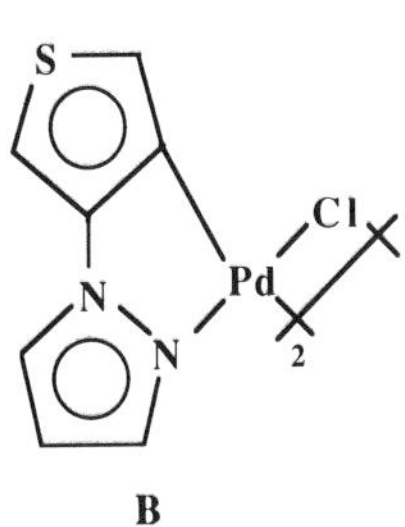

B

149

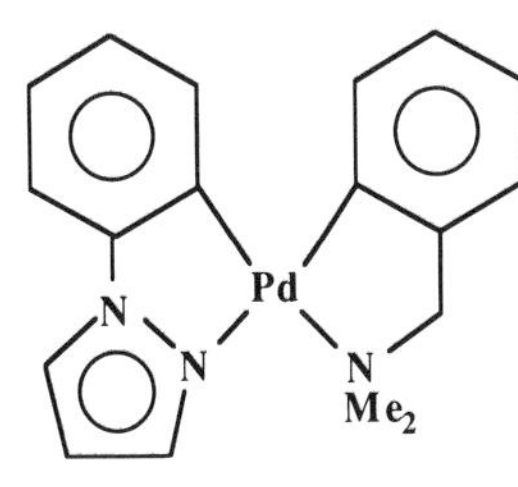

150

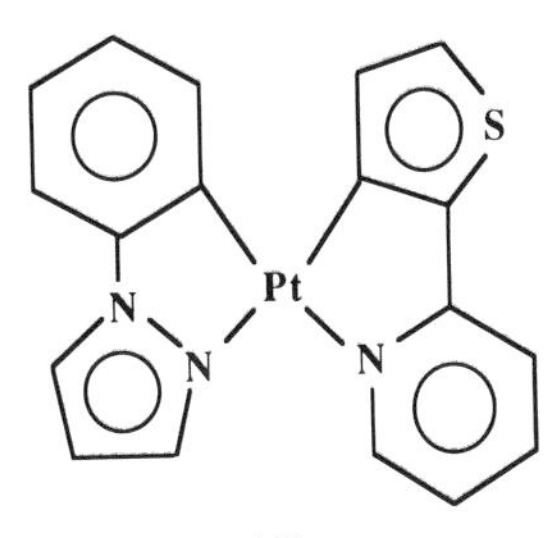

151

152 (M = Pt, X = Cl; R = R' = H, Me;

R = H, R' = Me; R = Me, R' = H)

153 (M = Pt, X = Br; R = R' = Me)

154 (M = Pd, X = Cl; R = H, R' = Me;

R = Me, R' = H)

complexes have been constructed in which both metal–carbon bonds are obtained by organolithium routes (*100,101*), e.g., **151**. An additional type of intramolecular coordination is illustrated by **152–154**, involving alkene and pyrazole coordination (*102*).

Ligand systems of the type **146–151** are referred to here as $[C{\sim}N]^-$ ligands; planar tridentate systems, $[N{\sim}C{\sim}N]^-$ and $[C{\sim}N{\sim}N]^-$, and tetradentate systems, $[C{\sim}N{\sim}N{\sim}C]^{2-}$, have also been obtained, and the complexes **155–159** fall within the scope of this review. The complexes

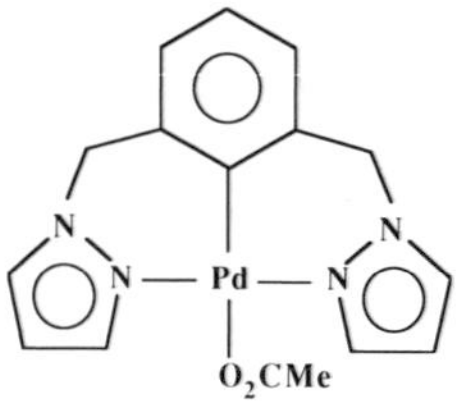

155

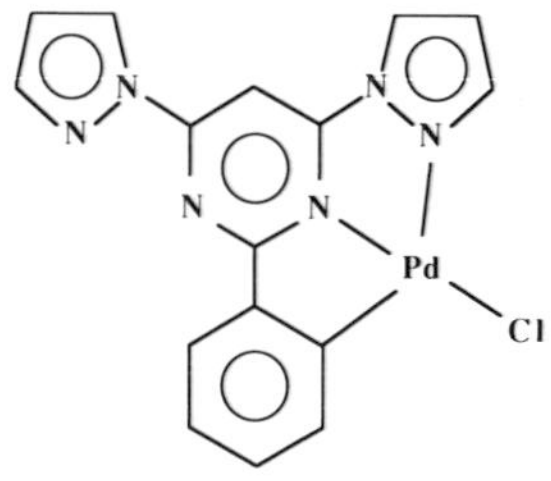

156

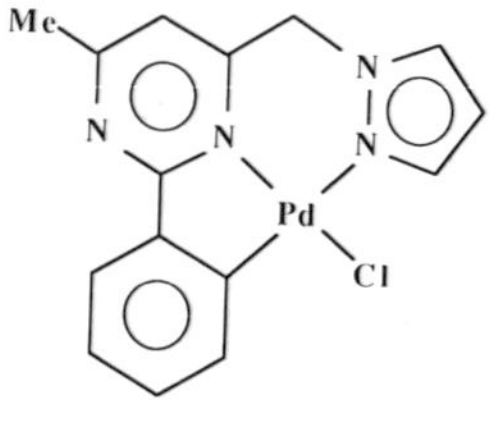

157

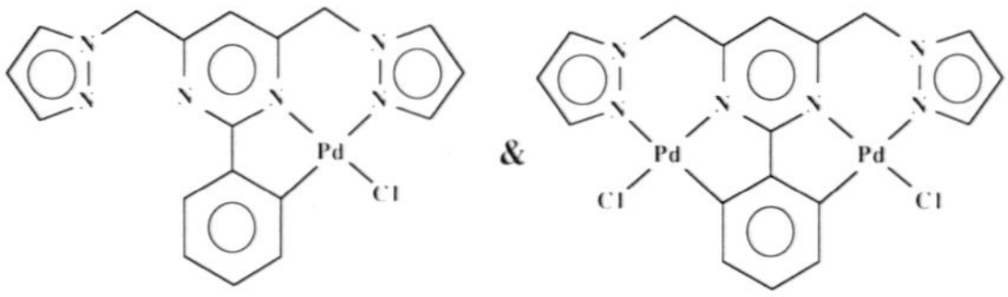

A & B

158

were formed by cyclometallation reactions of the appropriate LH or LH_2 reagent using $\{Pd(O_2CMe)_2\}_3$ (**155**) (*103*), or Li_2PdCl_4 [**156** and **157** and the mixture of **A** and **B** in **158** (*104*)], or by organolithium routes for **159** as indicated in Eq. (21) (*105*).

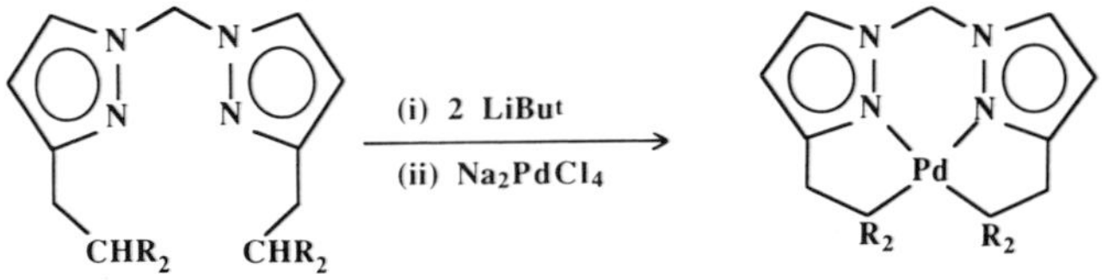

$$\xrightarrow[\text{(ii) } Na_2PdCl_4]{\text{(i) 2 LiBu}^t} \qquad (21)$$

159

When the aryl ring to be metallated is replaced by a donor ring, e.g., 2,2′-bipyridyl rather than 2-phenylpyridine, donor atom chelation is usually observed, although an important class of organometallic complexes involving cyclometallation of bpy and related ligands is emerging following the discovery of iridium(III) complexes such as [Ir(bpy-*N,N'*)$_2${$C_{10}H_7N(NH \cdots OH_2)$-*N,C*}$^{3+}$ (*106*). The synthesis of a related platinum(II) complex **160**, involving dimetallation, is shown in Eq. (22) (*107*). Extension of this approach

hot toluene/4-Bu^{t}pyridine; - PhH, - 1/2 bpy (22)

160

to ligands containing only donor rings connected by a bridging atom(s) has been little explored, although the poly(pyridin-2-yl)alkanes $(py)_2CH_2$ and $(py)_3CH$ are reported to not undergo cyclopalladation under the same conditions that **145** forms **146** (*92*). However, the complex $PtMe_2${$(pz)_3CH$}, containing the polydentate ligand $(pz)_3CH$, undergoes an unexpected cyclometallation in pyridine to form **161** and methane [Eq. (23)] (*58*).

solvent (S); - MeH (23)

66

161 (S = pyridine)

162 (S = 4-methylpyridine)

163 (S = 3,5-dimethylpyridine)

164 (S = *N*-methylimidazole)

Complex **161** (Fig. 12A) appears to be the first example of cyclometallation occurring for a donor ring that is connected to another donor ring via a bridging atom rather than directly as in bpy and related ligands (*57–59*). In view of this, and the "isoelectronic" relationship between $(pz)_3CH$, $(pz)_3BH^-$, and $(pz)_2(C_3H_2N_2)CH^-$, the chemistry of cyclometallation of $(pz)_3CH$ and related poly(pyrazol-1-yl)alkanes has been explored in detail, resulting in isolation of complexes such as [PtMe{$(pz)_2(C_3H_2N_2)CH$}]$_n$

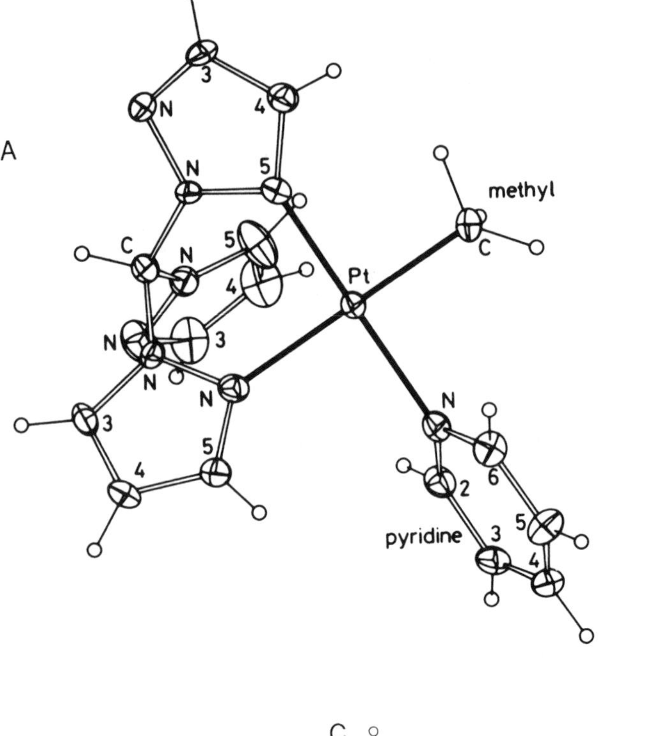

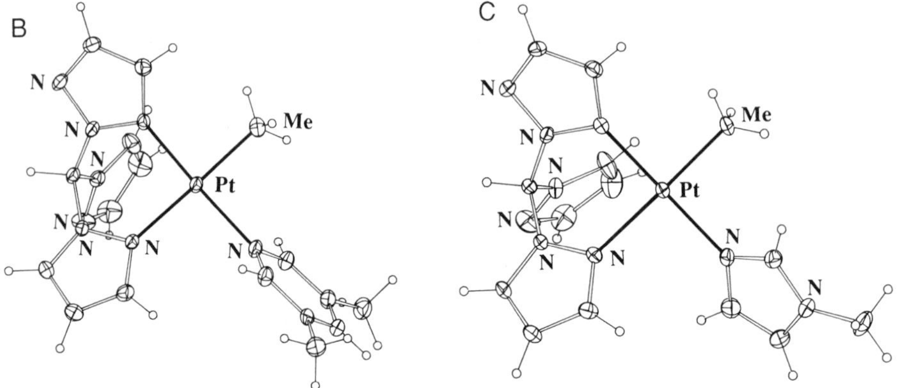

FIG. 12. Structures of (A) PtMe{$(pz)_2(C_3H_2N_2)CH$}(py) (**161**) (with permission from Ref. *59*; copyright Royal Society of Chemistry), (B) PtMe{$(pz)_2(C_3H_2N_2)CH$}(3,5-Me_2py) (**163**), and (C) PtMe{$(pz)_2(C_3H_2N_2)CH$}(*N*-methylimidazole) (**164**) [with permission from the authors (*108*)].

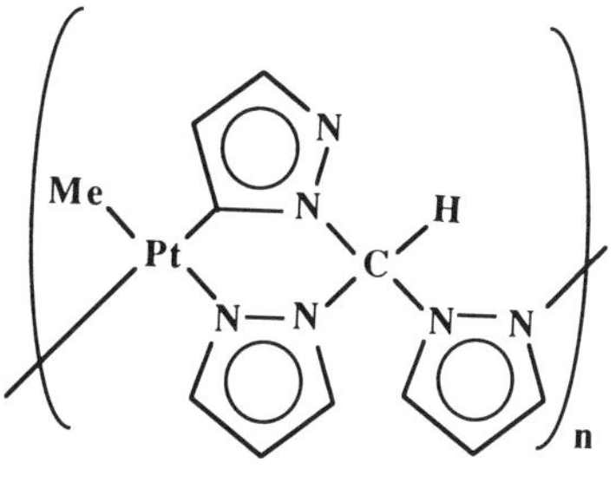

165

(**165**) (*57*), isoelectronic with $[PtMe\{(pz)_3BH\}]_n$ (**45**) and $[PtMe\{(pz)_3CH\}]_n(PF_6)_n$ (**47**).

Cyclometallation of a range of $Pt^{II}R_2(L)$ (R = Me, Ph) complexes (**71–80** and **82–89**) has been attempted in pyridine, and the reaction was found to occur at ambient temperature or on gentle warming for $PtMe_2(L)$-[L = $(pz)_3CH$, $(pz)_2(mim)CH$, $(pz)_2CH_2$, and $(pz)_2CHPh$] only (**161** and **166–168**) (*57*). The complexes of platinated $(pz)_2(mim)CH$, $(pz)_2CH_2$, and $(pz)_2CHPh$ formed as oils and were isolated as PPh_3 derivatives (**166–168**).

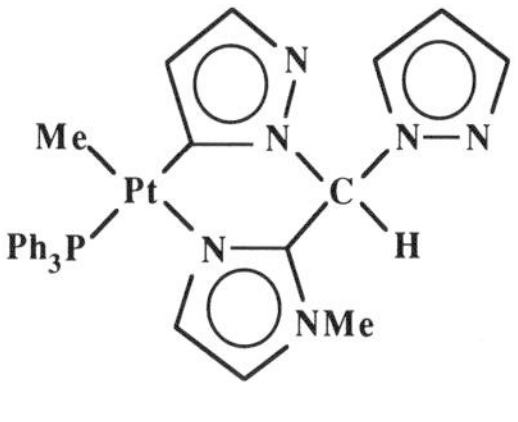

166

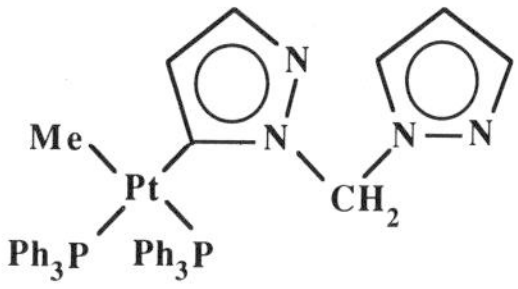

167

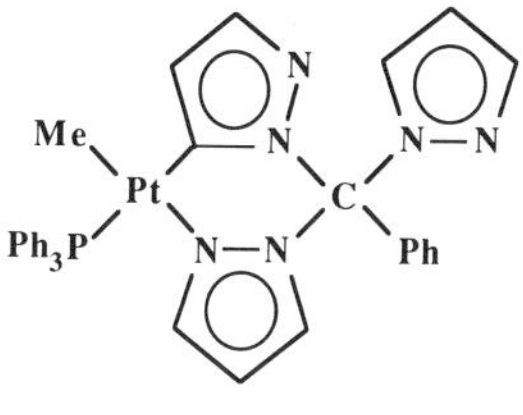

168

The remaining complexes dissolved in pyridine to form $PtR_2(py)_2$. The occurrence of cyclometallation for two complexes containing both pz groups initially N-coordinated to platinum, $PtMe_2\{(pz)_2CHR\}$ (R = H, Ph), indicates that, at least for these complexes, the reaction occurs via an intermediate with the ligand as a unidentate N-donor. A donor solvent is also necessary, e.g., $PtMe_2\{(pz)_3CH\}$ undergoes cyclometallation in pyridine (**161**), 4-methylpyridine (**162**), 3,5-dimethylpyridine (**163**) (Fig. 12B) (*108*) or *N*-methylimidazole (**164**) (Fig. 12C) at ambient temperature, but it does not undergo metallation in refluxing xylene. These reactivity patterns suggest that an oxidative addition (C–H) and reductive elimination (Me–H) mechanism occurs, with coordination of the stronger donor pyridine in a square planar intermediate **169** enhancing the nucleophilic character of platinum(II). A similar cyclometallation of a phenyl ring in **170** to form **172** has been recently reported, and this reaction is also considered to

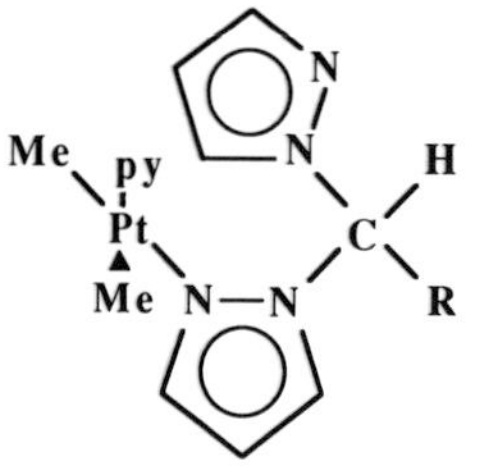

169 (R = H, pz, Ph; L = pz, mim)

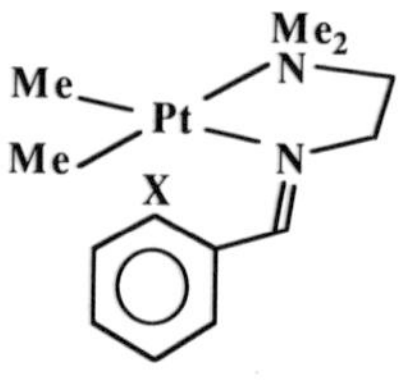

170 (X = H)

171 (X = Br)

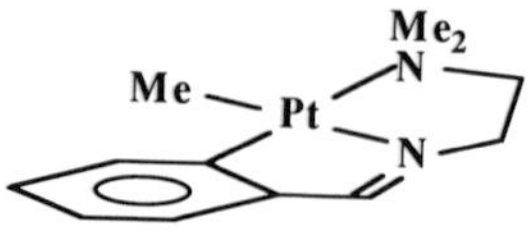

172

involve oxidative addition of an aryl–H bond followed by rapid reductive elimination of methane (*109*). The palladium complex $PdMe_2\{(pz)_3CH\}$ does not undergo cyclometallation in pyridine, forming instead $PdMe_2(py)_2$, consistent with the lower reactivity of palladium(II) toward oxidative addition (*83*) and the general occurrence of electrophilic mechanisms for cyclopalladation (*110*). The $Pt^{II}Ph_2$ complexes form $PtPh_2(py)_2$ rather than undergo cyclometallation (*60*), consistent with an expected lower reactivity of $Pt^{II}Ph_2$ complexes toward C–H oxidative addition, because $PtPh_2(bpy)$ reacts more slowly than $PtMe_2(bpy)$ with iodomethane (*111*).

The cyclometallated complexes react with a range of phosphines in acetone to form **166–168** and **173–175** and, in an early synthesis of PtMe $\{(pz)_2(C_3H_2N_2)CH\}$ $(PPh_3)_2$ (**174**) carried out in pyridine, a single crystal taken from the product and examined by X-ray crystallography was found

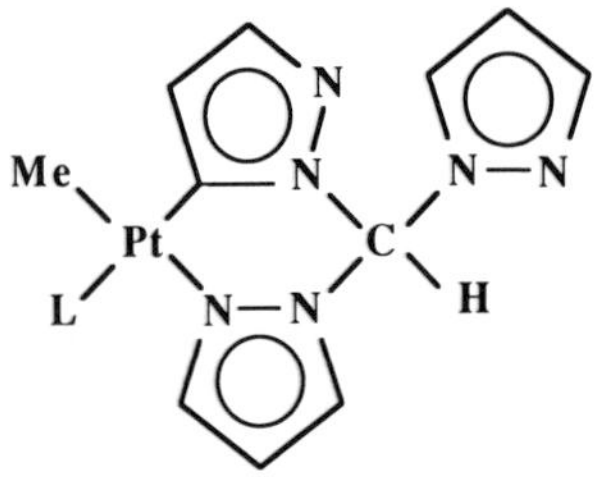

173 [L = PPh_2(*o*-tolyl), PPh_2(*o*-$MeOC_6H_4$)]

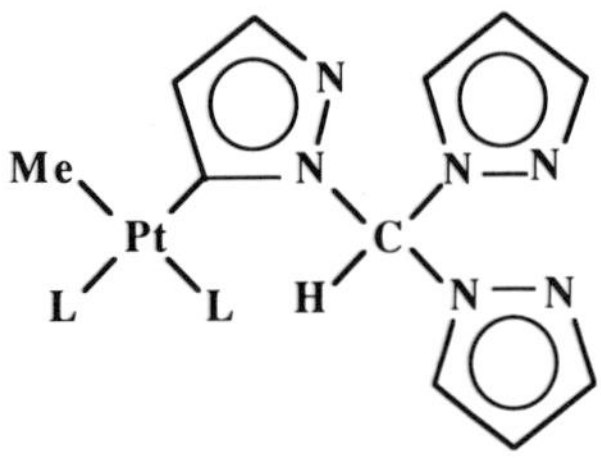

174 [L = PPh_3, $PMePh_2$, $PEtPh_2$, $PPh_2(CH_2Ph)$, $P(OPh)_3$]

175 [L_2 = $(PPh_2)_2CH_2$, $(PPh_2CH_2)_2$]

to have the formulation $PtMe\{(pz)_2(C_3H_2N_2)CH\text{-}C\}(PPh_3)(py) \cdot 2py$ (**176**) (Fig. 13A). The structures of two other phosphine complexes are illustrated in Fig. 13. The phosphine complexes illustrate the flexibility of

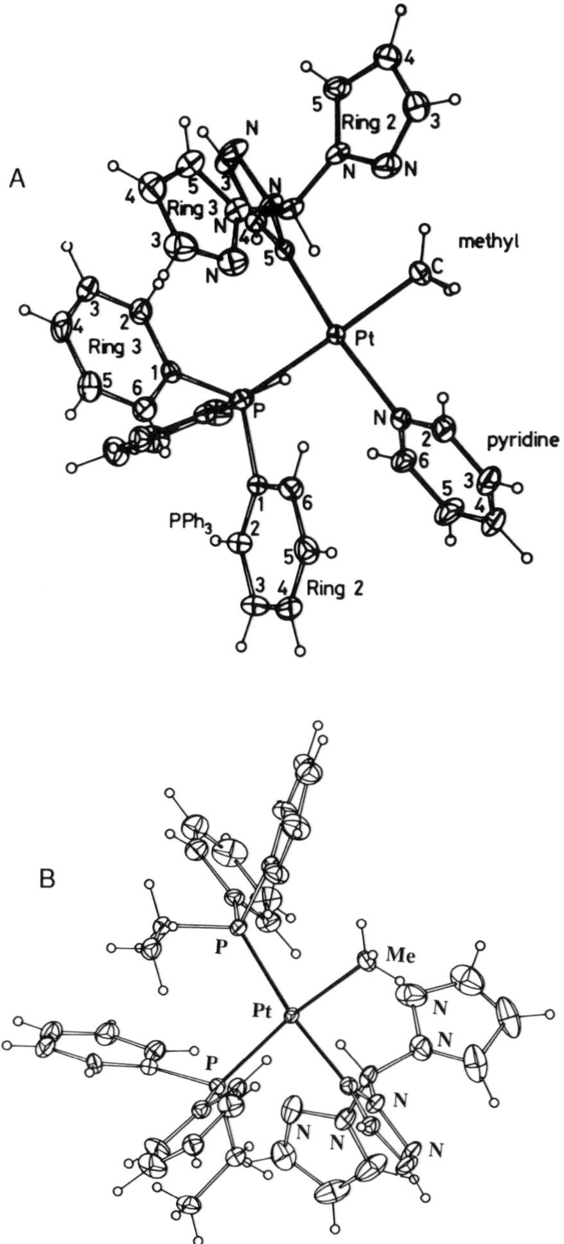

FIG. 13. Structures of phosphine complexes of platinated tris(pyrazol-1-yl)methane: (A) the complex in PtMe{$(pz)_2(C_3H_2N_2)CH$} $(PPh_3)(py) \cdot 2py$ (**176**) (with permission from Ref. 59; copyright Royal Society of Chemistry), (B) PtMe{$(pz)_2(C_3H_2N_2)CH$}$(PEtPh_2)_2$ (**174**), and (C) PtMe{$(pz)_2(C_3H_2N_2)CH$}$(PPh_2(o\text{-}MeOC_6H_4))$ (**173**) [with permission from the authors (*108*)].

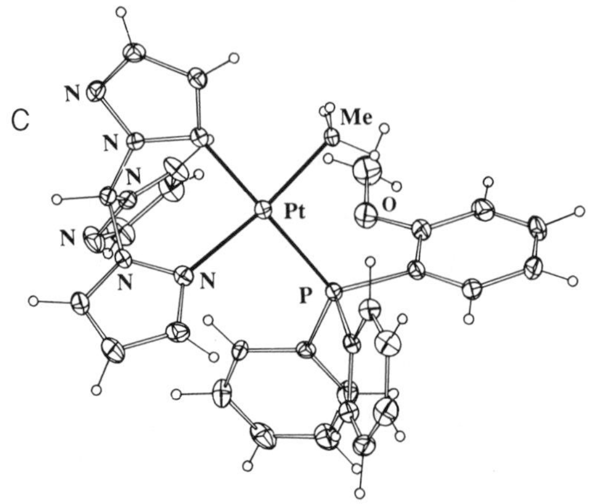

FIG. 13. (*continued*)

the metallated ligands to act as $[C{\sim}N]^-$ and $[C]^-$ donors. The complex $PtMe\{(pz)_2(C_3H_2N_2)CH\}(PPh_3)_2$ (**174**) undergoes an additional cyclometallation on melting to form $Pt\{(pz)_2(C_3H_2N_2)CH\}\{PPh_2(C_6H_4)\}$ (**177**),

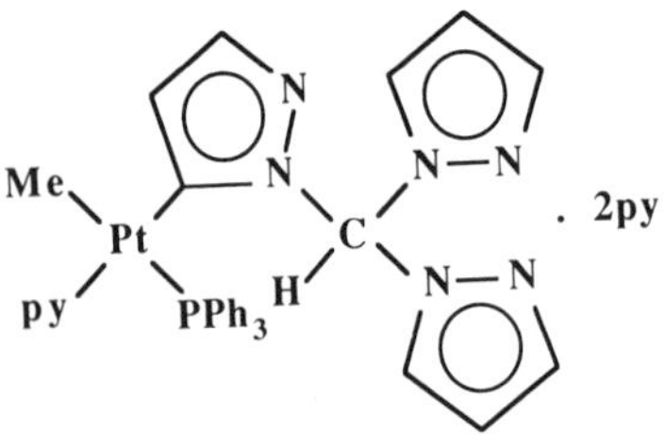

176

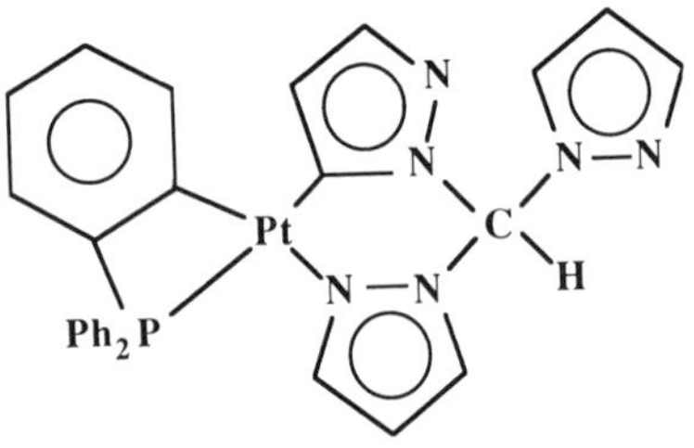

177

containing both nitrogen and phosphorus donor ligands metallated to form six- and four-membered rings, respectively. Mass spectral and thermogravimetric analysis studies of the other phosphine complexes are consistent with cyclometallation reactions for some of the complexes, with subsequent decomposition preventing characterization of products.

Pyridine is removed from **161** on reflux in benzene to form $[PtMe\{(pz)_2(C_3H_2N_2)CH\}]_n$ (**165**), assigned a polymeric structure by analogy with **45** and **47** and its low solubility. The complex **161** also dissolves at reflux in acetone, and reaction of this solution at ambient temperature with carbon monoxide gives the complex $PtMe\{(pz)_2(C_3H_2N_2)CH\}(CO)$ (**178**) (Fig. 14), a cyclometallated analog of $PtMe\{(pz)_3BH\}(CO)$ (**50**) (Fig. 3).

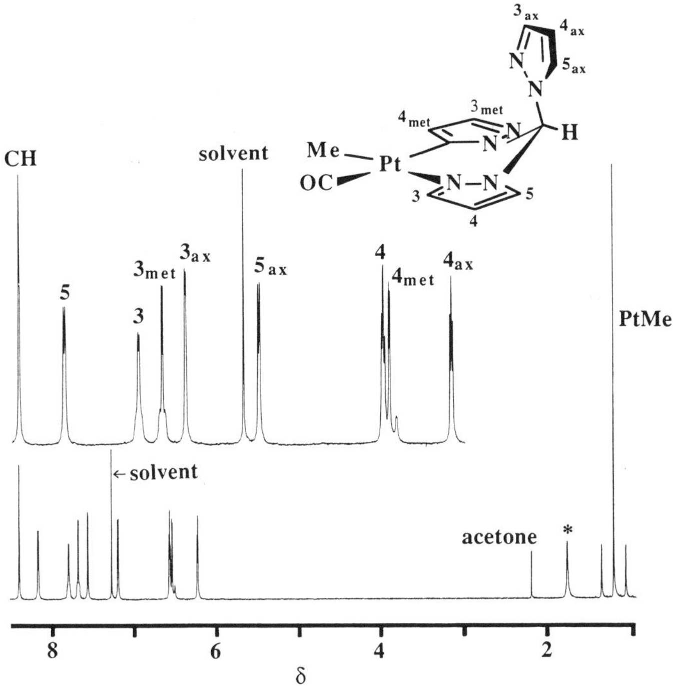

FIG. 14. ^{1}H NMR spectrum of PtMe{(pz)$_2$(C$_3$H$_2$N$_2$)CH}(CO) (**178**) in CDCl$_3$, illustrating ^{1}H–^{195}Pt coupling for protons of the metallated ring [^3J(HPt) 15.0, ^5J(HPt) 9.1 Hz] typical for the complexes with cyclometallated ligands, and unresolved ^{1}H–^{195}Pt coupling for H-3 of the *N*-coordinated ring (the asterisk denotes an impurity) (with permission from Ref. 57; copyright Elsevier Sequoia S.A.).

In contrast to the borate complex, **178** is not fluxional and is assumed to be square planar as found in the solid state for the borate analog.

B. *Platinum(IV)*

Although planar [C~D~D]$^-$ and [D~C~D]$^-$ ligands (D = N, P, etc.) have been widely studied for a range of transition metals and ligand systems, including the complexes **155–158** and **172**, tripodal systems have been obtained only recently. The synthesis of the first example of this type of complex [Eq. (24)] (*112*) was also prompted by the realization that a metallated tripodal ligand [(pz)$_2$(C$_3$H$_2$N$_2$)C-*N,N,C*]$^-$ may be regarded as isoelectronic with tridentate (pz)$_3$CH and (pz)$_3$BH$^-$.

$$\underset{\textbf{161}}{PtMe\{(pz)_2(C_3H_2N_2)CH\}(py)} + MeI \rightarrow \underset{\textbf{179}}{[PtMe_2\{(pz)_2(C_3H_2N_2)CH\}(py)]I} \qquad (24)$$

Several related complexes (**180–182**) were subsequently obtained using the oxidative addition strategy, including the reaction of $HC{\equiv}C{-}CH_2Br$ to form the allenyl complex **182** (*57*). Oxidative addition of MeI was

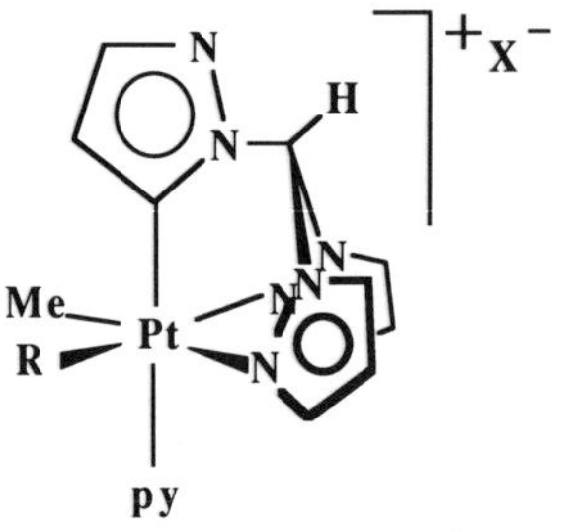

179 (R = Me, X = I)

180 (R = $PhCH_2$, X = Br)

181 (R = $CH_2{=}CH{-}CH_2$, X = Br)

182 (R = $CH_2{=}C{=}CH_2$, X = Br)

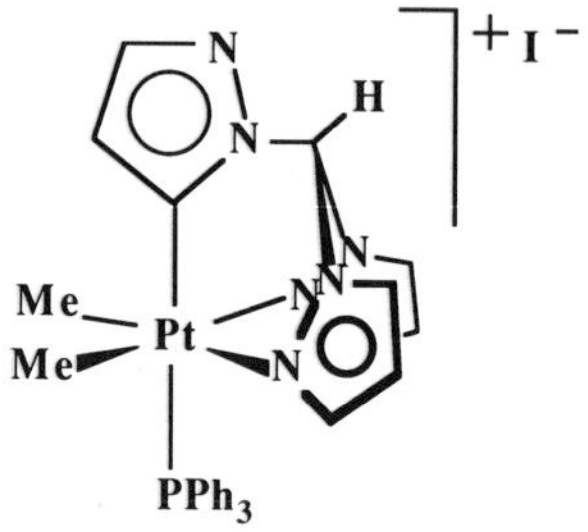

183

extended to include a PPh_3 derivative (**183**) by addition of PPh_3 to **161** prior to oxidative addition, and both py and PPh_3 derivatives of cyclometallated $(pz)_2(mim)CH$ were also isolated (**184**). Two neutral complexes (**185,186**) (Fig. 15) were obtained by oxidative addition to $[PtMe\{(pz)_2(C_3H_2N_2\}]_n$ (**165**), and oxidative addition of $pzCH_2CH_2Br$ to

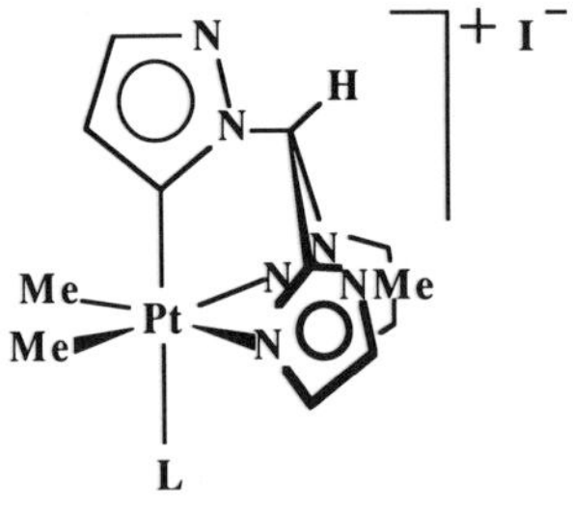

184 (L = py, PPh_3)

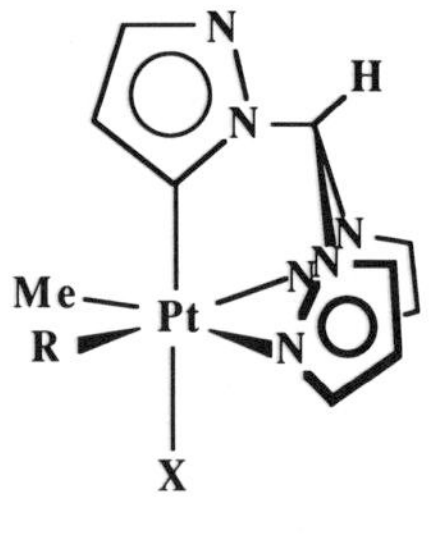

185 (R = Me, X = I)

186 (R = $PhCH_2$, X = Br)

187 (R = Et, X = I)

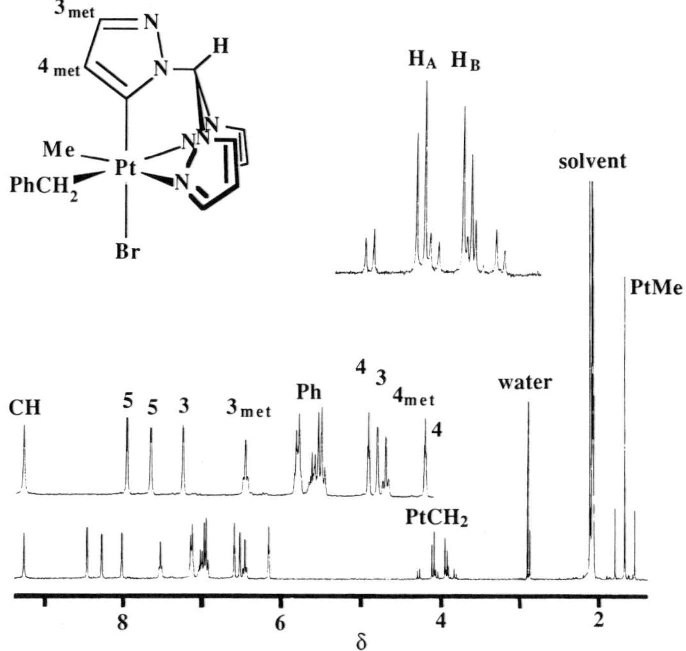

FIG. 15. ^{1}H NMR spectrum of PtBrMe(CH$_2$Ph){(pz)$_2$(C$_3$H$_2$N$_2$)CH} (**186**) in (CD$_3$)$_2$CO, illustrating three ring environments, ^{1}H–^{195}Pt coupling for the protons of the metallated ring [4_{met}, ^3J(HPt) 12.3; 3_{met}, ^5J(HPt) 9.0 Hz], and inequivalent benzylic protons with ^2J(HPt) 107.3 Hz for H$_A$ and 69.8 Hz for H$_B$; ^2J(HH) 8.8 Hz (with permission from Ref. 57; copyright Elsevier Sequoia S.A.).

either **161** or **165** gave **188**, containing both $[N{\sim}C]^-$ and $[N{\sim}C{\sim}N]^-$ ligands in the same complex (*57*).

188

189

As part of an X-ray study comparing structural parameters for organopalladium(IV) and platinum(IV) complexes (Fig. 10), the complex $[PtMe_3\{(pz)_3CH\}]I$ (**140**) was isolated by the reaction of $(pz)_3CH$ with a $Pt^{IV}Me_3$ substrate (*79*). An attempted synthesis of **140** by oxidative addition of MeI to $PtMe_2\{(pz)_3CH\}$ gave instead the cyclometallated complex **185** [Eq. (25)]. Similar results were obtained on reaction of $PtMe_2$-$\{(pz)_3CH\}$ with ethyl iodide (*60*) and benzyl bromide (*57*) to give **187** and **186**. Although $PtMe_2\{(pz)_4C\}$ reacted with MeI in acetone to form $PtIMe_3\{(pz)_4C\}$ (**126**), it also reacted in neat MeI to form $PtIMe_2\{(pz)_3$-$(C_3H_2N_2)CH\}$ (**189**).

$$\underset{\mathbf{73}}{PtMe_2\{(pz)_3CH\}} + RX \rightarrow \underset{\mathbf{185\text{–}187}}{PtXMe(R)\{(pz)_2(C_3H_2N_2)CH\}} \tag{25}$$

These unexpected reactions may proceed via cyclometallation of $PtMe_2(L)$, e.g., to form $[PtMe\{(pz)_2(C_3H_2N_2)CH\}]_n$ (**165**) (perhaps $n = 1$ with coordinated acetone), prior to rapid oxidative addition. Under these conditions unidentate $(pz)_3CH$ is unlikely as a precursor to metallation as seems likely in the pyridine-mediated cyclometallation of $PtMe_2\{(pz)_3CH\}$ (**169**). Consistent with this, a cyclometallation reaction does not occur on oxidative addition of MeI to $PtMe_2\{(pz)_2CH_2\}$ because unidentate $(pz)_2CH_2$ is required for cyclometallation, but cyclometallation of this complex does occur in pyridine. If unidentate coordination is precluded in the cyclometallation of complexes observed on addition of RX in acetone, then cyclometallation of the axial uncoordinated ring by the "$PtC_2(pz)_2$" center presumably occurs, with the reactivity of platinum perhaps activated by the initial stages of oxidative addition or coordination of RX.

None of the complexes **179–189** was obtained in a form suitable for crystallographic studies, and in an attempt to expand the range of tripodal $[N{\sim}C{\sim}N]^-$ donors and to include crystalline complexes a new strategy was adopted relying on the facile oxidative addition reactivity of dimethylplatinum(II). In this approach poly(pyrazol-1-yl)alkanes and related reagents **190–194** were synthesized and reacted with $[PtR_2(SEt_2)]_2$, and oxidative addition occurred to give **195–197** (*103, 112*), **198** and **200** (*60*), and **199** (*103*) [Eqs. (26)–(28)]. Complexes **195–198** react with pyridine to form cations **201**.

Structural studies of two of the complexes have been completed (Fig. 16) The $[N{\sim}C{\sim}N]^-$ tripod ligands form two five-membered $\overline{PtNNCC}$ rings and one six-membered $\overline{PtNNCNN}$ ring, with chelate angles at platinum ~7–12° less than 90°.

$[PtR_2(SEt_2)]_2$ (26)

190 (R = H, R' = Me)

191 (R = H, R' = CH_2Cl)

192 (R = Cl, R' = Me)

195 (R = Me, R' = H, R" = Me)

196 (R = Me, R' = H, R" = CH_2Cl)

197 (R = Me, R' = Cl, R" = Me)

198 (R = Ph, R' = H, R" = Me)

$[PtMe_2(SEt_2)]_2$ (27)

193 **199**

$[PtMe_2(SEt_2)]_2$ (28)

194 **200**

Complexes **195–200** are assumed to be formed via initial *N*-donor coordination to $Pt^{II}Me_2$ prior to oxidative addition, as recently demonstrated in a closely related study (*109*). In this report, $Me_2NCH_2CH_2N{=}CH$-*o*-C_6H_4Br reacted with $[PtMe_2(SMe_2)]_2$ to form isolable **171**, and this complex underwent intramolecular oxidative addition to form **202**. This reaction, and that of Eqs. (27) and (28), appear to be the first reported

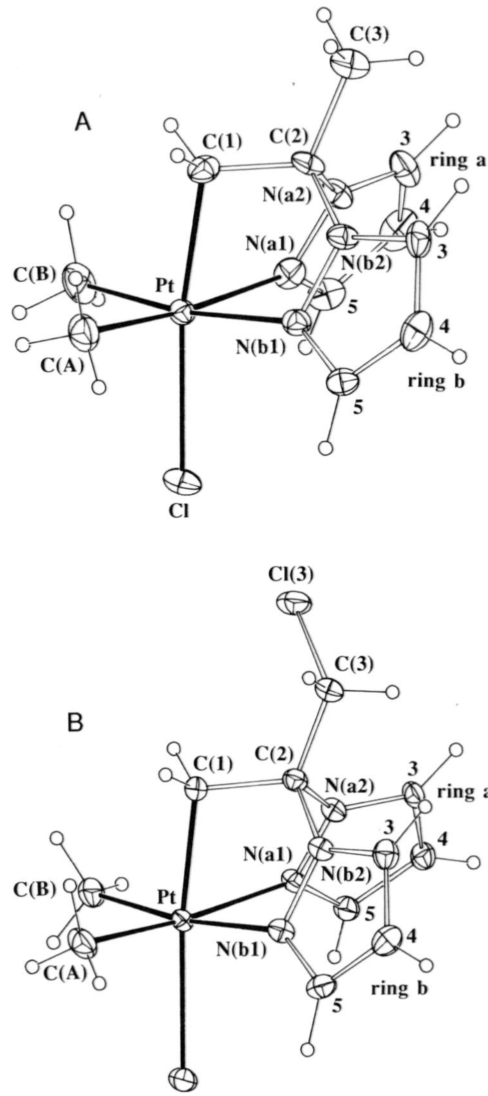

FIG. 16. Structures of $PtClMe_2\{(pz)_2C(R'')CH_2\}$. (A) $R'' = Me$ (**195**); (B) $R'' = CH_2Cl$ (**196**) (with permission from Ref. 103; copyright Elsevier Sequoia S.A.). Different projections are given in Ref. 112.

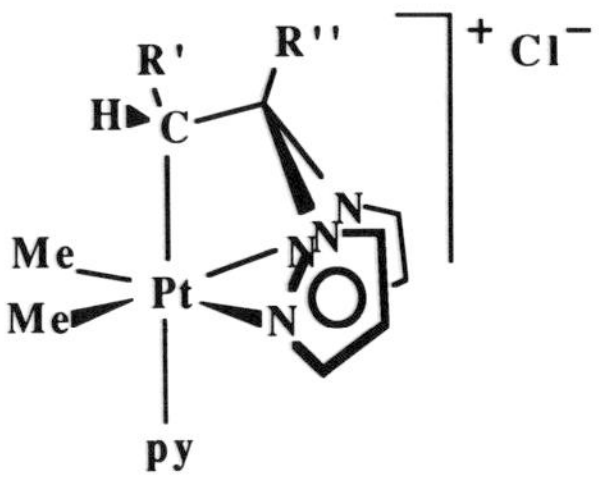

201

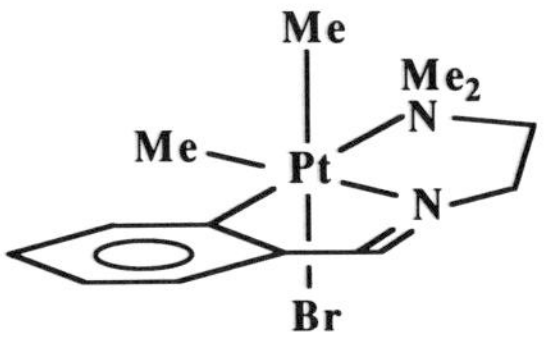

202

examples of oxidative addition of an aryl–bromine bond to platinum(II). The activation parameters for the reaction of **171** to form **202** are $\Delta H^{\ddagger} = 97 \pm 2$ kJ mol^{-1} and $\Delta S^{\ddagger} = 11 \mp 10$ J K^{-1} mol^{-1}, and the low value of $\Delta S^{\ddagger}$ precludes an S_N2 mechanism of oxidative addition for which large, negative values of $\Delta S^{\ddagger}$ are characteristic (*83,87,111,113*). A concerted mechanism, as proposed for C–H oxidative addition (*114*), is consistent with the experimental data.

The tripodal $[N{\sim}C{\sim}N]^-$ ligand in **199** is identical to that in the planar palladium(II) complex **152**, illustrating the ability of this ligand to coordinate in either fac or mer fashion, respectively. This flexibility distinguishes it from the closely related and widely studied system *mer*-$[1,3\text{-}(Me_2NCH_2)_2C_6H_3]^-$, which has provided a range of interesting advances in organometallic and coordination chemistry (*115*).

IV

CONCLUDING REMARKS

Palladium and platinum were involved in the early exploration of poly(pyrazol-1-yl)borates and poly(pyrazol-1-yl)alkanes as ligands, and these ligands have subsequently played key roles in the development of organopalladium and platinum chemistry. It is interesting to note, for example, that among the more recent reports are accounts of allylpalladium chemistry [**2** (*9*); **16** (*12*)] directly related to the initial applications of these ligands [**2** (*1*), **15** (*1*)]. This review has attempted to encompass all reported applications of these ligands in the organometallic chemistry of palladium and platinum and, in particular, has reviewed for the first time the role of poly(pyrazol-1-yl)alkanes in stabilizing organopalladium(IV) complexes and in roles as platinated intramolecular coordination systems.

V

APPENDIX

Methylpalladium(II) and trimethylpalladium(IV) complexes of poly(pyrazol-1-yl)borates have been obtained as indicated in Eqs. (29) and (30), and the structures of the complexes have been determined (Fig. 17) (*116*).

$$[PdIMe(SMe_2)]_2 \xrightarrow[\text{(ii) } PPh_3]{\text{(i) } K[(pz)_3BH] \text{ or } K[(pz)_4B]} \quad (29)$$

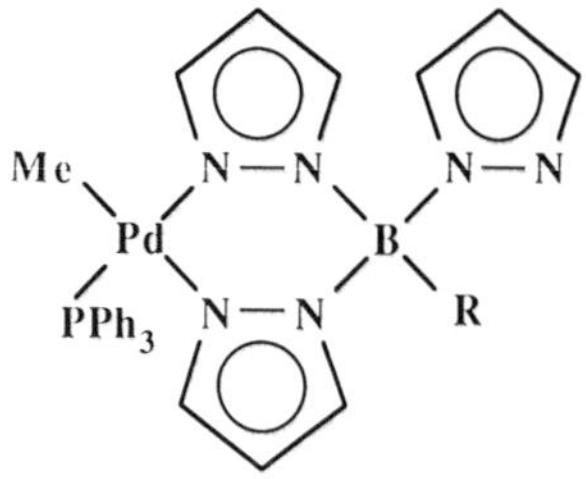

203 (R = H, pz)

$$[PdMe_2(pyridazine)]_n \xrightarrow[\text{(ii) MeI}]{\text{(i) } K[(pz)_3BH] \text{ or } K[(pz)_4B]} \quad (30)$$

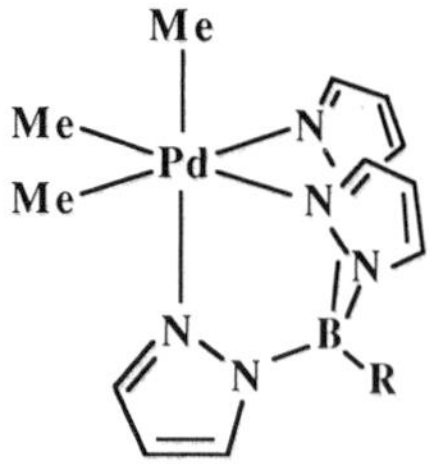

204 (R = H, pz)

Structural studies of complexes **108–110** have been completed (Figs. 18 and 19) (*117*). The complexes exhibit disorder in the solid state. The complex $[PtIPh_2\{(pz)_3CH\}]_2[I][I_3]$ (**108**) has two cations in the asymmetric unit, with one cation well ordered and the other showing disorder between the coordinated iodide and one phenyl position. As discussed in Section II,D, ^1H NMR spectra of **109** and **110** indicate the presence of isomers, with coordinated iodide trans to pz and to mim or py, and the structural results confirm this, with the coordinated iodide disordered with both methyl groups in both **109** and **110**. In **109** (Fig. 17B) the iodine atom is predominantly trans to the pz groups, and in **110** (Fig. 17C) the iodine atom is predominantly trans to the pyridine group.

A structural study of **133** has been completed, and a projection is shown in Fig. 20 (*108*). Additional structural studies of complexes containing

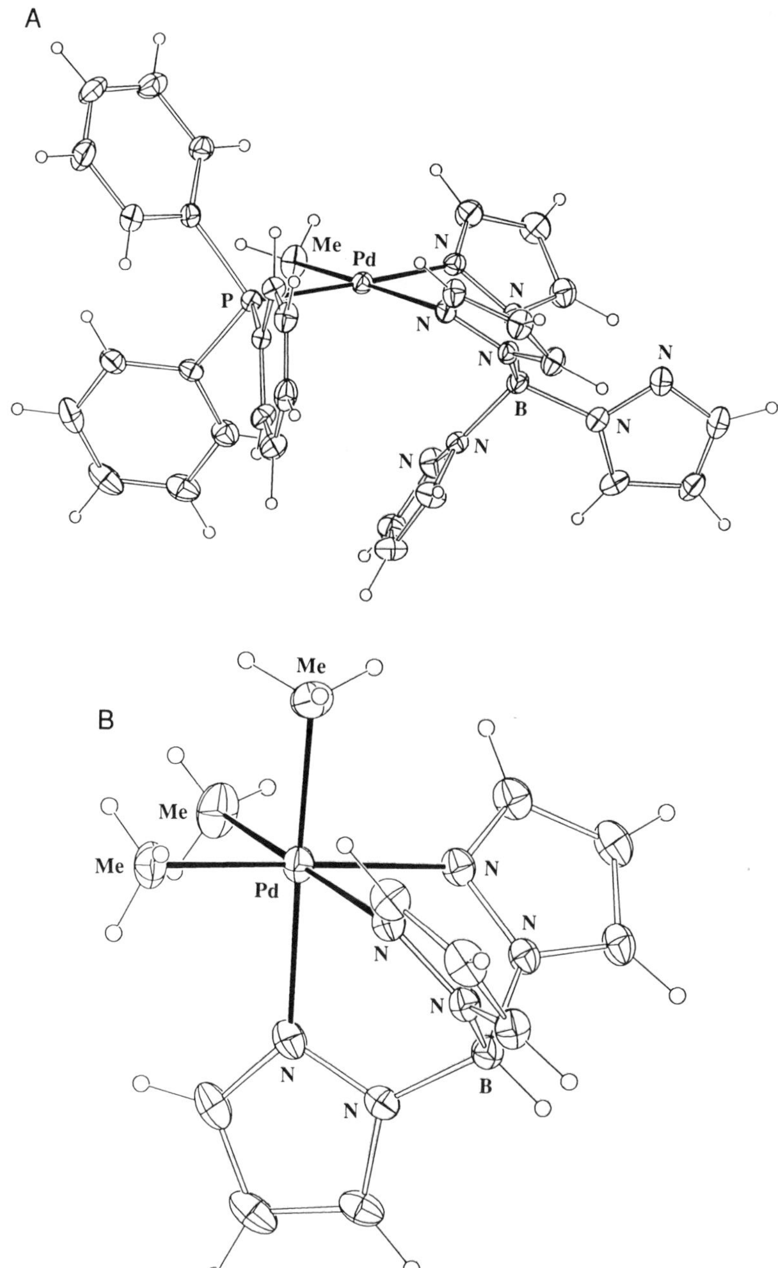

FIG. 17. Structures of methylpalladium(III) and trimethylpalladium(IV) complexes of poly(pyrazol-1-yl)borates. (A) PdMe{(pz)$_4$B}(PPh$_3$) (**203**); (B) PdMe$_3${(pz)$_3$BH} (**204**, R = H), shown in a similar orientation to the "isoelectronic" cation [PdMe$_3${(pz)$_3$CH}]$^+$ in Fig. 10A; (C) PdMe$_3${(pz)$_4$B} (**204**, R = pz) [with permission from the authors (*116*)].

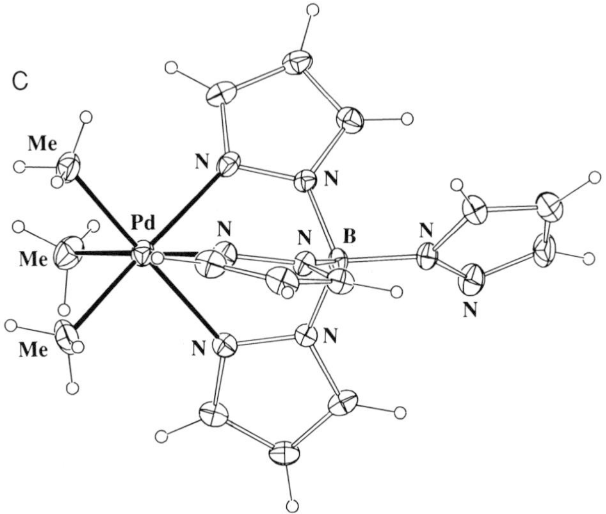

FIG. 17. (*continued*)

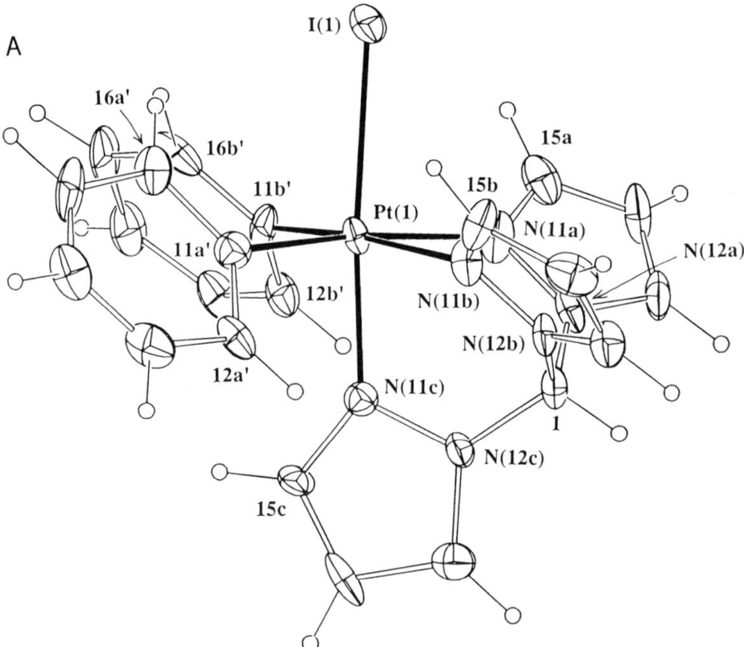

FIG. 18. Structures of the cations in $[PtIPh_2\{(pz)_3CH\}]_2[I][I_3]$ (**108**), showing (A) the well-ordered cation that occurs together with (B), the cation showing disorder in the position of the coordinated iodide and one of the phenyl groups (with permission from Ref. 117; copyright Elsevier Sequoia S.A.).

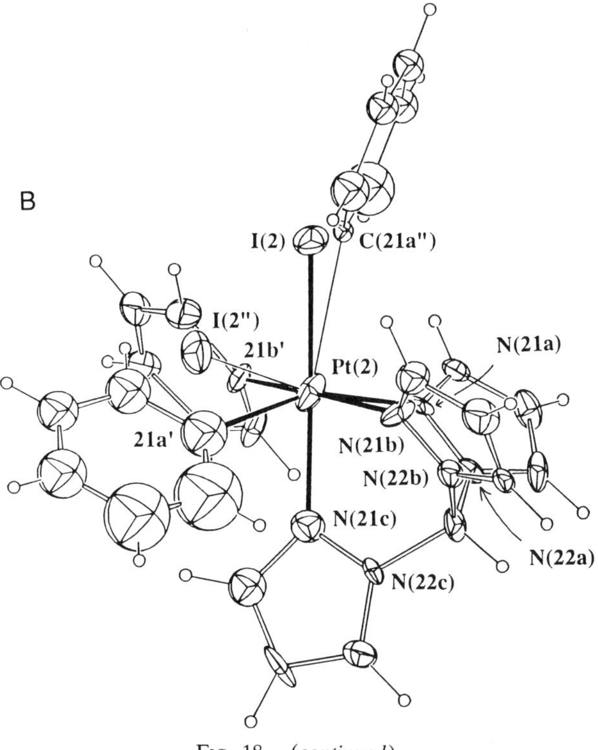

FIG. 18. *(continued)*

platinated $(pz)_3CH$ have been completed (Fig. 21) (*108*). Complexes **174** ($L = PPh_3$) and **175** have coordination geometries similar to that of $PtMe\{(pz)_2(C_3H_2N_2)CH\}(PEtPh_2)_2$ (**174**, $L = PEtPh_2$), with the ligand bonded to platinum via a carbon atom of one pyrazole ring and having two uncoordinated rings. In these three complexes, there are short Pt···H contacts of ~2.7 Å, involving the methine proton of the ligand, and these may represent agostic (*118*) interactions. However, a clear demonstration of the presence or absence of agostic interactions in these complexes has not been obtained. For example, ^1H NMR spectra of the complexes in $CDCl_3$ show $^4J(^1H-^{195}Pt)$ coupling for the resonances of the methine proton (δ 8.84–9.25, J_{HPt} 8.2–9.0 Hz), in contrast to complexes containing platinated $(pz)_3CH$ as a bidentate ligand (Fig. 12), which do not exhibit coupling. Although coupling could be attributed to an agostic interaction, it may also result from other factors, e.g., the quite different configuration of the "Pt···H–C–N–C" group compared to the boat configuration for the chelate ring of the complexes illustrated in Fig. 12.

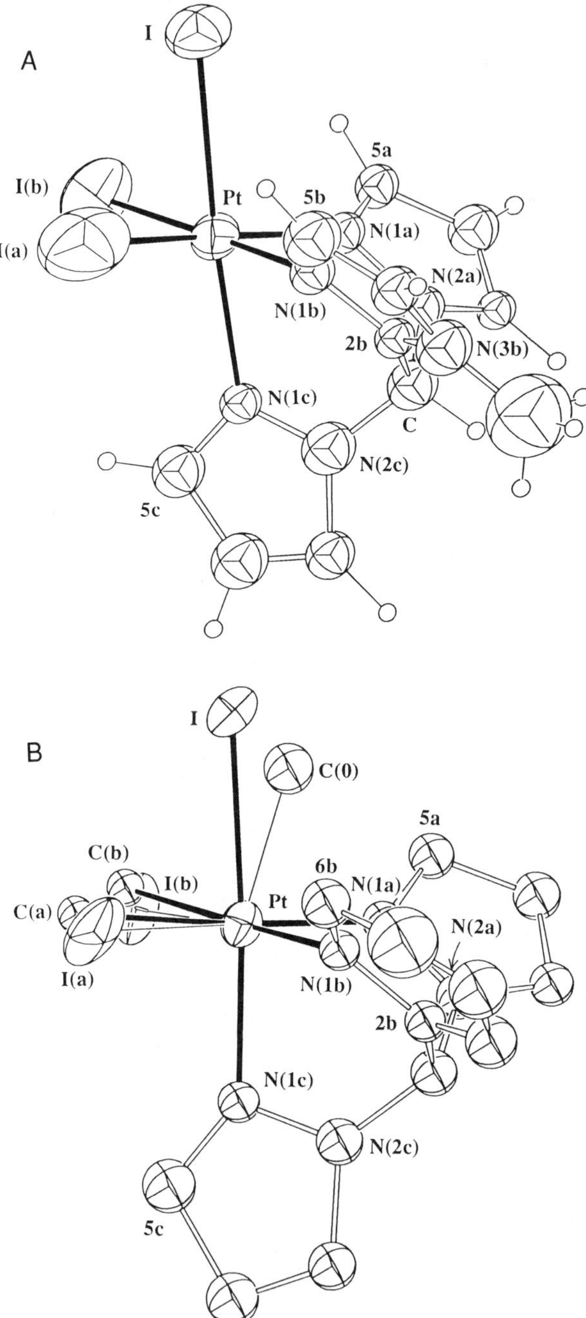

FIG. 19. Structures of the disodered cations in (A) $[PtIMe_2\{(pz)_2(mim)CH\}]I$ (**109**) and (B) $[PtIMe_2\{(pz)_2(py)CH\}]I$ (**110**). In complex **109**, disordered methyl and iodine groups are shown as I, I(a), and I(b), and in **110** the methyl and iodine groups are resolved (with permission from Ref. 117; copyright Elsevier Sequoia S.A.).

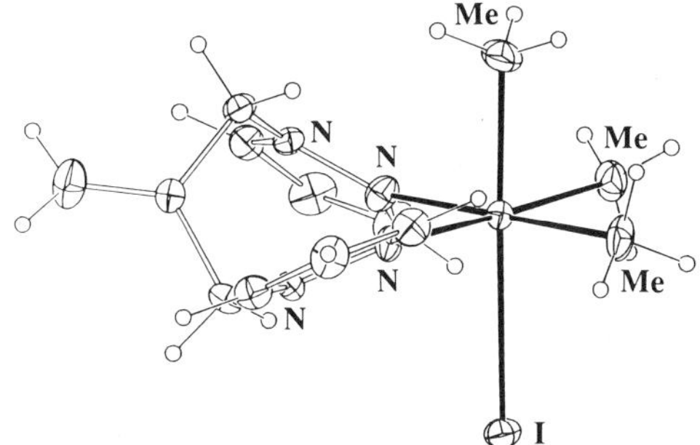

FIG. 20. The structure of $PtIMe_3\{(pzCH_2)_2C{=}CH_2\}$ (**133**) [with permission from the authors (*108*)].

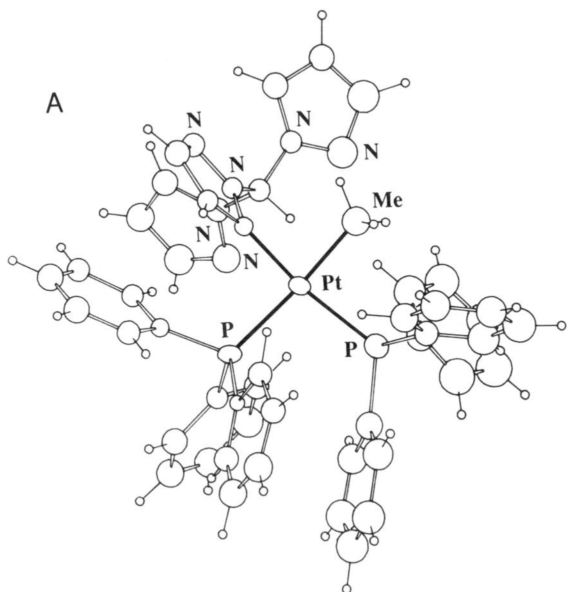

FIG. 21. Structures of (A) $PtMe\{(pz)_2(C_3H_2N_2)CH\}(PPh_3)_2$ (**174**) and (B) $PtMe\{(pz)_2(C_3H_2N_2)CH\}(PPh_2CH_2CH_2PPh_2)$ (**174**) [with permission from the authors (*108*)].

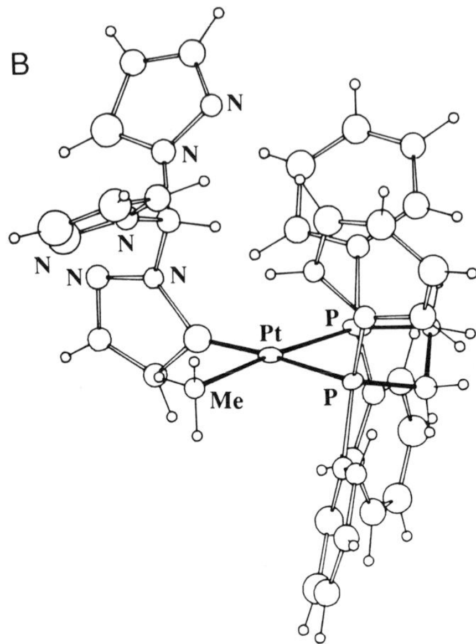

FIG. 21. *(continued)*

ACKNOWLEDGMENTS

We extend thanks to Dr. Allan H. White and his colleagues for X-ray crystallographic contributions to parts of the work discussed here, and to the Australian Research Council and The University of Tasmania for financial support.

REFERENCES

1. S. Trofimenko, *J. Am. Chem. Soc.* **92,** 5118 (1970).
2. S. Trofimenko, *Chem. Rev.* **72,** 497 (1972).
3. S. Trofimenko, *J. Am. Chem. Soc.* **88,** 1842 (1966).
4. S. Trofimenko, *Acc. Chem. Res.* **4,** 17 (1971).
5. S. Trofimenko, *Prog. Inorg. Chem.* **34,** 115 (1986).
6. A. Shaver, *in "Comprehensive Coordination Chemistry"* (G. Wilkinson, R. D. Gillard, and J. A. McLeverty, eds.), Vol. 2, p. 245. Pergamon (1987).
7. S. Trofimenko, *J. Am. Chem. Soc.* **91,** 588 (1969).
8. L. Komorowski, W. Maringgele, A. Meller, K. Niedenzu, and J. Serwatowski, *Inorg. Chem.* **29,** 3845 (1990).
9. L. Komorowski, A. Meller, and K. Niedenzu, *Inorg. Chem.* **29,** 538 (1990).
10. K. Niedenzu, J. Serwatowski, and S. Trofimenko, *Inorg. Chem.* **30,** 524 (1991).
11. S. Trofimenko, *J. Coord. Chem.* **2,** 75 (1972).
12. D. G. Brown, P. K. Byers, and A. J. Canty, *Organometallics* **9,** 1231 (1990).
13. G. Minghetti, M. A. Cinellu, A. L. Bandini, G. Banditelli, F. Demartin, and M. Manassero, *J. Organomet. Chem.* **315,** 387 (1986).

14. P. K. Byers and A. J. Canty, *Organometallics* **9,** 210 (1990).
15. S. Trofimenko, *J. Am. Chem. Soc.* **91,** 3183 (1969).
16. G. Paiaro and A. Musco, *Tetrahedron Lett.*, 1583 (1965).
17. P. K. Byers, A. J. Canty, P. R. Traill, and A. A. Watson, *J. Organomet. Chem.* **390,** 399 (1990).
18. L. S. Hegedus, B. Akermark, D. J. Olsen, O. P. Anderson, and K. Zetterberg, *J. Am. Chem. Soc.* **104,** 697 (1982).
19. C. P. Brock, M. K. Das, R. P. Minton, and K. Niedenzu, *J. Am. Chem. Soc.* **110,** 817 (1988).
20. G. Calvin and G. E. Coates, *J. Chem. Soc.*, 2008 (1960).
21. M. Onishi, Y. Ohama, K. Sugimura, and K. Hiraki, *Chem. Lett.*, 955 (1976).
22. M. Onishi, K. Sugimura, and K. Hiraki, *Bull. Chem. Soc. Jpn.* **51,** 3209 (1978).
23. M. Onishi, K. Hiraki, M. Shironita, Y. Yamaguchi, and S. Nakagawa, *Bull. Chem. Soc. Jpn.* **53,** 961 (1980).
24. G. van Koten, A. L. Spek, and J. M. Valk, to be published.
25. P. K. Byers, A. J. Canty, N. J. Minchin, J. M. Patrick, B. W. Skelton, and A. H. White, *J. Chem. Soc.*, Dalton Trans., 1183 (1985).
26. A. J. Canty, N. J. Minchin, P. C. Healy, and A. H. White, *J. Chem. Soc., Dalton Trans.*, 1795 (1982).
27. A. J. Canty, N. J. Minchin, J. M. Patrick, and A. H. White, *Aust. J. Chem.* **36,** 1107 (1983).
28. M. Onishi, K. Hiraki, T. Itoh, and Y. Ohama, *J. Organomet. Chem.* **254,** 381 (1983).
29. M. Onishi, T. Ito, and K. Hiraki, *J. Organomet. Chem.* **209,** 123 (1981).
30. M. Onishi, H. Yamamoto, and K. Hiraki, *Bull. Chem. Soc. Jpn.* **53,** 2540, (1980).
31. M. Onishi, K. Hiraki, H. Konda, Y. Ishida, Y. Ohama, and Y. Uchibori, *Bull. Chem. Soc. Jpn.* **59,** 201 (1986).
32. H. C. Clark and L. E. Manzer, *J. Am. Chem. Soc.* **95,** 3812 (1973).
33. H. C. Clark and L. E. Manzer, *Inorg. Chem.* **13,** 1291 (1974).
34. H. C. Clark and M. A. Mesubi, *J. Organomet. Chem.* **215,** 131 (1981).
35. H. C. Clark, G. Ferguson, V. K. Jain, and M. Parvez, *Organometallics* **2,** 806 (1983).
36. H. C. Clark, C. R. Jablonski, and K. von Werner, *J. Organomet. Chem.* **82,** C51 (1974).
37. H. C. Clark and K. von Werner, *J. Organomet. Chem.* **101,** 347 (1975).
38. H. C. Clark and L. E. Manzer, *Inorg. Chem.* **13,** 1996 (1974).
39. L. E. Manzer and P. Z. Meakin, *Inorg. Chem.* **15,** 3117 (1976).
40. P. E. Rush and J. D. Oliver, *J. Chem. Soc., Chem. Commun.*, 996 (1974).
41. J. D. Oliver and P. E. Rush, *J. Organomet. Chem.* **104,** 117 (1976).
42. J. D. Oliver and N. C. Rice, *Inorg. Chem.* **15,** 2741 (1976.
43. B. W. Davies and N. C. Payne, *J. Organomet. Chem.* **102,** 245 (1975).
44. S. Nussbaum and A. Storr, *Can. J. Chem.* **63,** 2550 (1985).
45. A. J. Canty, N. J. Minchin, L. M. Engelhardt, B. W. Skelton, and A. H. White, *J. Chem. Soc., Dalton Trans.*, 645 (1986).
46. P. K. Byers and A. J. Canty, *Inorg. Chim. Acta* **104,** L13 (1985).
47. P. K. Byers, A. J. Canty, B. W. Skelton, and A. H. White, *J. Chem. Soc., Chem. Commun.*, 1093 (1987).
48. P. K. Byers, A. J. Canty, B. W. Skelton, and A. H. White, *J. Organomet. Chem.* **336,** C55 (1987).
49. P. K. Byers, A. J. Canty, and R. T. Honeyman, *J. Organomet. Chem.* **385,** 417 (1990).
50. P. K. Byers, A. J. Canty, R. T. Honeyman, and A. A. Watson, *J. Organomet. Chem.* **363,** C22 (1989).

51. H. C. Clark and L. E. Manzer, *J. Organomet. Chem.* **59,** 411 (1973).
52. T. G. Appleton, J. R. Hall, and M. A. Williams, *J. Organomet. Chem.* **303,** 139 (1986).
53. J. Kuyper, R. van der Laan, F. Jeanneaus, and K. Vrieze, *Trans. Metal Chem.* **1,** 199 (1976).
54. J. D. Scott and R. J. Puddephatt, *Organometallics* **5,** 1538, 2522 (1986).
55. D. P. Bancroft, F. A. Cotton, L. R. Falvello, and W. Schwotzer, *Inorg. Chem.* **25,** 763 (1986).
56. P. K. Byers, A. J. Canty, R. T. Honeyman, and A. A. Watson, *J. Organomet. Chem.* **385,** 429 (1990).
57. A. J. Canty and R. T. Honeyman, *J. Organomet. Chem.* **387,** 247 (1990).
58. A. J. Canty and N. J. Minchin, *J. Organomet. Chem.* **226,** C14 (1982).
59. A. J. Canty, N. J. Minchin, J. M. Patrick, and A. H. White, *J. Chem. Soc., Dalton Trans.*, 1253 (1983).
60. R. T. Honeyman, Ph.D. Thesis, University of Tasmania, Hobart, Australia (1989).
61. T. Astley, A. J. Canty, M. A. Hitchman, G. L. Rowbottom, B. W. Skelton, and A. H. White, *J. Chem. Soc., Dalton Trans.*, 1981 (1991).
62. P. Meakin, S. Trofimenko, and J. P. Jesson, *J. Am. Chem. Soc.* **94,** 5677 (1972).
63. H. C. Clark and L. E. Manzer, *J. Chem. Soc., Chem. Commun.*, 870 (1973).
64. L. E. Manzer, *Inorg. Chem.* **15,** 2354 (1976).
65. B. W. Davies and N. C. Payne, *Inorg. Chem.* **13,** 1843 (1974).
66. N. C. Rice and J. D. Oliver, *Acta Crystallogr., Sect. B: Crystallogr. Cryst. Chem.* **B34,** 3748 (1978).
67. H. C. Clark and L. E. Manzer, *Inorg. Chem.* **11,** 2749 (1972).
68. J. R. Hall and G. A. Swile, *Aust. J. Chem.* **24,** 423 (1971).
69. A. J. Canty, R. T. Honeyman, B. W. Skelton, and A. H. White, *J. Organomet. Chem.* **396,** 105 (1990).
70. R. B. King and A. Bond, *J. Am. Chem. Soc.* **96,** 1338 (1974).
71. S. Roth, V. Ramamoorthy, and P. R. Sharp, *Inorg. Chem.* **29,** 3345 (1990).
72. H. C. Clark, G. Ferguson, V. K. Jain, and M. Parvez, *J. Organomet. Chem.* **270,** 365 (1984).
73. P. K. Byers, A. J. Canty, R. T. Honeyman, B. W. Skelton, and A. H. White, *J. Organomet. Chem.*, in press (1992).
74. W. de Graaf, J. Boersma, and G. van Koten, *Organometallics* **9,** 1479 (1990).
75. J. P. Collman, J. N. Cawse, and J. W. Kang, *Inorg. Chem.* **8,** 2574 (1969).
76. B. E. Mann, B. L. Shaw, and N. I. Tucker, *J. Chem. Soc. A*, 2667 (1971).
77. C. J. Elsevier, H. Kleijn, J. Boersma, and P. Vermeer, *Organometallics* **5,** 716 (1986).
78. P. K. Byers, A. J. Canty, B. W. Skelton, and A. H. White, *J. Chem. Soc., Chem. Commun.*, 1722 (1986); A. J. Canty, *Accs. Chem. Res.* **25,** 83 (1992).
79. P. K. Byers, A. J. Canty, B. W. Skelton, and A. H. White, *Organometallics* **9,** 826 (1990).
80. R. Uson, J. Fornies, and R. Navarro, *J. Organomet. Chem.* **96,** 307 (1975); R. Uson, J. Fornies, and R. Navaro, *Synth. React. Inorg. Met–Org. Chem.* **7,** 235 (1977).
81. P. K. Byers, A. J. Canty, B. W. Skelton, P. R. Traill, A. A. Watson, and A. H. White, *Organometallics* **9,** 3080 (1990).
82. P. K. Byers and A. J. Canty, *J. Chem. Soc., Chem. Commun.*, 639 (1988).
83. K.-T. Aye, A. J. Canty, M. Crespo, R. J. Puddephatt, J. D. Scott, and A. A. Watson, *Organometallics* **8,** 1518 (1989).
84. S. Otsuka, T. Yoshida, M. Matsumoto, and K. Nakatsu, *J. Am. Chem. Soc.* **98,** 5850 (1976).
85. J. M. Wisner, T. J. Bartczak, and J. A. Ibers, *Organometallics* **5,** 2044 (1986).

86. J. M. Wisner, T. J. Bartczak, J. A. Ibers, J. J. Low, and W. A. Goddard III, *J. Am. Chem. Soc.* **108,** 347 (1986).
87. P. K. Byers, A. J. Canty, M. Crespo, R. J. Puddephatt, and J. D. Scott, *Organometallics* **7,** 1363 (1988).
88. A. C. Cope and R. W. Siekman, *J. Am. Chem. Soc.* **87,** 3272 (1965).
89. A. Kasahara, *Bull. Chem. Soc. Jpn.* **41,** 1272 (1968).
90. M. A. Gutierrez, G. R. Newkome, and J. Selbin, *J. Organomet. Chem.* **202,** 341 (1980).
91. K. Hiraki, Y. Fuchita, and K. Takechi, *Inorg. Chem.* **20,** 4316 (1981).
92. A. J. Canty, N. J. Minchin, B. W. Skelton, and A. H. White, *J. Chem. Soc., Dalton Trans.*, 2205 (1986).
93. S. Trofimenko, *Inorg. Chem.* **12,** 1215 (1973).
94. M. I. Bruce, B. L. Goodall, and I. Matsuda, *Aust. J. Chem.* **28,** 1259 (1975).
95. M. Nonoyama and H. Takayanagi, *Trans. Met. Chem.* **1,** 10 (1976).
96. M. Nonoyama, *J. Organomet. Chem.* **229,** 287 (1982).
97. A. A. Watson, D. A. House, and P. J. Steel, *J. Organomet. Chem.* **311,** 387 (1986).
98. G. B. Caygill and P. J. Steel, *J. Organomet. Chem.* **327,** 115 (1987).
99. P. J. Steel and G. B. Caygill, *J. Organomet. Chem.* **327,** 101 (1987).
100. L. Chassot and A. von Zelewsky, *Inorg. Chem.* **26,** 2814 (1987).
101. C. Deuschel-Cornioley, R. Luond, and A. von Zelewsky, *Helv. Chim. Acta* **72,** 377 (1989).
102. T. Miyamoto, K. Fukushima, T. Saito, and Y. Sasaki, *Bull. Chem. Soc. Jpn.* **49,** 138 (1976).
103. A. J. Canty, R. T. Honeyman, B. W. Skelton, and A. H. White, *J. Organomet. Chem.* **389,** 277 (1990).
104. G. B. Caygill and P. J. Steel, *J. Organomet. Chem.* **395,** 375 (1990).
105. O. Juanes, J. de Mendoza, and J. C. Rodriguez-Ubis, *J. Organomet. Chem.* **363,** 393 (1989).
106. W. Wickramasinghe, P. H. Bird, and N. Serpone, *J. Chem. Soc., Chem. Commun.*, 1284 (1981).
107. A. C. Skapski, V. F. Sutcliffe, and G. B. Young, *J. Chem. Soc., Chem. Commun.*, 609 (1985).
108. A. J. Canty, R. T. Honeyman, B. W. Skelton, and A. H. White, to be published.
109. C. M. Anderson, R. J. Puddephatt, G. Ferguson, and A. J. Lough, *J. Chem. Soc.*, Chem. Soc., Chem. Commun., 1297 (1989).
110. A. D. Ryabov, *Chem. Rev.* **90,** 403 (1990).
111. J. K. Jawad and R. J. Puddephatt, *J. Chem. Soc., Dalton Trans.*, 1466 (1977).
112. A. J. Canty, R. T. Honeyman, B. W. Skelton, and A. H. White, *Inorg. Chim. Acta* **114,** L39 (1986).
113. P. K. Monaghan and R. J. Puddephatt, *J. Chem. Soc., Dalton Trans.*, 595 (1988).
114. R. H. Crabtree, *Chem. Rev.* **85,** 245 (1985).
115. G. van Koten, *Pure Appl. Chem.* **61,** 1681 (1989).
116. A. J. Canty, B. W. Skelton, P. R. Traill, and A. H. White, to be published.
117. A. J. Canty, R. T. Honeyman, B. W. Skelton, and A. H. White, *J. Organomet. Chem.* **424,** 381 (1992).
118. M. Brookhart, M. L. H. Green, and L.-L. Wong, *Prog. Inorg. Chem.* **36,** 1 (1988).

ADVANCES IN ORGANOMETALLIC CHEMISTRY, VOL. 34

Organometallic Chemistry of the N=N Group[1]

HORST KISCH and PETER HOLZMEIER

Institut Für Anorganische Chemie der Universität Erlangen-Nürnberg
D-8520 Erlangen, Germany

I
INTRODUCTION

The interest in the interaction of 1,2-diazenes (RN=NR) with transition metal complexes in low oxidation states stems from its relevance to the reduction of dinitrogen. Diimine, 1,2-diazene (HN=NH), is assumed to be an intermediate in biological N_2 fixation[2] (*1a,b*) but it is an open question whether the further transformation to ammonia occurs with or without catalytic action of iron or molybdenum (*1b*). It therefore seems important to study the coordination behavior of diimine toward transition metals. Up to now this has not been investigated because diimine starts to decompose at −180°C (*2*). For this reason the only diimine complexes known, $L_nM(HN{=}NH)ML_n$, were obtained through oxidation of the corresponding hydrazine complexes (*3*). To overcome these difficulties, cyclic diazenes were used as model compounds for *cis*-diimine (*4*).

[1] This article is dedicated to Professor G.N. Schrauzer on the occasion of his 60th birthday.
[2] This is supported by the observation that formation of HD in enzymatic and inorganic model systems only occurs when D_2 and N_2 are present.

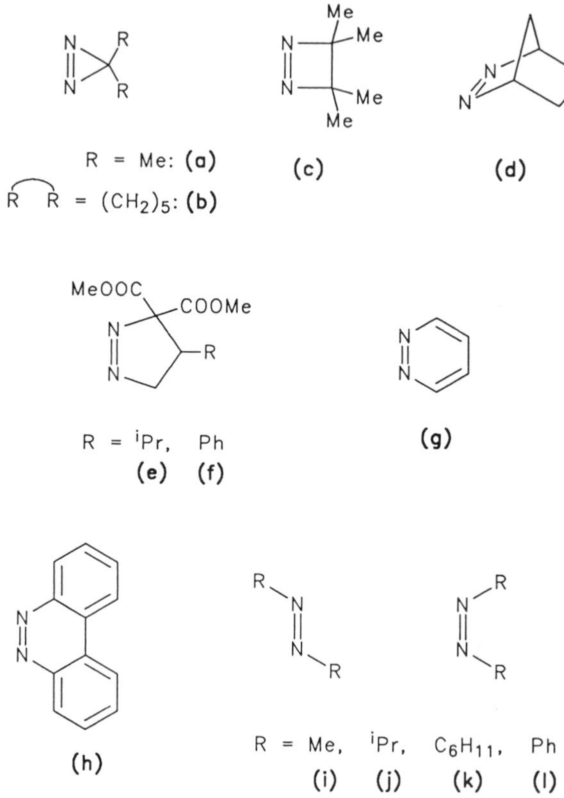

R = Me: (a)

R R = $(CH_2)_5$: (b)

(c)

(d)

R = ^{i}Pr, Ph

(e) (f)

(g)

(h)

R = Me, ^{i}Pr, C_6H_{11}, Ph

(i) (j) (k) (l)

FIG. 1

A further point of more general interest is the question of whether low-valent transition metal fragments activate the N═N group toward C–N bond formation with unsaturated compounds such as olefins or alkynes. It is known from the organic chemistry of 1,2-diazenes that such reactions occur only if electron-withdrawing groups are present in the α,α'-positions. An example is the reaction of bis(ethoxycarbonyl)diazene with cyclopentadiene, one of the first Diels–Alder reactions (*5*). Recently, an intramolecular $[2 + 2]$ photocycloaddition was observed between parallel C═C and N═N bonds (*6*). If a metal complex can induce similar but intermolecular transformations, it may be possible to synthesize nitrogen heterocycles from dinitrogen, because the latter, when coordinated to manganese, can be alkylated to yield dimethyldiazene (*7*).

According to these two points of general interest, the first part of this review contains the types of complexes obtained from the reactions of metal carbonyls with cyclic diazenes of different steric and electronic properties (Fig. 1); the second part describes thermal and photochemical reactions of some diazene iron complexes with alkynes and 1,3-dienes.

Complexes of diimine (*3*) and bis(trimethylsilyl)diazene (*8*) with metal carbonyls and cyclopentadienyl metal fragments, respectively, have been recently reviewed. 1,2-Diphenyldiazenes (azobenzenes) usually react with metal carbonyls by orthometallation, which has been discussed elsewhere (*9*). Similarly, coordination compounds of the classical type [as obtained from dialkyl- or diaryldiazenes and metal halides (*10*)] and azo dyes containing a transition metal (*11*) are not treated in this review. There are only few cases in which 1,2-diaryldiazenes form π complexes with transition metals; they have been summarized in an earlier review (*4*).

II
ELECTRONIC STRUCTURE OF DIAZENES

In the classical description of the N=N bond, the two electron lone pairs are assumed to occupy sp^2 atomic orbitals that are localized on the nitrogen atoms (Fig. 2a). Molecular orbital (MO) calculations and photoelectron spectra (*12*) indicate that through-space (*13*) and through-bond (*14*) interactions lead to the molecular orbitals n_1 and n_2 (Fig. 2b) (*15*). The highest occupied molecular orbital (HOMO) n_1 of a *trans*- and *cis*-diazene originates from the symmetric and antisymmetric combination of the lone pairs, respectively. The energy of this MO decreases in the ligand sequence **a–k** (Fig. 1) from 9.76 eV in **a** to 8.24 eV in *cis*-**j**; the latter value is 8.98 eV in the trans isomer (*12*).

The energy separation between the n_1 and n_2 orbital depends on the NNC bond angle and therefore on the ring size of the cyclic diazene. It amounts to 3–3.6 eV for three- and five-membered diazenes, and only about 1.5 eV for the four-membered, 1,2-diazetines (*12*). This is the reason why the usual orbital sequence n_1, π, n_2 becomes n_1, n_2, π in the latter case. MO theory further reveals that the n_1 and n_2 orbitals may be considerably delocalized onto the neighboring C–N bonds (*16,17*). Thus, the lone pair character of the n_1 orbital is 67, 42, and 32% in *cis*-dimethyldiazene, 1,2-diazetine, and diazirine, respectively. In the same sequence the N–C bond character of the HOMO amounts to 29, 50, and 68% (*18*). Therefore, the HOMO in the case of diazirines is a bonding σ(C–N) rather than a nonbonding lone pair orbital. The N=N bond character of the n_1 orbital remains zero whereas for the n_2 orbital it increases from 9 to 12 and 16%, respectively, in the diazene sequence

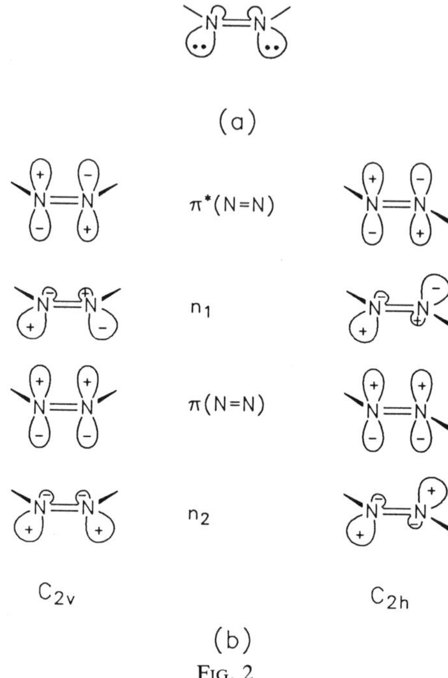

Fig. 2

mentioned above; in contrast, the N═N character decreases from 85 to 80 and 74%, respectively, for the π MO but remains almost constant (96–100%) for the lowest unoccupied molecular orbital (LUMO), i.e., the $\pi^*_{(N=N)}$ orbital.

From these electronic properties one expects a weakening of the C–N bond when the diazene becomes coordinated to one metal atom *via* the n_1 orbital; coordination of a second metal through the n_2 orbital in addition should result in a lesser weakening of the N═N bond. Both effects are expected to decrease in the *cis*-diazene sequence **a–h**.

III

MODES OF COORDINATION AND GENERAL SYNTHESES

The initial interaction of a diazene with a metal is mainly determined by the nature of the HOMO (n_1), which has a more or less pronounced lone pair character. This explains the good ability of diazenes to form mononuclear σ complexes of types **I** and **II** (Fig. 3) wherein the diazene ligand acts as two-electron donor. Additional involvement of the n_2 orbital generates

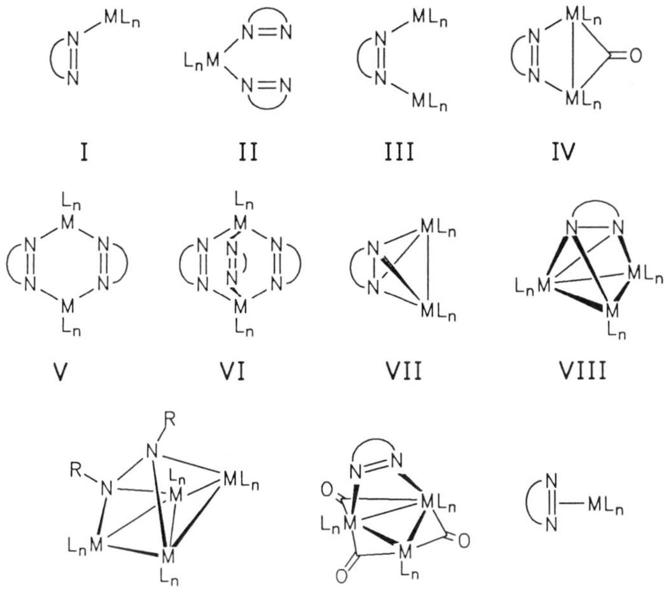

I II III IV

V VI VII VIII

IX X XI

FIG. 3

a four-electron ligand that is able to form the binuclear complexes of types **III–VI** and the cluster **X**. When the π-electrons interact with the metal, the N═N group becomes a six-electron σ/π ligand inducing the formation of binuclear (type **VII**) and tri- and tetranuclear (**VIII** and **IX**) complexes. M–M bond formation from type **III** compounds is observed only in the case of M = Fe and depends on the ring size of the cyclic diazene. The M–M distance increases with decreasing ring size due to a widening of the angle N–N–M (*19*). For this reason the three-membered diazirines are unable to form type **IV** and **VII** complexes. Whereas the doubly bridged complex **V** has been isolated with a three- to six-membered cyclic diazene (*20a–22b*), the stability of triply bridged complex **VI** again depends on the ring size of the cyclic diazene; **VI** cannot be isolated with a diazirine as ligand. The formation of π complexes (type **XI**) is restricted to cases where the diazene is a poor σ-donor, for example, a diaryldiazene, and the metal fragment is a poor σ-acceptor (*4*).

For the synthesis of diazene complexes there are generally three main pathways. The most common is thermal or photochemical ligand substitution according to Eq. (1).

$$\text{RN{=}NR} + \text{M}_x(\text{CO})_y \rightarrow \text{M}_x(\text{CO})_{y-n}(\text{RN{=}NR}) + n\text{CO} \tag{1}$$

Insertion of aryldiazonium groups into the M–H bond of some hydride complexes (*23a–f,24*) yields monosubstituted aryldiazene complexes [Eq. (2)]. A further route is the protonation of aryldiazenido complexes (*23g*).

$$\mathrm{Ar{-}N_2}^{+} + [\mathrm{L}_n\mathrm{M{-}H}]^{m+} \rightarrow [\mathrm{L}_n\mathrm{M(HN{=}NAr)}]^{(m+1)+} \qquad (2)$$

A third way of synthesis is the oxidation of coordinated hydrazine [Eq. (3)].

$$\mathrm{L}_n\mathrm{M(N_2H_4)} \xrightarrow{\mathrm{H_2O_2}} \mathrm{L}_n\mathrm{M(HN{=}NH)} \qquad (3)$$

This method was used by Sellmann *et al.* (*3,25*) to obtain the first complex of unsubstituted *trans*-diazene. In the case of methylhydrazine, oxidation with lead tetraacetate leads to a *cis*-methyldiazene ligand (*26*).

IV

COMPLEXES FORMED WITH RETENTION OF THE N–N BOND

A. *Iron and Ruthenium*

1. *Synthesis and Physical Properties*

Iron carbonyls form a wide variety of different types of compounds, including the first detected metal carbonyl diazene complex $Fe_2(CO)_6L$ (L = benzo[*c*]cinnoline (**h**, Fig. 1; type **VII**, Fig. 3), reported by Bennett (*27*) in 1970. A convenient synthesis is the thermal reaction of $Fe_2(CO)_9$ with a cyclic diazene. Mononuclear complexes of type **I** (Fig. 3) are synthesized using an excess of diazene to prevent further reaction to the more favored complexes **IV**, **VII**, and **VIII** in the case of four- or higher-membered cyclic diazenes (*18,28,29*). An exception would be the highly unstable $Fe(CO)_4$(diazirine) complexes (*19*), which react easily to isolable $[Fe(CO)_4]_2$(diazirine) compounds (type **III**) and which do not tend to form an Fe–Fe bond (see Section III). Another possible way to prepare **I** is the photochemical reaction of $Fe(CO)_5$ with a cyclic diazene (*28*).

Acyclic diazenes do not afford stable mononuclear σ complexes with iron carbonyls; *trans*-dimethyldiazene (*30*) or azobenzene (*31a,b*) yields binuclear complexes **VII** or, via N=N bond cleavage, bi- and trinuclear *o*-semidine complexes, respectively. *cis*-Diazenes are catalytically isomerized to the *trans* isomers, affording the same binuclear products (*32*). The same isomerization occurs in the presence of MH_2Cp_2 (M = Mo, W) (*33*).

As X-ray crystallographic studies on the pyrazoline complex $Fe(CO)_4$(**f**) (*34*) and $Fe(CO)_4$(pyridazine) (*35*) show, the diazene ligand occupies an apical position in the trigonal bipyramid. Dynamic NMR measurements indicate that this apical isomer is preferred by at least 3 kcal mol^{-1} over the equatorial isomer even in solution (*35*). For the corresponding pyridine and pyrazine complex, the same coordination geometry is observed (*36*). In the IR spectrum these latter two $Fe(CO)_4$(amine) compounds show three carbonyl stretching modes ($2\ A_1 + E$) as expected for local C_{3v} symmetry. Contrary to that, all $Fe(CO)_4$(diazene) complexes reveal four ν_{CO} bands (*4*). Because the analogous 2-methylpyridine complex (*37*) likewise shows only three ν_{CO} bands, it can be assumed that the appearance of four bands in the diazene complexes is not caused by a steric effect but rather by a partial double-bond character of the Fe–N bond (*4*). This interpretation is supported by the slight decrease of the Fe–N distance in $Fe(CO)_4L$ from 205 to 198 pm when the ligand π^* orbital is lowered in energy [as estimated from the charge-transfer metal-to-ligand (CTML) transition energy of the corresponding complexes] by replacing pyridine with the pyrazoline (**f**, Fig. 1) (Table I). Whereas steric factors should favor the formation of a staggered rotamer ($\delta = 60°$), an eclipsed conformation seems to be preferred when the N^1FeC^4 angle increases.

From Mössbauer studies on several $Fe(CO)_4L$ complexes (*38–41*) it is known that the isomer shift δ for complexes containing tertiary phosphine ligands PR_3 is significantly lower than in cases in which L represents a nitrogen donor such as a diazene or pyridine, indicating a higher *s*-electron density at the iron nucleus in the former cases. Herberhold *et al.* (*38*) have interpreted this behavior in terms of a higher π-acceptor capacity and therefore stronger p_π–d_π interaction of the phosphine ligands compared to the nitrogen donors. This assumption is further supported by the fact that the $A_1(1)$ ν_{CO} vibration in these complexes, representing the stretching mode of the axial CO ligand, is shifted to lower wavenumbers when going from phosphine to diazene ligands or pyridine (*38*). A comparison of the IR spectra of the pyridine and of the diazene complexes indicates that the latter have a slightly lower electron density at the metal atom pointing to a better π-acceptor capability of the diazene ligands due to their energetically low $\pi^*_{(N=N)}$ orbital.[3] A further comparison shows that the isomer

[3] Note that the absence of N=N bond lengthening is not necessarily a criterion for the presence of π bonding in these complexes: σ-donation through the n_1 orbital, which is antibonding with respect to the N=N bond, is expected to slightly strengthen this linkage, whereas back-donation should cause the opposite and therefore no net effect may be observed.

TABLE I
STRUCTURAL PARAMETERS OF $Fe(CO)_4L$ COMPLEXES

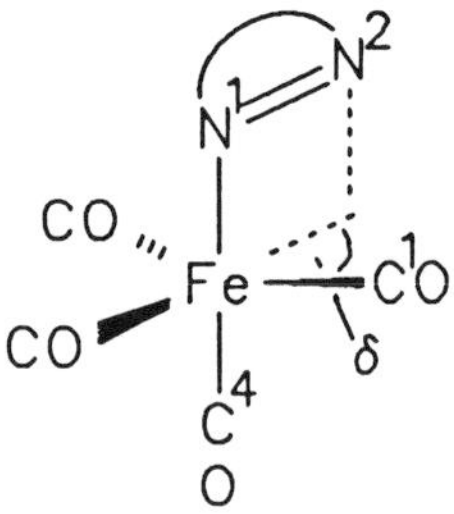

L	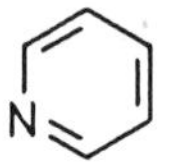 pyridine	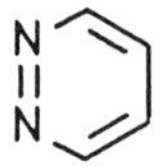 pyridazine	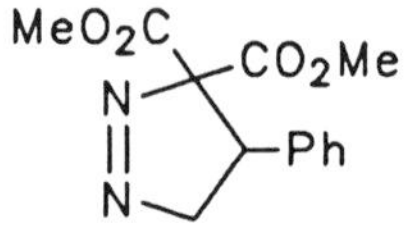 MeO_2C, CO_2Me, Ph pyrazoline
Fe-N	205 pm	201 pm	198 pm
N^1N^2Fe		118°	126°
δ	5.5°	23.8°	0°
N^1FeC^4	176.2°	175.6°	178.0°

shift in $Fe(CO)_4L$ complexes with an aliphatic diazene L (L = **d** and **f**, Fig. 1) (*41*) is about 0.06 mm sec^{-1} higher than in complexes of the aromatic ligands pyridazine and phthalazine (*38*) (Table II). This points to a poorer σ-donor or π-acceptor capability for the former ligands.

The Mössbauer quadrupole coupling constants ΔE_Q of $Fe(CO)_4L$ compounds are 2.3, 2.58, and 3.0 mm sec^{-1} for L = PR_3, CO, and nitrogen donors, respectively (*38*). This enhancement can be correlated with the decreasing σ-donor strength of L, which effects the negative contribution of the axial *d*- and *p*-electron density to the electric field gradient at the iron nucleus.

In the UV–VIS spectrum all $Fe(CO)_4L$ complexes containing a cyclic diazene ligand L show a distinct absorption in the range of 430–480 nm that is assigned to a CTML transition populating the antibonding $\pi^*_{(N=N)}$ orbital (*4*). This band is reversibly shifted to lower wavelength on cooling the compound to −60°C (*29, 42*). Together with a temperature-dependent ^1H NMR spectrum of the 2,3-diazanorbornene complex (L = **d**, Fig. 1), this effect may be explained by a conformational equilibrium due to hindered rotation around the Fe–N bond (*4*).

Mononuclear aryldiazene iron(II) complexes, $[Fe(CO)(HN{=}NAr)L^2_4]^{2+}$ and $[Fe(HN{=}NAr)L^1L^2_4]^{2+}$ [Ar = 4-MeC_6H_4, L^1 = a nitrile, L^2 = $P(OEt)_3$], were prepared by Albertin *et al.* (*23a*) by insertion of an aryldiazonium group into the Fe–H bond of $[Fe(CO)(H)L^2_4]^+$ and $[Fe(H)L^1L^2_4]^+$, respectively. Because the monosubstituted aryldiazene proved to be a good leaving group, the complex was used as a precursor in the synthesis of new iron(II) compounds (*23b*). The corresponding bisdiazene iron(II) complex, $[Fe(HN{=}NAr)_2L^2_4]^{2+}$, was synthesized in a similar manner using the dihydride $FeH_2L^2_4$ (*23c*). The insertion reactions of the ruthenium dihydrides $RuH_2(CO)(PPh_3)_3$ and $RuH_2L^2_4$ afford ruthenium(II) aryldiazene complexes. Whereas only one insertion step can be performed with the former compound (*24*), both mono- and bisdiazene complexes can be prepared using $MH_2L^2_4$ (M = Ru, Os) (*23d*). The synthesis and structural analysis of the mixed aryldiazene/arylisocyanide complex $[Ru(4\text{-}MeC_6H_4N{=}NH)(4\text{-}MeC_6H_4NC)L^2_4]^{2+}$ has been recently reported (*23f*).

The binuclear σ complexes $LFe_2(CO)_7$ (type **IV**, Fig. 3) are obtained by thermal reaction of the diazene ligand with an excess of $Fe_2(CO)_9$ below or at room temperature (*43, 44*). Except the cases in which L represents an aromatic diazene such as pyridazine or phthalazine, wherein the $\pi_{(N=N)}$ orbital is part of the aromatic ring system, they are easily decarbonylated to yield the σ/π complexes $Fe_2(CO)_6L$ (*43*) (type **VII**, Fig. 3).

The molecular structures of $Fe_2(CO)_7$(pyridazine) (*45*) and $Fe_2(CO)_7$(3,3,4,4-tetramethyl-1,2-diazetine) (*18*) have been established by X-ray analysis. Both complexes belong to structure type **IV** [Fig. 3; $ML_n = Fe(CO)_3$] with Fe–N and Fe–Fe distances of 196 and 257 pm and 191 and 262 pm, in the former and latter case, respectively. The longer metal–metal bond in the diazetine complex most likely reflects the larger distance of the lone pairs imposed by the smaller four-membered ring (see Section III). Although the diazene ring in both compounds is coplanar with the Fe_2N_2 ring, the Fe_2CO triangle is folded out from this plane (interplanar angle = 101°). The Fe–Fe bond distances are 257 and 262 pm, respectively. Dynamic ^{13}C NMR studies on the pyridazine complex (*45*) reveal a fast CO scrambling. Between −140 and −30°C the averaging of the seven CO ligands occurs in two distinct steps, the first one including only the five coplanar CO ligands.

Binuclear iron diazene complexes $[Fe(CO)Cp]_2L$ (Cp = η^5-C_5H_5; L = **d** and **g**, Fig. 1, and 2,3-diazabicyclo[2.2.2]oct-2-ene) containing two bridging CO ligands have been synthesized by refluxing a solution of $[Fe(CO)_2Cp]_2$ and L in benzene or toluene for 2–5 days (depending on the ligand L) (*46*). The 2,3-diazanorbornene complex (**1**) shown in Fig. 4 was obtained through thermal decarbonylation of $Fe_2(CO)_3Cp_2$(2,3-diazanorbornene). The latter compound was prepared photochemically from $[Fe(CO)_2Cp]_2$ (*47*).

In the reaction of iron carbonyls with cyclic diazenes, the σ/π complexes $Fe_2(CO)_6L$ (type **VII**, Fig. 3) are the thermally most stable compounds. They are available through thermal or photochemical reaction of $Fe_2(CO)_9$ or $Fe(CO)_5$ with a diazene ligand. Due to their easy formation they are

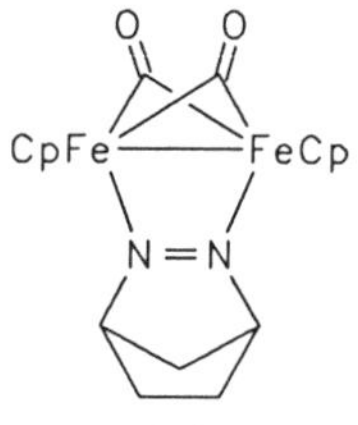

(1)

FIG. 4

often observed as by-products in the synthesis of **I** and **IV** (Fig. 3), e.g., in the reaction of 3,3,4,4-tetramethyl-1,2-diazetine or its mono-*N*-oxide with $Fe_2(CO)_9$ (*18*). In the latter reaction the ratio of the products $Fe_2(CO)_7L$ $Fe_2(CO)_6L$ depends on the solvent used. Argo and Sharp (*48*) obtained complexes **VII** with several alkyl-substituted 3*H*-1,2-diazepines. As a result of coordination, the rate of the [1,5] sigmatropic hydrogen migration across the seven-membered diazepine ring is slowed down significantly.

In contrast to the σ complexes **I** and **IV** (Fig. 3), compounds of type **VII** are also isolable with acyclic diazene ligands, in a few cases. Reaction of $Fe_2(CO)_9$ with dimethyldiazene yields $Fe_2(CO)_6(CH_3N{-}NCH_3)$ (*30*); with phenyl azide the corresponding azobenzene complex is obtained (*31b*).

X-Ray analysis has been carried out for several complexes of type **VII** (*49–52*); structural data have been reviewed elsewhere (*53*). The basic geometry of this type of complex can be regarded as a combination of two distorted tetragonal pyramids, the basal plane of each consisting of the two nitrogen atoms and two carbonyl groups at an iron center (*54*). The coordination sphere around each iron atom corresponds to a distorted octahedron. For the complexes $LFe_2(CO)_6$ with the diazene ligands 2,3-diazanorbornene (**d**), benzo[c]cinnoline (**h**), and *cis*-dimethyldiazene (*cis*), the Fe–Fe distances are 249, 251, and 250 pm, respectively, showing that the nature of the diazene ligand has little impact on the metal–metal bond. The fact that this bond distance is about 10 pm higher than in related complexes, such as $Fe_2(CO)_6[(CH_3N)_2CO]$ (*55*) and $Fe_2(CO)_6(H_2N)_2$ (*56*), which lack N–N bond, whereas at the same time the Fe–N bonds (188–191 pm) are shorter by this amount, was interpreted by Doedens (*50*) as a stereochemical effect caused by strain at the bridging nitrogen atoms. MO calculations for this type of complex following the Fenske–Hall model (*54*) have provided, however, an electronic explanation for this effect. Beyond this, the further outcome of this study is a bonding model for $Fe_2(CO)_6X_2$-type complexes (with and without an X–X bond) in which the Fe–Fe bond is composed of hybrid metal orbitals that have 20% d_{z^2}, 10% p_z, and 5% d_{yz} character at each iron atom. Because the d_{z^2} and the p_z

atomic orbitals are colinear with the *trans*-axial carbonyl ligands, this brings about some "bent" bond character for the metal–metal bond. The Fe–Fe bonding and antibonding orbitals are separated by about 5 eV and correspond closely to the HOMO and the LUMO of the molecule.

Each of the three complexes $Fe_2(CO)_6$(**d**), $Fe_2(CO)_6$(**h**), and Fe_2-$(CO)_6$(*cis*-**i**) shows a small but significant deviation from its ideal symmetry, which would be C_s for $Fe_2(CO)_6$(**d**) and C_{2v} for both of the other compounds. The coordinated benzo[*c*]cinnoline is distorted from planarity by a dihedral angle of 7° between the two phenyl rings (*50*). The nonequivalency of the two iron atoms in complexes of type **VII** is also expressed by their Mössbauer data, summarized in Table II (which lists complexes **2–13**).

Within the three $Fe_2(CO)_6L$ complexes listed, the quadrupole splittings ΔE_Q, measured at 77 K (*41,57*), are different for the two $Fe(CO)_3$ groups.

TABLE II

MÖSSBAUER DATA OF SOME SELECTED IRON CARBONYL COMPLEXES

Complex	L	δ (mm sec^{-1})	ΔE_Q (mm sec^{-1})	Γ (mm sec^{-1})	Ref.
$LFe(CO)_4$[a]					
2	**g**	−0.087	2.97	—	*38*
3	Phthalazine	−0.104	2.93	—	*38*
4	**d**	−0.03	2.83	0.26	*41*
5	**f**	−0.01	3.04	0.28	*41*
$LFe_2(CO)_7$[a]					
6	**d**	−0.029	1.64	—	*38*
7	**h**	−0.020	1.99	—	*38*
8	Phthalazine	−0.034	2.01	—	*38*
9	**g**	−0.031	2.00	—	*38*
$LFe_2(CO)_6$[b]					
10	**f**	0.08	1.12	0.30	*41*
		0.07	0.85		
11	**h**	0.10	1.08	0.25	*41*
		0.09	0.94		
12	**d**	0.05	1.01	0.28	*57*
		0.05	0.81		
$LFe_3(CO)_9$[b]					
13	**d**	0.07	1.23	0.24	*57*
		0.06	0.78		
		0.05	0.62		

[a] Room temperature.
[b] 77 K.

As an effect of the unsymmetrical pyrazoline and 2,3-diazanorbornene ligand this difference is higher for complexes **10** and **12** than for the benzo[*c*]cinnoline complex (**11**). In contrast to these findings, Mössbauer and NMR measurements indicate two equivalent iron centers for the related maleic anhydride adduct of $Fe_2(CO)_6$(3,6-diphenylpyridazine) (*58*). The same result was obtained by Trusov *et al* (*59*) for the ketiminato complexes **14** and **15**. In these compounds the nature of the substituents R at the methyleneimino groups is insignificant for the equivalency of the metal centers. The closely related 3,5,7-triphenyl-4*H*-1,2-diazepine complex (**15**) and its 5,6-dihydro derivative, however, clearly show two different $Fe(CO)_3$ moieties in their Mössbauer spectra (*60*). Note however, that **14** and **15** do not contain a N–N bond (Fig. 5). The ΔE_Q values of the $Fe_2(CO)_6L$ complexes listed in Table II are about 2 mm sec^{-1} lower than those for $Fe(CO)_4L$ compounds, representing the difference in symmetry between a distorted octahedral and a trigonal bipyramidal coordination sphere.

The trinuclear cluster compounds $M_3(CO)_9L$ (M = Fe, Ru) of type **VIII** (Fig. 3) have been synthesized with the five- and six-membered diazenes **f** and **d** (Fig. 1) (*29, 61*). They can be prepared by the thermal reaction of the free diazene with $Fe(CO)_5$ or $Ru_3(CO)_{12}$ as well as photochemically from the corresponding $Fe_2(CO)_6L$ compound and $Fe(CO)_5$. The complexes contain one seven-coordinated and two six-coordinated $M(CO)_3$ groups. As can be seen from the Mössbauer data of $Fe_3(CO)_9$(2,3-diazanorbornene) (**13**) (*57*) in Table II, the ΔE_Q values are in agreement with these two types of coordination spheres; the value of 1.23 mm sec^{-1} can be attributed to the seven-coordinated iron center. There is also a slight difference between the two six-coordinated iron atoms, indicating three nonequivalent $Fe(CO)_3$ groups in this molecule. The molecular structure of the complex has been established by X-ray analysis and is schematically depicted in Fig. 6 (*41*). It possesses a triangular $Fe_3(CO)_9$ core with two Fe–Fe distances of 257 and 258 pm, which are within the range of a single bond, and a longer one (261 pm) between the two six-coordinated iron atoms. The elongation of this bond may be due to its integration in a

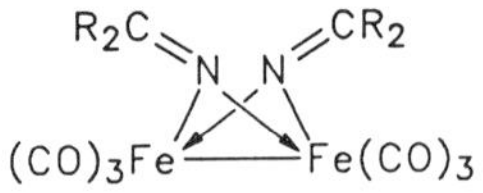

(14) 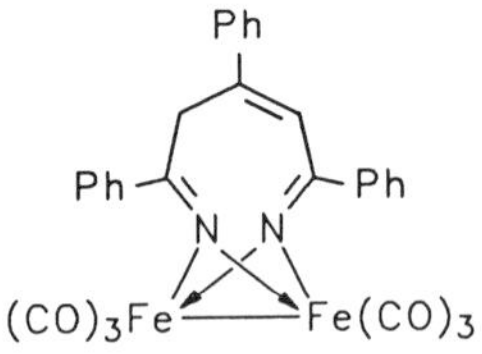 (15)

Fig. 5

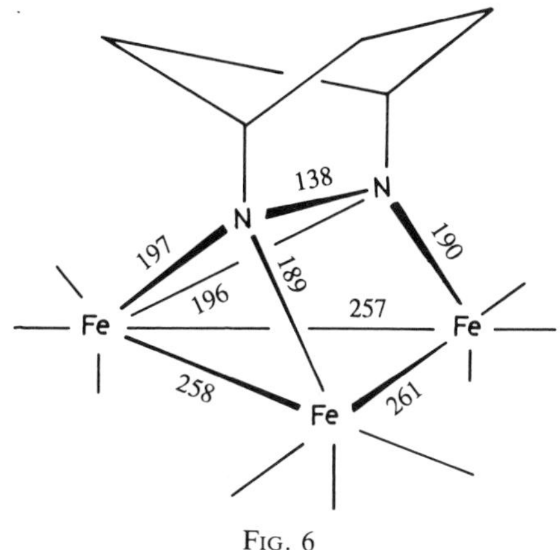

FIG. 6

strained four-membered Fe_2N_2 ring (*41*). Fe–N bond lengths within this ring are 190 pm and therefore shorter by 6 pm as compared to those outside.

Dynamic ^{13}C NMR studies (*61*) reveal that there is no internuclear carbonyl scrambling within the cluster compound at room temperature. For localized CO exchange the activation energy differs significantly for the seven- and the six-coordinated centers. Whereas at the latter two iron atoms scrambling is slow up to −40°C, a fast local CO exchange can be observed at the seven-coordinated $Fe(CO)_3$ group even at −130°C.

Cotton *et al.* (*62*) have shown that internuclear carbonyl scrambling occurs in different types of trinuclear diazene complexes. In $Ru_3(CO)_{10}$(pyridazine) (type **X**, Fig. 3), in which the pyridazine ligand acts as a four-electron donor, a fast averaging process affecting the three bridging and the three equatorial CO ligands occurs within the temperature range of −156 to −86°C. At higher temperatures the axial CO ligands become involved in additional scrambling processes.

A tetranuclear diazene cluster, $Ru_4(CO)_{12}(Et_2N_2)$ (type **IX**, Fig. 3) was obtained by Bantel *et al.* (*63*) by irradiating the corresponding trinuclear cluster **VIII** (*64*) in the presence of $Ru(CO)_5$. The N–N bond distance of 146 pm in this compound closely corresponds to a single bond.

2. *Reactivity*

Among the reaction pathways that can be followed by ruthenium and iron diazene complexes, the most important type is the reactivity of the complexed diazene group toward alkynes and 1,3-dienes yielding metallacycles and organic N-heterocycles. Reactions of this type will be discussed in Section VI.A separate discussion will also be given for the reactions of iron and ruthenium diazirine complexes leading to cleavage of the diazene group.

Substitution chemistry with iron and ruthenium diazene complexes with retention of the N–N bond is restricted to a few examples. Thus, in the trinuclear cluster compound $Fe_3(CO)_9$(2,3-diazanorbornene) up to three carbon monoxide ligands can be substituted on electrochemical reduction in the presence of trimethylphosphite (*65*). Whereas a stoichiometric amount of charge is consumed during the third substitution step, the formation of the mono- and the disubstitution product was found to follow an electron-transfer chain (ETC)-catalyzed mechanism.

Oxidation of the binuclear complexes $[Fe(CO)Cp]_2L$ (L = **d** and **g**, Fig. 1 or 2,3-diazabicyclo[2.2.2]oct-2-ene) with iodine or silver perchlorate affords the mononuclear cationic complex $[Fe(CO)_2Cp(L)]^+$ wherein L acts as a two-electron donor (*46*). In contrast to the oxidation behavior of the isostructural diphosphine complexes $[Fe(CO)Cp]_2[Ph_2P(CH_2)_nPPh_2]$ ($n = 1, 2$, or 3) (*66*), there is no indication for the formation of a binuclear cationic species in this reaction.

b. *Chromium, Molybdenum, and Tungsten*

1. *Synthesis and Physical Properties*

The different types of complexes observed in the reaction of Group VIB carbonyls with cyclic diazenes have been established primarily by Herberhold *et al.* (*22a, 67, 70, 72a–d*) and are shown in Scheme 1. Mononuclear diazene and bisdiazene complexes (types **I** and **II**) are conveniently prepared by ligand displacement reactions (*20a–22b, 67–72d*).

$$M(CO)_5L + RN{=}NR \rightarrow M(CO)_5(RN{=}NR) + L \tag{4}$$

$$M(CO)_4(nbd) + 2RN{=}NR \rightarrow M(CO)_4(RN{=}NR)_2 + nbd \tag{5}$$

where L = tetrahydrofuran (THF), CH_3CN; M = Cr, Mo, W; and nbd = nor-C_7H_8). Type **I** complexes have been isolated with all types of cyclic diazenes (shown in Fig. 1). Contrary to the corresponding $Fe(CO)_4$-(diazene), they are also accessible with acyclic diaryl- and dialkyldiazenes. Thus, whereas in the case of azobenzene only the *cis* isomer gives rise to a mononuclear chromium complex (*70*), Ackermann *et al.* obtained complexes $M(CO)_5(RN{=}NR)$ with both the *cis* and *trans* isomers of diisopropyldiazene (*71a*), dimethyldiazene (*71b*), and several unsymmetrical dialkyldiazenes RN=NR′ (*71c*) by reaction with $M(CO)_5$THF [Eq. (4)]. These dialkyldiazene complexes can also be prepared by oxidation of the corresponding hydrazine complexes with activated MnO_2. The overall

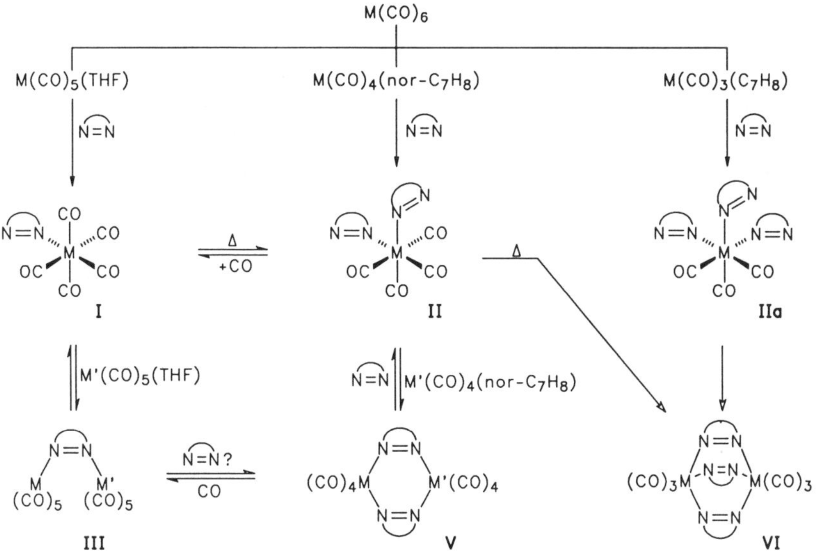

SCHEME 1

yield as well as the ratio of *cis* and *trans* product of this reaction show a significant increase with decreasing size of the alkyl substituents R; in addition, the course of the reaction is dependent on the nature of the central metal (*71c*). Reaction of $Cr(CO)_5THF$ with *trans*-bis(cyclohexyl)-diazene affords the corresponding mononuclear σ complex, which also been prepared by a unique dehalogenation of $C_6H_{11}NCl_2$ with $Na_2[Cr_2(CO)_{10}]$ (*68*).

Out of the several complexes $M(CO)_5(RN{=}NR)$ with acyclic dialkyldiazenes investigated so far (*71a–c*), only those of dimethyldiazene exhibit further reactivity to form binuclear complexes of type **III** (Fig. 3). This is interpreted by Ackermann *et al* (*71b*) in terms of steric hindrance in the case of larger substituents R. The high instability of the bimetallic complex $(CO)_5Cr$(*trans*-azobenzene)$Cr(CO)_5$ (*70*), which decomposes in solution above −40°C, is in agreement with this interpretation.

In the case of cyclic diazenes, a high instability is observed also for the mononuclear σ complexes of the three-membered diazirines. The easy conversion of these compounds to the singly bridged complexes $(CO)_5M$(diazirine)$M(CO)_5$ at temperatures above −30°C in solution and even in the solid state (*20a*) can be explained by the steric and electronic properties of the diazirine HOMO (vide supra).

The stability of the bisdiazene complexes **II** (Scheme 1) likewise is a function of the ring size of the cyclic diazene ligand; the decreasing lone pair character of the diazene HOMO (n_1, Fig. 2) with decreasing ring size should weaken the M–NN bond in the case of small rings due to weaker overlap with an empty metal orbital. Accordingly, complexes of type **II** have been prepared with four-membered and higher membered cyclic diazenes (*21,22,a,b*) but not with three-membered ones.

Starting from $Mo(CO)_3(C_7H_8)$ and pentamethylenediazirine (**b**) Beck *et al.* (*20b*) have obtained $Mo(CO)_3$(**b**)$_2$ in which the diazirine ligand is assumed to act as a two- and four-electron donor. *fac*-$Cr(CO)_3$-(pyridazine)$_3$ (type **IIa**) was obtained from $Cr(CO)_3(C_7H_8)$ and pyridazine (*22a*). Several monometallic diazene complexes containing an arene ligand have been prepared. Thus, the reaction of pentamethylenediazirine (**b**) (*20b*), 2,3-diazanorbornene (*d*) (*72a, b*), benzo[*c*]cinnoline (**h**), and phthalazine and pyridazine (**g**) (*72c*) with photochemically generated $Cr(CO)_2$-($Ar = C_6H_6$ and substituted benzenes), $Mn(CO)_2(Cp)THF$, or $Mn(CO)_2$-$(MeC_5H_4)THF$ yields the corresponding σ complexes.

Smith and Hillhouse (*73*) recently reported the stereoselective synthesis of monosubstituted *cis*-aryldiazenes by displacement of coordinated *cis*-ArN=NH from the tungsten complex [*trans,trans*-$W(CO)_2$(ArN=NH)-$(NO)(PPh_3)_2]^+$. In analogy to the already mentioned monoaryldiazene complexes of iron and ruthenium (*23a–g*), the latter was prepared by reaction of the aryldiazonium salt ArN_2^+ with the corresponding hydrido complex. This study is the first report on a monosubstituted *cis*-diazene ligand.

As dynamic ^{13}C and ^1H NMR spectroscopy reveals, some of the $M(CO)_5$(diazene) complexes exhibit fluxional behavior. A metal 1,2-shift, resulting in a coordination site exchange of the $M(CO)_5$ unit between the lone pairs of the two nitrogen atoms, has been demonstrated to occur in the *cis*-dialkyldiazene compounds. In contrasts, no exchange occurs in the complexes of the *trans*-diazenes but the two isomeric complexes can be isolated in the case of unsymmetrical diazenes (*71a–c*). The same dynamic process was found for $M(CO)_5$(benzo[*c*]cinnoline) (M = Cr, Mo, W) (*69*), $M(CO)_5$(phthalazine) (*74*), and the isolobal arene complexes Cr(Ar)-$(CO)_2$(2,3-diazanorbornene) (*72b*). Concerning this fluxionality there is a significant difference between complexes of cyclic and acyclic diazenes. Whereas the activation free energy $\Delta G^\ddagger$ calculated from the temperature of coalescence amounts to 12–15 kcal mol^{-1} for the diisopropyldiazene complexes (*71a*), values between 17 and 20 kcal mol^{-1} have been reported for $Cr(Ar)(CO)_2$(2,3-diazanorbornene) (*72b*). This difference has been interpreted as an effect of the geometrical properties of the diazene ligand, because the smaller C–N=N angle of the cyclic diazenes as compared to the acyclic ones leads to a longer distance between the lobes of the n_1 diazene orbital (Fig. 2). Thus, the energy of the symmetrical transition state is raised in the case of cyclic diazenes due to poorer overlap between the n_1 and the metal d_{xz} (or d_{yz}) orbitals (*71a*). Because within the series of cyclic diazenes the spatial separation between the lobes of n_1 increases with decreasing ring size, this interpretation may also explain the fact that the metal 1,2-shift does not occur in complexes of three- to five-membered cyclic diazenes. However, extended Hückel calculations on the pentacarbonylchromium complexes of *cis*-HN=NH and pyridazine (*75*) have revealed that the transition state of lowest energy corresponds to a π complex. On the basis of these results, fluxionality should be possible also in complexes of *trans*-diaryldiazenes. In fact, $Re(CO)_2Cp$(*trans*-PhN=NPh) in solution consists of isomers which contain η^1- and η^2-bound azobenzene; the former undergoes a 1,2-metal shift (*72d*).

Within the different $Cr(Ar)(CO)_2$(2,3-diazanorbornene) complexes the activation energy of the migration process shows a clear dependence on the nature of the arene ligand. The coalescence temperatures are lower for electron-rich arenes (52–77°C) than for electron-poor ones (90–95°C), indicating an electronic interaction between the π-bonded arene ligand and the diazene. The linear relationship between $\Delta G^\ddagger$ and the Hammett constant σ_p (*72a*) also points to this type of interaction. The lower values of $\Delta G^\ddagger$ in the case of electron-rich arenes may be explained by destabilization of the metal orbitals, leading to a stronger interaction with n_1 and $\pi_{(N=N)}$ in the transition state.

Most of the $M(CO)_5$(diazene) compounds exhibit four or five ν_{CO} IR absorptions, which indicates that the local symmetry at the metal cannot be approximated by C_{4v}. This behavior parallels the situation in the $Fe(CO)_4$(diazene) compounds. Low-temperature IR spectroscopy (−30 to −50°C) performed with the pyridazine pentacarbonyl complexes of chromium and molybdenum has revealed a splitting of the E band (1944 and 1947 cm^{-1}, respectively) by 6–7 cm^{-1} and a slight shift of the $A_1(2)$ mode (1919 and 1922 cm^{-1}) of about 3 cm^{-1} (*76*). This behavior may be interpreted as the effect of an increasing asymmetry of the pentacarbonyl metal fragment, caused by slower rotation of the diazene ligand around the M–N bond on cooling. In addition, the observed shift of the $A_1(2)$ mode is indicative for a higher relative concentration of the staggered conformer in the cold solution (*76*). In the case of more bulky diazene ligands such as the pyrazolines **e** and **f** the rotation around the M–N bond is more restricted, causing a split E mode at room temperature. Thus, the pentacarbonyltungsten complex of 4-phenyl-3,3-bis(methoxycarbonyl)-1-pyrazoline (**f**) exhibits five carbonyl vibrations.

For the pentacarbonyl complexes of acyclic dialkyldiazenes (*71a–c*), the $\nu_{N=N}$ vibration in the region of 1500–1600 cm^{-1} is shifted to lower wavenumbers by 25 to 50 cm^{-1} relative to the free ligand. This reduction is caused by π back-donation from the metal d_{xz} or d_{yz} orbital into the diazene π^* orbital. However, no clear correlation between the magnitude of this shift and the π-acceptor capability of the diazene ligand can be drawn (*71c*), which may be due to a concomitant weak strengthening of the N=N bond via σ donation through the n_1 HOMO. The presence of back-bonding into the π^* orbital could also be shown within a series of $Cr(CO)_2L(Me_6C_6)$ complexes (L = amine and diazene ligands) by comparison of their ν_{CO} frequencies (*4, 72b*).

As an example for monometallic Group VIB pentacarbonyl complexes containing cyclic diazene ligands, the UV–VIS spectrum of $W(CO)_5$-(pentamethylenediazirine), obtained in solution at −125°C, is shown in Fig. 7, curve 1. As in the case of the $Fe(CO)_4$(diazene), the lowest energy absorption band of these compounds (391 nm, Fig.7) can be assigned to a CTML transition populating the diazene $\pi^*_{(N=N)}$ orbital. This can be deduced from the characteristic blue shift of this band with increasing solvent polarity and from a comparison with several $M(CO)_5$(amine) complexes (*76, 77*). Beyond this, for the same metal center the transition energy is strongly dependent on the nature of the diazene. Thus, in the case of the tungsten complexes $W(CO)_5L$, the transition energy continuously decreases when the diazene ligand L is varied in the sequence **d** to **h** (Fig.1); the corresponding wavelengths are 424, 427, 442, 490, and 538 nm for L = **d, e, f**, *cis*-**1**, and **h**, respectively. In agreement with these findings the high π^* level of the diazene-related 1,2,4-triazoles raises the CTML energy of the complexes $M(CO)_5$(1,2,4-triazole) (M = Cr,

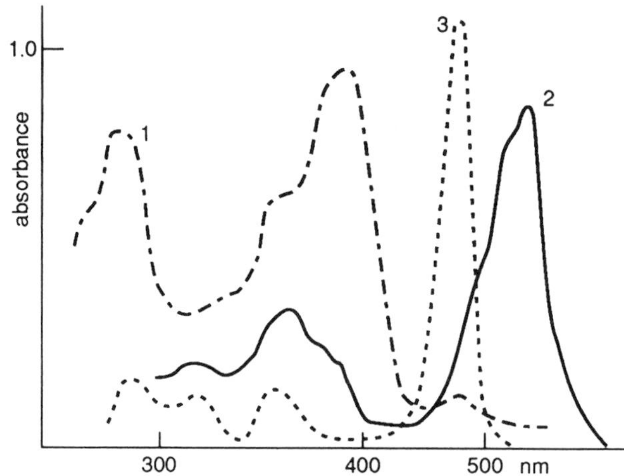

FIG. 7. Electronic absorption spectra of $W(CO)_5$(**b**) (curve 1), $(CO)_5W$(**b**)$W(CO)_5$ (curve 2), and $(CO)_4W$(**b**)$_2W(CO)_4$ (curve 3) in *n*-hexane.

W) into the ultraviolet region (*78*); their long wavelength absorption near 400 nm therefore possesses ligand field (LF) character. However, this correlation between the diazene π^* level and the CTML transition energy of the corresponding complex may not be generalized, because the energies of both the occupied *d* and empty ligand π^* orbitals in the complex are subjected to variations, depending on the effectiveness of the metal–diazene bonding as well as on steric factors (*71a*).

The next higher absorption band in the spectra of $M(CO)_5$(diazene) complexes arises from an LF transition; for $W(CO)_5$(dimethyldiazirine) and $W(CO)_5$(benzo[*c*]cinnoline) it is located at 352 and 390 nm, respectively (*n*-hexane) (*20a, 69*). Additionally, a weak band can be observed in some cases [476 nm for $W(CO)_5$(pentamethylendiazirine), Fig. 7], which was assigned to singlet–triplet LF transition (*20a, 71*). Due to overlap between the CTML and the LF band, several compounds, for example, $M(CO)_5$(diisopropyldiazene) (M = Cr, Mo, W) (*71a*) and $M(CO)_5$(2,3-diazanorbornene) (M = Cr, Mo) (*76*), exhibit a single broad absorption maximum.

Dynamic UV–VIS spectroscopy has been performed on several $M(CO)_5L$ complexes (M = Cr, Mo or W; L = **d, e, g, h, l** in Fig. 1) (*76*); on cooling to −185°C the maximum of the LF absorption remains relatively unchanged, but the CTML band experiences a reversible hypsochromic shift of 15 to 50 nm. As the splitting of the *E* mode in the carbonyl IR spectrum, this temperature-dependent behavior may be thought to arise from the relative concentration changes of the two possible conformers in these compounds (vide supra). For some complexes of the type $M(CO)_5L$, emission is observed at 77 K. In Fig. 8 the absorption and emission spectrum of $W(CO)_5$(*cis*-azobenzene) is shown; the emitting state most likely has triplet LF character because no emission occurs in the case of the corresponding chromium and molybdenum compounds (*76*).

Marabella *et al* (*79*) have reported the structure of another type of mononuclear diazene complex that is obtained as 1:1 adduct of oxobis-(*N,N*-dimethyldithiocarbamato) molybdenum(IV) and dibenzoyldiazene. In the resulting compound, $MoO(PhCON{=}NCOPh)(S_2CNMe_2)_2$, the chelating diazene ligand is coordinated through nitrogen and oxygen.

Since the first bis(pentacoarbonylmetal) complexes of the unsubstituted diazene were synthesized and characterized by Sellmann *et al.* (*25*), a number of isostructural compounds (type **III**, Fig. 3) containing cyclic and acyclic diazenes have been prepared. As already mentioned, Group VIB complexes $(CO)_5M(RN{=}NR)M(CO)_5$ of acyclic diazenes can be obtained only in the case of small substituents R. Thus, besides R = H, *cis*- and *trans*-$H_3CN{=}NCH_3$ are so far the only acyclic dialkyldiazenes to form this type of complex (*71c*). Additionally, in the case of diaryldiazenes, only

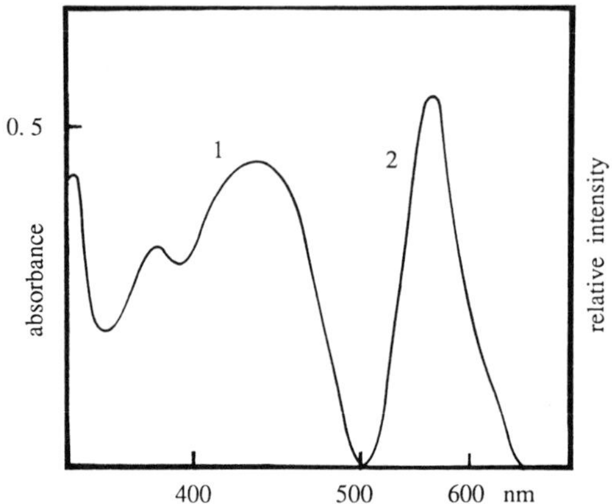

FIG. 8. Electronic absorption (curve 1) and emission spectrum (curve 2) of $W(CO)_5$(*cis*-azobenzene) in 2,3-dimethylbutane/*n*-hexane = 8/3 (v/v) at 77 K.

trans-azobenzene gives rise to a highly unstable singly bridged chromium complex (*70*).

Complexes **III** containing the *cis*-diazene group fixed in a three- or four-membered ring are obtained by reaction of photochemically generated $M(CO)_5THF$ with the free ligand. They have been prepared with the two diazirines, **a** and **b** (M = Cr, Mo, W) (*20a, b*), as well as with the diazetine, **c** (M = W) (*21*). In contrast to the synthesis of the corresponding monometallic complexes **I**, wherein a high excess of diazene is required, stoichiometric amounts are used for the synthesis of **III** (*20a*). Heterobinuclear diazirine complexes $(CO)_5M$(**b**)$M'(CO)_nL$ (M = Cr, Mo, W; M = M′ = Cr, Mo, W, Fe, Mn; L = CO, MeC_5H_4) have been obtained by reaction of $M(CO)_5$(**b**) with $M'(CO)_5THF$, $Fe_2(CO)_9$ or $Mn(CO)_2$-$(MeC_5H_4)THF$ (*80*). As a consequence of the directional properties of the diazene HOMO (n_1, Fig. 2), causing the metal–metal distance within a binuclear complex to decrease with increasing ring size of the diazene (vide supra), compounds of type **III** cannot be obtained with five-membered or higher membered cyclic diazenes. Accordingly, only the centrosymmetric isomer (**16**) was found to be present in the bimetallic *s*-tetrazine complexes $[M(CO)_5]_2(C_2H_2N_4)$ (M = Cr, Mo, W) (*81*) (Fig. 9); these and the corresponding monometallic compounds were synthesized by addition of the ligand to a photolyzed solution of $M(CO)_6$ in diethyl ether.

As in the case of the monometallic complexes **I**, the appearance of five ν_{CO} bands in complexes **III** of cyclic diazenes suggests a deviation from the ideal C_{4v} symmetry. Therefore, restricted rotation around the M–N bonds with steric interference of the two *cis*-fixed $M(CO)_5$ groups can be

(16)

FIG. 9

assumed. In contrast, the bis(pentacarbonyl metal) complexes of *trans*-HN=NH (*25*) and *trans*-$H_3CN{=}NCH_3$ (*71b*), in agreement with C_{4v} symmetry, exhibit three carbonyl vibrations. In the IR spectra of the mono- and bimetallic complexes **I** and **III**, the νCO bands are shifted by 10–20 cm^{-1} to lower energy for complexes of four-membered cyclic diazenes (**c**) as compared to three-membered ones (**a** and **b**). This points to a weaker σ-donor or stronger π acceptor capability of the diazirine ligands. The pattern of the ν_{CO} bands is more complicated than the superimposition of the spectra of the corresponding monometallic complexes. This indicates the presence of different conformers in solution (vide supra). Thus, the IR spectrum of $(CO)_5Cr$(**b**)$Fe(CO)_4$ has 12 bands in the region of terminal ν_{CO} vibrations (*80*).

The molecular structures of $(CO)_5Cr$(**b**)$Mo(CO)_5$ (*80*) (Fig. 10) and $(CO)_5W$(**c**)$W(CO)_5$ (*82*) have been established by X-ray analysis. As a

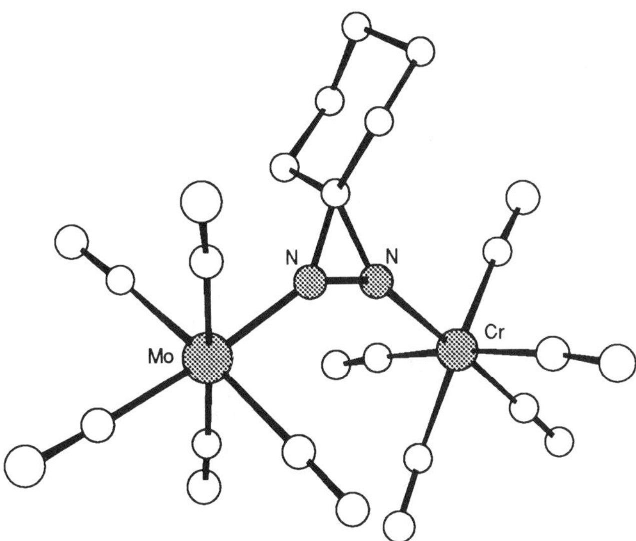

FIG. 10

consequence of the weaker π-acceptor capability of the diazetine ligand as compared to the diazirine, a distinct trans effect can be observed in the diazetine complex; the W–C bonds *trans* to nitrogen have a length of 201 pm as compared to 204 pm measured for the *cis* positions. Out of the three possible conformers ss, ee, and se that arise from rotation around both M–N bonds [where s and e denote staggered and eclipsed conformation of one $M(CO)_5$ fragment with respect to the relative orientation of the M–C and N=N bonds], structural analysis has revealed the se arrangement to be present in both complexes. This result is not surprising, because a minimum of steric strain between the two $M(CO)_5$ groups can be assumed in this conformer (Fig. 10).

The UV–VIS spectrum of $(CO)_5W(\mathbf{b})W(CO)_5$ shows the typical pattern of this type of complex (Fig. 7, curve 2). Irrespective of a *cis* or *trans* configuration of the bridging diazene, this can be regarded as a general feature of the M(RN=NR)M moiety (*83*). Whereas the metal-centered absorptions are located at about the same energy as compared to the monometallic compounds **I**, the CTML band is significantly red-shifted—about 150 nm for the displayed $(CO)_5W(\mathbf{b})W(CO)_5$. Due to the three possible conformers in the complexes **III** (vide supra), their low-temperature spectra exhibit shoulders on both major bands (Fig. 7). The presence of two different metal centers in the heterobimetallic complexes does not change the pattern of the UV–VIS spectra; only the CTML band of the mixed-metal compound $(CO)_5Cr(\mathbf{b})Mn(CO)_2(MeC_5H_4)$ (*80*), located at 494 nm, is blue-shifted by about 40 nm with respect to the dichromium complex.

The UV–VIS spectra of the dimetallacycles **V** (Fig. 3) closely resemble those of **III**. Thus, in the case of $(CO)_4W(\mathbf{b})_2W(CO)_4$ only the low-energy CTML absorption is affected by the introduction of a second diazene ligand, as is evident by a hypsochromic shift of about 70 nm (Fig. 7, curve 3). An additional indication for the charge-transfer character of this absorption is given by its high absorptivity; thus, for the pentamethylenediazirine compounds $(CO)_4M(\mathbf{b})_2M(CO)_4$, the ε values are 50.5×10^3, 70.2×10^3 and 58.0×10^3 M^{-1} cm^{-1} in the case of M = Cr, Mo, and W, respectively. In contrast to complexes of types **I–III** (Scheme 1), the rigid structure of the dimetallacycles prevents the formation of conformers, and the UV–VIS and IR spectra therefore lack temperature-dependent behavior. For the same reason the half-width of the CTML band is significantly lowered as compared to the singly bridged compounds **III** (*20a*).

Complexes **V** are prepared by reaction of $M(CO)_4(\text{nor-}C_7H_8)$ with the free diazene ligand; the reaction can be assumed to proceed via

$M(CO)_4(diazene)_2$ as an intermediate (Scheme 1). Unlike the situation in the singly bridged compounds **III**, the ring size of the cyclic diazene ligand seems to have little impact on the stability of **V**. Thus, they have been isolated with the diazirines (**a** and **b**) (*20a, b*), the diazetine (**c**) (*21*), and the pyridazine (**g**) and 2,3-diazanorbornene (*d*) (*22a, b*). In addition, Ackermann and Kou (*22b*) obtained heterobimetallic compounds **V** from $M(CO)_4(\mathbf{d})_2$ and $M'(CO)_4(nor\text{-}C_7H_8)$ ($M' = M = Cr, Mo, W$).

It may be speculated that electron delocalization within the six-membered bisdiazene dimetal ring, besides a chelate effect, is one of the reasons for the stability of the dimetallacycles. Further support for this assumption is given by the X-ray analysis data of the molybdenum pentamethylenediazirine compound $(CO)_4Mo(\mathbf{b})_2Mo(CO)_4$ (*84*) that revealed a planar M_2N_4 ring for this complex. Beyond this, the elongated N–N bond distance of 126 pm as compared to 123 pm in the free ligand (*85*) points to some back-donation into the diazene π^* orbital. On the contrary, a non-planar M_2N_4 arrangement was found to be present in the corresponding tungsten complex of the four-membered diazetine (**c**) (*86*). The different geometries of the two compounds are also reflected by the IR spectra; whereas all diazirine dimetallacycles exhibit four carbonyl stretching modes, in agreement with local C_{2v} symmetry at the $M(CO)_4$ units (*20a*), five carbonyl vibrations are observed in the case of the diazetine complex (*21*).

A rhodium complex of type **V**, Rh_2Cl_2(benzo[*c*]cinnoline)$_2(CO)_4$, and the corresponding monomeric compound were synthesized by Nixon and Kooti (*87*) through the reaction of $[RhCl(CO)_2]_2$ with the free ligand in refluxing benzene.

Group VIB metal complexes $(CO)_3M(diazene)_3M(CO)_3$(type **VI**) containing three bridging diazene ligands were first isolated by Herberhold *et al.* (*22a*) from the reaction of $Cr(CO)_6$ with 2,3-diazanorbornene or pyridazine in boiling dioxane. Since the order of stability within the different types of complexes is **VI** > **V** > **II** for these ligands (*22a,b*), compounds **VI** can also be obtained by heating or irradiating the corresponding complexes **V**, **II**, or **I** in the presence of the diazene. Whereas thermal reaction of the diazetine (**c**) with $M(CO)_6$ ($M = Cr, W$) analogously affords $(CO)_3M(\mathbf{c})_3M(CO)_6$ (*21*), this method cannot be employed in the case of the thermally unstable pyrazolines (*88*); the corresponding chromium complex of the pyrazoline (**e**) therefore is obtained from the more reactive $Cr(CO)_4(nor\text{-}C_7H_87)$ and the diazene at 45°C (*89*). The reaction of $M(CO)_3(C_7H_8)$ with an excess of diazene offers a more specific method for the synthesis of **VI** (Scheme 1); *fac*-$M(CO)_3(diazene)_3$ can be isolated as an intermediate in some cases (*22a, b*).

The electronic spectra of the triply bridged complexes **VI** exhibit a broad absorption around 550 nm; thus, there is a bathochromic shift of the low-energy band by about 70 nm as compared to the dimetallacycles **V**. Additionally, a high-energy shoulder at about 500 nm can be observed in the spectrum of the pyrazoline complex; this absorption is shifted to 478 nm when diazetine (**c**) is employed as a bridging ligand (*21*).

In the IR spectrum, all triply bridged complexes **VI** show two sharp ν_{CO} absorptions in the range of 1900–1810 cm^{-1}, consistent with local C_{3v} symmetry at the $M(CO)_3$ moieties; a considerable extent of back-donation into the π^* orbitals of the three diazene ligands is indicated by the low wavenumbers. The high symmetry of the molecule (D_{3h}) has been confirmed by X-ray analysis in the case of the diazetine complex (*90*).

2. *Reactivity*

Despite the large number of papers that have appeared concerning the synthesis and spectroscopic properties of Group VIB diazene complexes, there is little knowledge about their chemical reactivity. Kooti and Nixon (*69*) have made attempts to probe the thermal reactivity of $M(CO)_5$-(benzo[*c*]cinnoline) (M = Cr, Mo). Heating the chromium compound with an excess of PPh_3 results in the formation of $Cr(CO)_4(PPh_3)_2$. Irradiation of the singly bridged $(CO)_5W(\mathbf{b})W(CO)_5$ in the presence of PPh_3 within the 370-nm region entails substitution of carbon monoxide; on the contrary, no reaction is observed upon excitation of the CTML state (*20a*). This behavior reflects the known difference in reactivity between the LF and CTML excited states (*91*).

Albini and Fasani (*92*) have investigated the photochemistry of $M(CO)_5$(2,3-diazanorbornene) (M = Cr, W). On irradiation within the lowest excited state, CO as well as the diazene ligand are substituted in both complexes. In the tungsten complex the photoreactive state has triplet CTML character whereas the chromium compound reacts from a singlet LF state. Beyond this, the labilization of the diazene ligand is clearly favored in the tungsten complex and can be sensitized by benzophenone, probably via electron transfer (*92*).

Both diazene ligand labilization and substitution of CO ligands have also been observed on photoexcitation of the dimetallacycles $(CO)_4M(\mathbf{b})_2M$-$(CO)_4$ (**17**, Scheme 2; M = Cr, Mo, W) (*93*). Laser flash experiments suggest that the primary photoproduct is $(CO)M(\mathbf{b})_2M(CO)_3$(solvent), which has an absorption maximum at 520 nm and a half-life above 1 second in argon-saturated solution. For the chromium compound the quantum yields of educt disappearance are 0.056 and 0.0063 for LF and CTML excitation, respectively. In contrast, the corresponding quantum

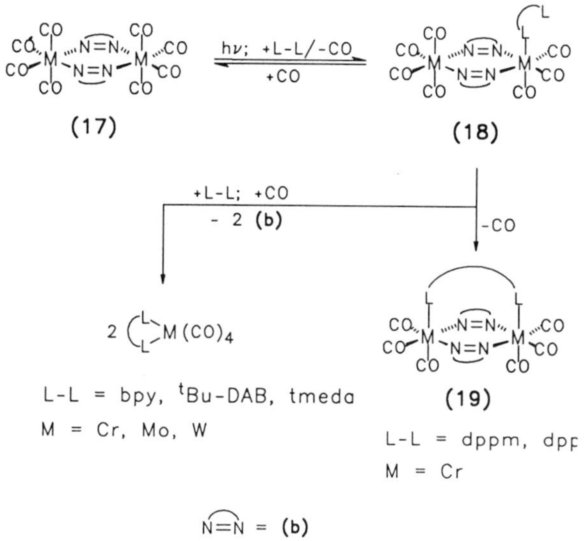

SCHEME 2

yields for the molybdenum and tungsten complex are about 60 times lower for CTML excitation. When irradiated within its LF or CTML band, each of the three complexes undergoes elimination of one CO ligand, whereupon a monosubstituted dimetallacycle is formed in the presence of an added phosphine ligand. Substitution of CO by better σ-donor ligands such as phosphines entails significant alterations in the UV–VIS spectra of the dimetallacycles. Thus, with increasing extent of substitution the CTML band shifts from 475 to 503 and 539 nm in the sequence from $(CO)_4$-$Cr(\mathbf{b})Cr(CO)_4$ to $(CO)_4Cr(\mathbf{b})_2Cr(CO)_3PPh_3$ and **19**, respectively (Fig. 11). This is agreement with an increasing destabilization of metal *d* orbitals on replacing CO by phosphines.

Irradiation of the chromium compound in the presence of the ditertiary phosphine $Ph_2PCH_2CH_2PPh_2$ (dppe) yields the triply bridged complex $(CO)_3Cr(\mathbf{b})_2(dppe)Cr(CO)_3$ (**19**) in which the planar geometry of the M_2N_4 ring is retained (*93*). Contrary to that, X-ray analysis of the analogous complex $(CO)_3Cr(\mathbf{b})_2(dppm)Cr(CO)_3$ (dppm = $Ph_2PCH_2PPh_2$) reveals a "folded book" shape of the Cr_2N_4 ring, which is expressed by an angle of 152.8° between the two planes defined by Cr(1), Cr(2), N(1), N(2) and Cr(1), Cr(2), N′(1), N′(2) (*89*). This deviation from planarity can be attributed to steric strain caused by the shorter P–P distance of the dppm ligand. The doubly bridged compound $(CO)_4Cr(\mathbf{b})(dppm)Cr(CO)_4$ could be isolated as a by-product of the photochemical reaction of $(CO)_4Cr(\mathbf{b})_2$-$Cr(CO)_4$ and dppm.

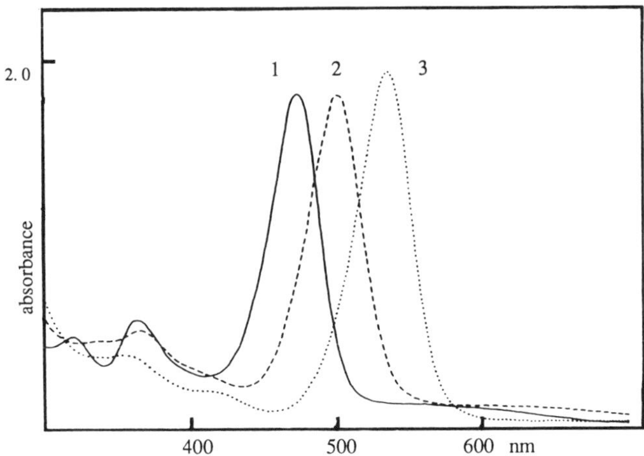

FIG. 11. Electronic absorption spectra of $(CO)_4Cr(\mathbf{b})_2Cr(CO)_4$ (curve 1), $(CO)_4Cr(\mathbf{b})_2Cr(CO)_3PPh_3$ (curve 2), and $(CO)_3Cr(\mathbf{b})_2(dppe)Cr(CO)_3$ (curve 3) in toluene.

However, irradiation of **17** in the presence of stronger σ-donors, such as 2,2′-bipyridine (bpy), 1,4-di-*tert*-butyl-1,4-diaza-1,3-diene (^{t}Bu-DAB), or tetramethylethylenediamine (tmeda), leads to destruction of the dimetallacycle, affording $M(CO)_4(L–L)$ as reaction product. Both observed reaction pathways proceed via the common intermediate **18** (Scheme 2) containing a η^1-coordinated ligand. The subsequent attack of the uncomplexed ligand part occurs at the same metal in the case of nitrogen ligands, with concomittant M–N cleavage, whereas the other metal is preferred by phosphorous ligands. This difference most likely is due to a more crowded transition state for attack at the same metal in the case of the latter ligands, which have the two large phenyl groups at the ligating atom, compared to only one or two small methyl substituents at the nitrogen atom of the other ligands.

In addition to photochemical activation, CO substitution in the diazirine dimetallacycles can also be induced by electrochemical reduction. The reaction follows an electron-transfer chain-catalyzed mechanism [Eqs. (6)–(8)] with the efficiency of the catalytic chain decreasing in the order $Mo \gg W > Cr$.

$$\mathbf{17} + e^- \rightarrow \mathbf{17}^- \quad (6)$$

$$\mathbf{17}^- + L \rightarrow \mathbf{18}^- \quad (7)$$

$$\mathbf{18}^- + \mathbf{17} \rightarrow \mathbf{18} + \mathbf{17}^- \quad (8)$$

Depending on the applied potential, mono- and disubstituted products can be selectively obtained. As in the photochemical reaction, the same difference in reactivity is observed with phosphine and nitrogen ligands. This indicates that **18** is also an intermediate in the ETC-catalyzed reaction (*94*).

V

COMPLEXES FORMED BY CLEAVAGE OF THE N–N BOND

The cleavage of the N=N bond of substituted 1,2-diazenes in the coordination sphere of a transition metal can be regarded as a model reaction for the transformation of diimine to ammonia in the course of biological nitrogen fixation. Further interest in this type of reaction stems from its close relationship to diazene metathesis, for which the reaction of *cis*-azobenzene with $(CO)_5Cr{=}C(OMe)Me$, yielding a mixture of heterocycles (1,2- and 1,3-diazetidinones) and the metathesis product PhN=C(OMe)Me (*95a*), is a rare example. Due to *trans–cis* isomerization of azobenzene, the reaction can also be performed with the *trans* isomer when conducted photochemically. In contrast to the chromium carbene complex, irradiation of *trans*-azobenzene in the presence of the analogous tungsten compound $(CO)_5W{=}C(OMe)Me$ does not give heterocycles; in this case only products of azo metathesis are formed. The reaction has been shown to proceed via a zwitterionic intermediate that could be isolated and characterized by its NMR data (*95b*). It must be emphasized that unlike olefin metathesis, both the described reactions as well as the formation of $[NbCl_2(Me_2S)(NC_6H_5)]_2(\mu\text{-}Cl)_2$ from azobenzene and $Nb_2Cl_6(Me_2S)_3$ (*96*) are stoichiometric examples of azo metathesis; a catalytic version involving substituted 1,2-diazenes has not yet been observed.

Most transition metal-induced N=N bond cleavages are known with aromatic diazenes. The reaction of azobenzenes with carbonylcyclopentadienyl complexes of molybdenum and cobalt as well as iron and ruthenium carbonyls yields mono- or polynuclear complexes containing arylimido or *o*-semidine ligands (*31,97,98a,b*). $Ru_3(NAr)(CO)_{10}$ reacts with diaryldiazenes under a CO atmosphere to form $Ru_3(NAr)_2(CO)_9$ and ArNCO (*98c*). Evans *et al.* (*99*) reported the insertion of two CO molecules into the N=N bond of $Cp_2Sm(\mu\text{-}PhN{=}NPh)SmCp_2$.

Due to the importance of titanium species in organic synthesis, e.g., for the reductive coupling of ketones (*100*), many attempts have been made to study the behavior of substituted 1,2-diazenes toward titanium complexes. Whereas the reaction of azobenzene with $Ti(CO)_2Cp_2$ yields the monomeric complex $TiCp_2(PhN{=}NPh)$, in which the diazene ligand is coordinated side-on (*101*), the reaction with $TiCl_2Cp$ partially leads to a cleavage of the N=N bond and the formation of the dinuclear μ-phenylimido complex (**20**) (*102*) (Fig. 12).

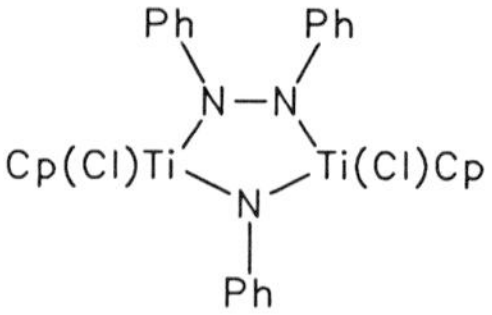

FIG. 12

The nature of the products formed by the reaction of $Ti(CO)_2Cp_2$ with several aliphatic and aromatic azodicarboxylic acid esters $RO_2CN{=}NCO_2R$ changes with the ratio of the educts (*103*). Using both components in a molar ratio of 1:1 yields the five-membered metallacycles (**21**, Scheme 3) through formal 1,4-addition of the diazene to the $TiCp_2$ fragment. When a two-fold excess of $Ti(CO)_2Cp_2$ is employed, the initially formed metallacycle reacts by cleavage of the N═N bond to give **22, 23**, and **24**. The same three complexes are formed upon slow thermal decomposition of **21** in solution or further reaction with titanocenedicarbonyl. Although the metallacycle (**21**) can also be obtained with the substituted

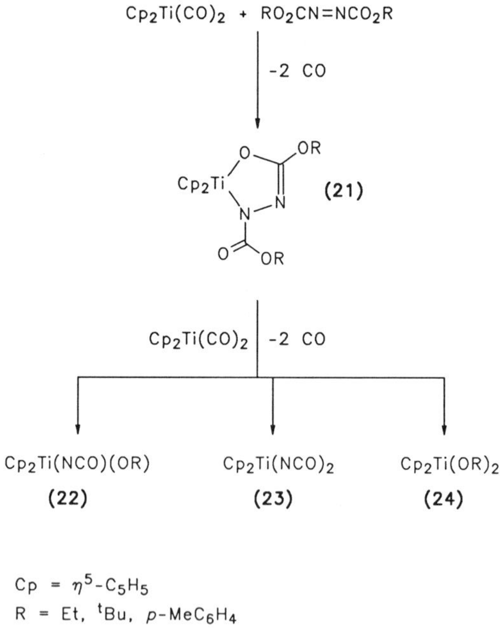

SCHEME 3

azobenzene $MeC_6H_4CO–N{=}N–COC_6H_4Me$, it does not afford analogous cleavage products in this case. Thus, it may be speculated that the formation of the strong Ti–OR bonds in the subsequent products is essential for the cleavage of the diazene group.

In agreement with this interpretation, steric strain in the three-membered diazirines (**a** and **b**) can be regarded as the driving force for N=N cleavage with titanocenedicarbonyl. This reaction can be assumed to proceed via initial σ-coordination of the diazirine (Scheme 4). As the diazirine HOMO is mainly a bonding C–N orbital (see Section II), this should facilitate C–N cleavage to the postulated $Ti(CO)Cp_2(N–N{=}CR_2)$; subsequent migration of carbon monoxide to the nitrenic nitrogen and concomitant N–N cleavage finally lead to the observed isocyanato–ketiminato complex (**25**).

Cleavage of the diazirine N=N bond also occurs during the reaction with $Fe_2(CO)_9$ or $Fe(CO)_5$ (*19*). The singly bridged bis(*N,N'*)tetracarbonyliron complex **III**, which is formed on thermal reaction of the diazirines (**a** and **b**) with $Fe_2(CO)_9$ in tetrahydrofuran slowly decomposes in solution to yield the ketiminato complexes (**26** and **27**, Scheme 5). Both complexes can be obtained directly by conducting the reaction in inert hydrocarbon solvents or photochemically from $Fe(CO)_5$ and the diazirines.

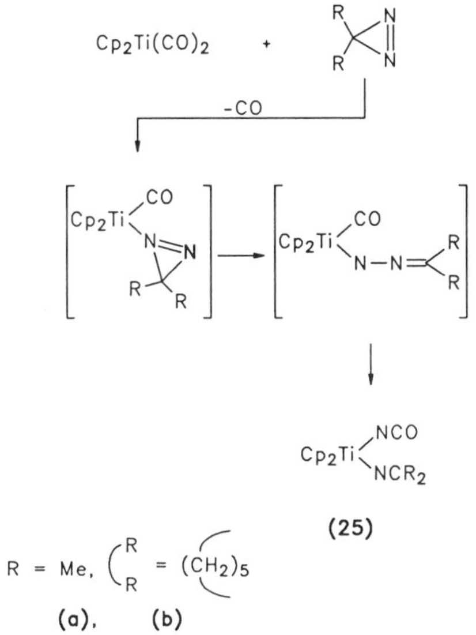

SCHEME 4

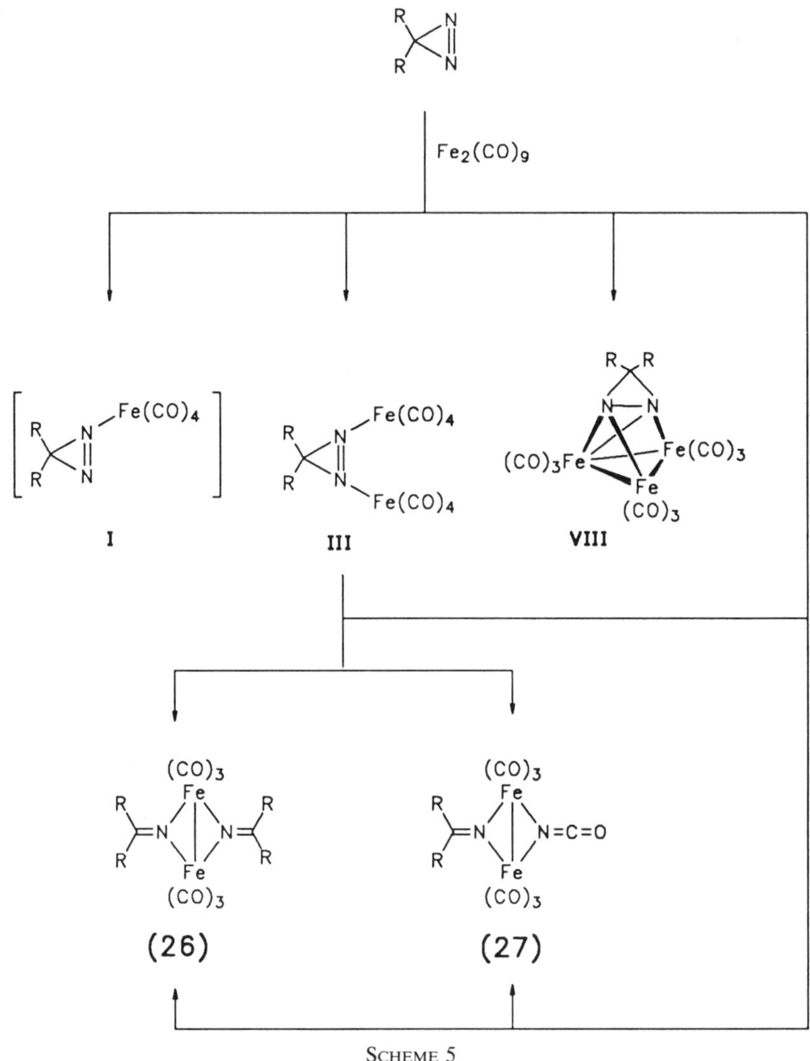

SCHEME 5

IR spectroscopic analysis of the reaction of $Fe_2(CO)_9$ with pentamethylenediazirine in *n*-hexane indicated the initial formation of the mononuclear species **I** (Scheme 5); however, this species could not be isolated. Due to the special geometrical properties of the diazirine HOMO (see Section III), Fe–Fe bond formation under retainment of the N–N bond, leading to stable complexes **IV** or **VII** (Fig. 3), is not possible in the case of singly bridged diazirine species **III** [accordingly, the cluster compound **VIII** that can be thought to arise from addition of a third $Fe(CO)_3$ fragment to the binuclear σ/π complex **VII** is observed only in small traces during this reaction]. This can be achieved only after opening of the diazirine ring. The formation of the isocyanate complex (**27**, Scheme 5) suggests the involvement of an nitrenic intermediate as postulated in reactions of organic azides with metal carbonyls (*104*). No molecular nitrogen is formed, although the composition of the bisketiminato complex (**26**) would imply that; the missing two nitrogen atoms are part of an insoluble by-product containing isocyanate groups.

Thermal reaction of pentamethylenediazirine (**b**) with $Ru_3(CO)_{12}$ in toluene affords the trinuclear isocyanato–ketiminato complex (**28**) (*105*) (Fig. 13). As X-ray analysis reveals, there is no bonding interaction between the two $Ru(CO)_3$ fragments that are linked by the isocyanato, the ketiminato, and one $Ru(CO)_4$ group. The ruthenium atom of the latter is 284 pm distant from both other metal atoms; Ru–N bond lengths are in the range of 210–218 pm and the interplanar angle between the planes defined by one of the nitrogen atoms and the two metals of the $Ru(CO)_3$ groups is 124°.

The cleavage of diethyldiazene coordinated to a trinuclear iron cluster of type **VIII** (Fig. 3) leading to the bicapped bis(μ_3-etylimido) species $Fe_3(CO)_9(\mu_3\text{-NEt})_2$ was reported by Wucherer and Vahrenkamp (*64*). The reaction is complete within 11 hours in boiling ligroine. The corresponding bis(μ_3-methylimido) complex has been prepared by the reaction of $Fe_2(CO)_9$ with methylazide (*31b*); the molecular structures of this compound (*106*) and the isostructural bis(μ_3-phenylimido) complex (*107*) have been evaluated.

The cleavage of the N=N bond of diazirines during the reaction with titanium, iron and ruthenium carbonyls stands in clear contrast to the behavior of cyclic diazenes toward group VIB carbonyls; in the latter case exclusively, complexes containing the intact diazenee group are formed. The cleavage with titanocenedicarbonyl may be due to the strong reducing capability of this compound; in the case of iron carbonyls the high tendency

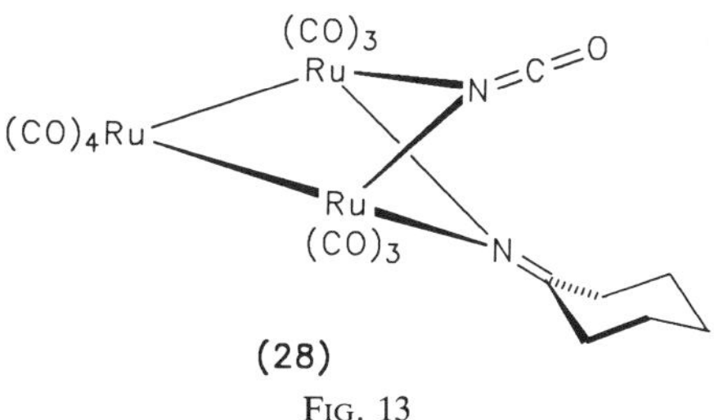

FIG. 13

to form metal–metal bonds is probably the main reason for the observed N═N cleavage; whereas an Fe–Fe bond can be achieved without encroaching on the N═N bond in the case of higher membered diazene rings, it is not possible in the singly bridged diazirine complexes of type **III** due to the larger metal–metal distance.

There are few other examples for N═N bond cleavage of an aliphatic diazene caused by coordination to a transition metal complex: 1,2-bis-(trifluoromethyl)diazene reacts with *trans*-$IrCl(CO)(PMePh_2)_2$ to give the nitrene complexes *cis*- and *trans*-$IrCl(CO)(NCF_3)(PMePh_2)_2$ (*108*). A dinuclear complex containing two bridging and two terminal trimethylsilylimido ligands, $[CrCp(NSiMe_3)(\mu_2\text{-}NSiMe_3)]_2$, is formed on reaction of $CrCl_2Cp$ or with $Me_3Si–N═N–SiMe_3$ (*109a*). X-Ray analysis of this compound reveals a single bond between the two metal centers and Cr–N distances of 164 and 165 pm for the terminal $NSiMe_3$ groups; these are values lying between a formal double and a formal triple bond. In contrast to this cleavage reaction, a 1,2-shift of the $NSiMe_3$ group occurs in the reaction of bis(trimethylsilyl)diazene with the metallocenes MCp_2 (M = V, Mn), affording complexes that contain terminal or bridging 1,1-bis(trimethylsilyl)diazene ligands (*8,109b*)].

VI

REACTIONS WITH ALKYNES AND 1,3-DIENES

A. *Iron-Assisted Synthesis of 1,2,3-Diazepinones*

As mentioned in Section I, the N═N double bond of 1,2-diazenes usually does not easily form C–N bonds with alkynes or olefins. Because some of the complexes described in the preceding discussions, for example the binuclear compounds of type **VII**, contain a formal N–N single bond that may be more reactive, we investigated their reactions with alkynes (*110–113*).

Irradiation of **VII** (Scheme 6) in the presence of an aliphatic or aromatic alkyne affords the adduct (**29**). Note that only one Fe_2N fragment of the tetrahedral Fe_2N_2 core is involved in the reaction and that the second one remains unchanged. Because carbon monoxide does not influence the rate of the reaction, it is likely that opening of one Fe–N bond is the primary photochemical process. The created coordination site is then occupied by the alkyne, which inserts into the second Fe–N bond, affording the final product (**29**). Another mechanistic possibility, direct insertion of the alkyne into one Fe–N bond followed by coordination of the generated

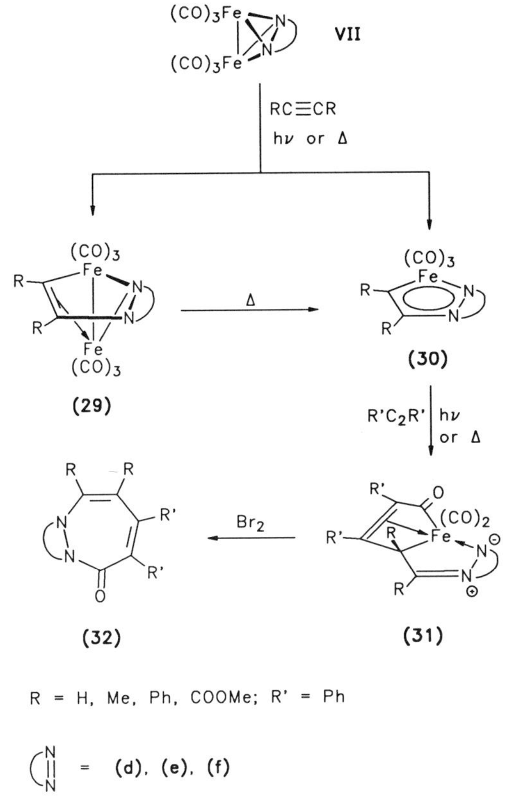

R = H, Me, Ph, COOMe; R' = Ph

= (d), (e), (f)

SCHEME 6

σ-vinyl group and concomitant opening of the other Fe–N bond, cannot be excluded, however. The reactivity of **29** depends on the nature of the cyclic diazene. When this is 2,3-diazanorbornene (**d**) or a pyrazoline (**e** and **f**) the compounds are converted while standing at room temperature into the 2,3,1-diazaferroles (**30**); in the case of benzo[*c*]cinnoline (**h**), even heating to 180°C with or without added carbon monoxide does not yield **30**. Therefore the benzo[*c*]cinnoline complexes (**29h**) can be also prepared thermally. Another route to (**30d**) is the reaction of $Fe_3(CO)_9$(**d**) with $MeO_2CC_2CO_2Me$ at 80°C (*111*).

Although the mechanism of the transformation of **29** → **30** is not known, it is most likely that loss of the $Fe^2(CO)_3$ group (Fig. 14) is initiated by nucleophilic attack of solvent or carbon monoxide at this metal center. The latter is much less sterically shielded in the case of **29f** because the lone pair at N-1 points away from it, whereas it is directed toward it in the complex

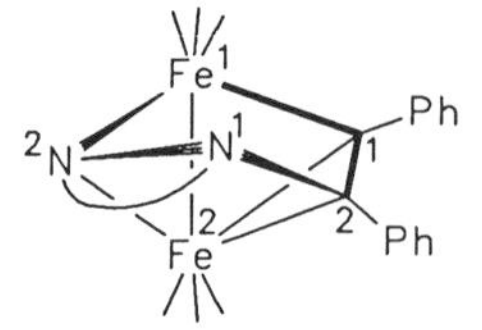

(29f)	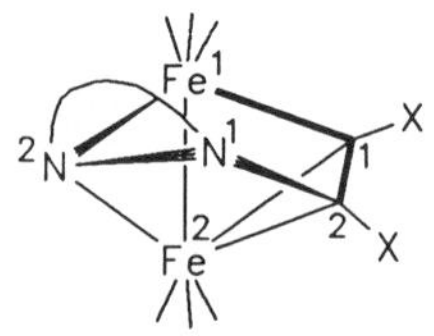	(29h) ($X = CO_2Me$)
239	Fe^2-C^1	207
217	Fe^2-C^2	207
250	Fe^1-Fe^2	246
199	Fe^1-C^1	194
197	Fe^1-N^2	199
201	Fe^2-N^2	198
143	N^1-C^2	145
149	N^1-N^2	146
138 pm	C^1-C^2	142 pm

FIG. 14

(**29h**). This difference arises from the different conformations of the diazene ring in both complexes, as schematically shown in Fig. 14. In addition, elimination of $Fe^2(CO)_3$ should be favored in **29f** because Fe-2 is much more weakly bound to C-1=C-2 than in **29h**, as indicated by the long iron–carbon distances of 239 and 217 pm as compared to 207 pm. The other characteristic bond lengths of these bicyclic ferracycles are similar for both diazene ligands. The structure of **30** was originally proposed to contain an $Fe(CO)_4$ group in an octahedral environment (*111,114*), as suggested by the similarity of the Mössbauer spectra of **30d** and $H_2Fe(CO)_4$, which exhibited chemical shifts of -0.12 and -0.18 mm sec^{-1} and quadrupole splittings of 0.60 and 0.55 mm sec^{-1} respectively (*41*). However, X-ray analysis of **30d** revealed the presence of a distorted square pyramidal $Fe(CO)_3$ group, as shown in Fig. 15, wherein the methoxycarbonyl groups at C-4 and C-5 were omitted (*115*). The small quadrupole splitting found for **30d** most likely arises from a delocalized bonding situation within the planar diazaferrole ring. This is also suggested by the short N–N distance of 133 pm and the elongated C-4–C-5 bond of 137 pm.

Irradiation of **30** in the presence of an alkyne affords the "double-addition" product **31** (Scheme 6). In contrast to the reaction of alkynes with **VII**, carbon monoxide strongly inhibits product formation (*114*), indicating that the primary photoreaction is loss of carbon monoxide. This is corroborated by detection of the intermediate (**34**, Fig. 16, where S

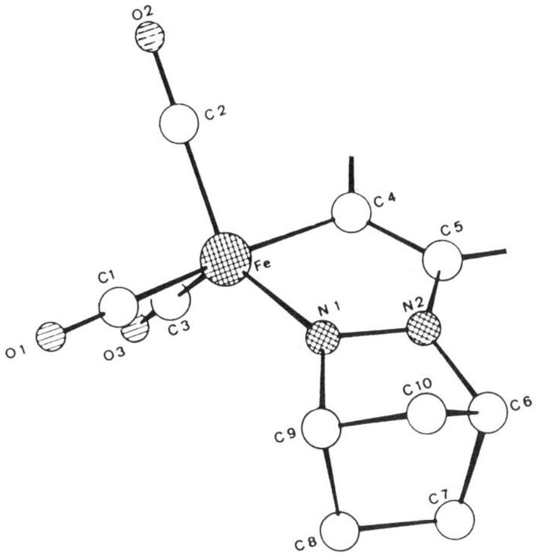

$Fe-C^4$: 194
$Fe-N^1$: 186
N^1-N^2 : 133
N^2-C^5 : 138
C^4-C^5 : 137 pm

FIG. 15

denotes solvent) following flash photolysis of **30e**. The transient absorbs at 562 nm whereas the starting material has two widely separated maxima at about 350 and 500 nm. In degassed benzene solution the rather long lifetime of 70 msec points to mesomeric stabilization as indicated in Fig. 16. The rate constant of recombination with carbon monoxide to **30e** is $2 \times 10^5\ M^{-1}\ sec^{-1}$. Cycloaddition of the alkyne to the C=Fe=CO fragment of **34e**′ followed by uptake of carbon monoxide was proposed as a likely mechanism (*114*). From the fact that the quantum yield is doubled when the wavelength of the absorbed light is varied from 350 to 500 nm, it was assumed that the excited states have metal-centered and charge-transfer-to-ligand character in the former and latter cases, respectively.

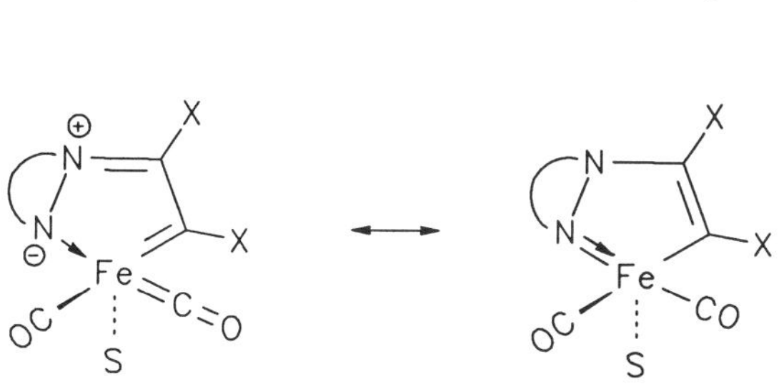

(34e′) **(34e″)**

τ = 70 ms, λ_{max} = 562 nm

FIG. 16

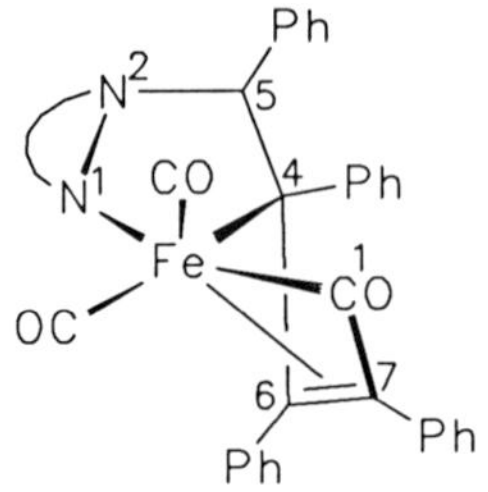

Fe $-C^4$	: 213
Fe $-N^1$	: 203
N^1 $-N^2$	: 135
N^2 $-C^5$	: 132
C^4 $-C^5$	: 147 pm

(31f)

FIG. 17

The basic results of an X-ray analysis of the tricyclic (**31f**) are schematically depicted in Fig. 17. The diazaferracyclopentane ring is planar like in **30d** but the sp^3-hybridized carbon atom (C-4) prevents cyclic electron delocalization. Accordingly, the quadrupole splitting of the distorted octahedral iron atom is increased to 1.04 mm sec^{-1} (*41*). The NNC^{-5} fragment resembles an azomethinimine as drawn in Scheme 6. In fact, the N–N and N═C distances of 135 and 132 pm compare very well with the corresponding values of 130 and 133 pm, respectively, found in an uncomplexed azomethinimine (*116*). The other ferracycle fragment present in **31f** has a distorted envelope conformation as indicated by the dihedral angle C-1,Fe,C-4/C-1,C-4-C-6,C-7 of 88° (*114*).

The double-addition product (**31**) is converted into the hitherto unknown 1,2,3-diazepinone (**32**) on oxidative degradation with bromine. This new heterocyclic ring system forms a boat conformation without any π-conjugation as indicated by the typical average $C(sp^2)$–$C(sp^2)$ bond length of 136 pm measured for **34f** (*113*). The N–N and N–$C(sp^2)$ bond distances of 143 and 144 pm are significantly longer than in **31f**; this further supports the formulation of an azomethinimine fragment structure for **31f** as shown in Scheme 6.

The results presented above demonstrate that the N═N group of cyclic diazenes is activated by complexation to undergo C–N bond formation with alkynes. This ultimately leads to an iron-assisted synthesis of a new class of heterocycles by a formal cyclooligomerization of the diazene with carbon monoxide and two molecules of an alkyne. In addition to the ability of **31** to produce **32**, **31f** ($R = R' = Ph$) was found to afford $Fe(CO)_3$-(tetraphenylcyclobutadiene) when heated in a vacuum to 150°C. This would offer an efficient route to unsymmetrically substituted cyclobutadiene complexes if **31f** also underwent the same reaction in the case when R and/or R′ are not phenyl groups. However, this does not occur under the experimental conditions mentioned. Attempts to convert **30** into 1,2-

diazacyclobutene-3 derivatives by oxidative or reductive methods were not successful.

B. *Photo-Diels-Alder Reactions of 2,3-1-Diazaferroles*

The diazaferrole (**30**), on irradiation in the presence of 1,3-butadiene or its 2,3-substituted derivatives, affords the Diels–Alder adduct (**33**) as summarized in Scheme 7 (*114,115*). No products are formed when the reaction is conducted thermally, even at the boiling point of the 1,3-diene. Introduction of substituents in the 1- and/or 4-position also prevents the reaction. When the cyclic diazene ligand is the pyrazoline (**e** or **f**), the Diels–Alder adducts in solution suffer decomposition *via* C–N and Fe–C bond cleavage into the pyrazoline, the corresponding 1,4-cyclohexadiene, and iron carbonyl fragments (*114*). In contrast, diazanorbornene as diazene ligand in **33d** leads to more stable adducts. Attempts to obtain tricyclic heterocycles by reductive or oxidative C–N bond formation failed.

The molecular structure of **33d** is depicted in Fig. 18 wherein hydrogen atoms and the two methoxycarbonyl groups at C-4 and C-5 are omitted for the sake of clarity. The diazaferracycle is not planar as in **30d** and **31f** because N-2 and C-5 are 27 and 75 pm below the FeN-1C-4 plane. The absence of electron delocalization is indicated by the long Fe–C-4 and N-2–C-5 bonds of 213 and 147 pm as compared to 194 and 138 pm, respectively, in **30d** (*115*).

Mechanistic investigations revealed that two photochemical steps are involved in product formation. In the first step, loss of carbon monoxide leads to the intermediate **34** in which the solvent **S** is then substituted by the 1,3-diene. When this intermediate complex is generated at −30°C and subsequently warmed up to room temperature, no Diels–Alder adduct is formed. This occurs only on irradiation, indicating that the complexed 1,3-diene can undergo the cycloaddition only photochemically. Uptake of carbon monoxide finally leads to **33**. In this proposed mechanism the transition metal formally acts as an intramolecular catalyst. As observed

CO_2Me CO_2Me N N Fe $(CO)_3$ + R R' $h\nu$ MeO_2C R' R' N N Fe $(CO)_3$ CO_2Me

(30d) (33)

R, R' = H, Me

SCHEME 7

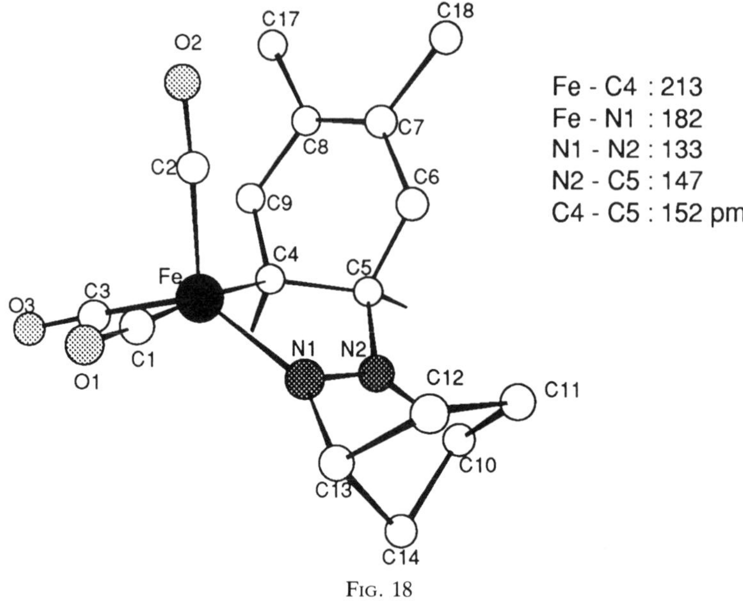

FIG. 18

for the diazirine dimetallacycles (**17** Section IV,B,2, substitution reactions of **30d** can be induced by electrochemical reduction, resulting in ETC catalysis (*117*).

VII

SUMMARY AND OUTLOOK

The reactions of metal carbonyls with cyclic 1,2-diazenes reveal several modes of coordination of the *cis*-N═N group. As expected from the electronic structure, up to six electrons may be formally donated to the metal fragment, resulting in the formation of mono- and polynuclear complexes, including metalla- and dimetallaheterocycles. Depending on the ring size of the diazene and on the metal carbonyl employed, cleavage of the N═N bond can occur, a reaction with relevance to dinitrogen fixation. Binuclear iron complexes contain an activated diazene ligand that is able to form C–N bonds on reaction with alkynes, resulting in bi- and tricyclic organoiron complexes. Oxidative removal of iron leads to 1,2,3-diazepinones, a new class of heterocycles. These reactions are of relevance for the direct synthesis of nitrogen heterocycles from dinitrogen and unsaturated hydrocarbons because coordinated dinitrogen and be alkylated to a 1,2-diazene ligand.

Future research should focus on easily accessible acyclic diazenes such as azobenzene derivatives and on catalytic versions of C–N bond formations through the use of carbon monoxide-free transition metal complexes.

References

1a. C. E. Hoch, K. C. Schneider, and R. Burris, *Biochim. Biophys. Acta* **37,** 273 (1960).

1b. G. N. Schrauzer, *Angew. Chem.* **87,** 579 (1975); *Angew. Chem., Int. Ed. Engl.* **14,** 514 (1975).

2. N. Wiberg, G. Fischer, and H. Bachhuber, *Angew. Chem.* **89,** 828 (1977); *Angew. Chem. Int. Ed. Engl.* **16,** 780 (1977).

3. D. Sellmann, *Angew. Chem.* **86,** 692 (1974); *Angew. Chem., Int. Ed. Engl.* **13,** 639 (1974). D. Sellmann, W. Soglowek, F. Knoch, and M. Moll, *Angew. Chem.* **101,** 1244 (1989); *Angew. Chem., Int. Ed. Engl.* **28,** 1271 (1989).

4. A. Albini and H. Kisch, *Top. Curr. Chem.* **65,** 105 (1976).

5. O. Diels, J. H. Blom, and W. Koll, *Liebigs Ann. Chem.* **443,** 242 (1925).

6. G. Fischer, D. Hunkler, and H. Prinzbach, *Tetrahedron Lett.* **25,** 2459 (1984); K. Beck and S. Hünig, *Angew. Chem.* **98,** 193 (1986); *Angew. Chem., Int. Ed. Engl.* **25,** 187 (1986).

7. D. Sellmann and W. Weiss, *J. Organomet. Chem.* **160,** 183 (1978).

8. N. Wiberg, *Adv. Organomet. Chem.* **23,** 131 (1984).

9. M. I. Bruce and B. L. Goodall, *in* "The Chemistry of Hydrazo, Azo and Azoxy Groups" (S. Patai, ed.), Vol. 1, p. 259. Wiley, New York, 1975.

10. D. Nicholls, B. A. Warburton, and D. H. Wilkinson, *J. Inorg. Nucl. Chem.* **32,** 1075 (1970).

11. H. Baumann and H. R. Henschel, *Top. Curr. Chem.* **7,** 643 (1967); R. Price, *in* "The Chemistry of Synthetic Dyes" (K. Venkataraman, ed.), Vol. 3, p. 303. Academic Press, New York, 1970.

12. F. Brogli, W. Eberbach, W. Haselbach, E. Heilbronner, V. Hornung, and D. M. Lemal, *Helv. Chim. Acta* **56,** 1933 (1973).

13. R. Hoffmann, *Acc. Chem. Res.* **4,** (1971).

14. B. M. Gimarc, *J. Am. Chem. Soc.* **92,** 266 (1969).

15. K. N. Houk, Y. M. Chang, and P. S. Engel, *J. Am. Chem. Soc.* **97,** 1824 (1975); see also Ref. 4.

16. R. Hoffmann, *Tetrahedron* **22,** 539 (1966).

17. E. Haselbach and E. Heilbronner, *Helv. Chim. Acta* **53,** 684 (1970).

18. A. Albini, J. Kisch, C. Krüger, and A. P. Chiang, *Z. Naturforsch. B: Anorg. Chem. Org. Chem.* **37b,** 463 (1982).

19. A. Albini and H. Kisch, *J. Organomet. Chem.* **94,** 75 (1975).

20a. E. Battaglia, H. Matthäus, and H. Kisch, *J. Organomet. Chem.* **193,** 57 (1980).

20b. W. Beck and W. Danzer, *Z. Naturforsch. B: Anorg. Chem. Org. Chem.* **30b,** 716 (1975).

21. A. Albini and H. Kisch, *Z. Naturforsch. B: Anorg. Chem.. Org. Chem.* **37b,** 486 (1982).

22a. M. Herberhold, W. Golla, and K. Leonhard, *Chem. Ber.* **107,** 3209 (1974).

22b. M. N. Ackermann and K. J. Kou, *J. Organomet. Chem.* **86,** C7 (1975); M. N. Ackermann and K. J. Kou, Inorg. Chem. **15,** 1423 (1976).

23a. G. Albertin, S. Antoniutti, and E. Bordignon, *J. Chem. Soc., Chem. Commun.*, 1688 (1984); G. Albertin, S. Antoniutti, and E. Bordignon, *J. Am. Chem. Soc.* **111,** 2072 (1989); G. Albertin, S. Antoniutti, M. Lafranchi, G. Pelizzi, and E. Bordignon, *Inorg. Chem.* **25,** 950 (1986).
23b. S. Antoniutti, G. Albertin, and E. Bordignon, *Inorg. Chem.* **26,** 2733 (1987).
23c. G. Albertin, S. Antoniutti, G. Pelizzi, F. Vitali, and E. Bordignon, *J. Am. Chem. Soc.* **108,** 6627 (1986).
23d. G. Albertin, S. Antoniutti, G. Pelizzi, F. Vitali, and E. Bordignon, *Inorg. Chem.* **27,** 829 (1988), and references therein.
23e. G. Albertin, S. Antoniutti, and G. Bordignon, *J. Chem. Soc., Dalton Trans.*, 2353 (1989).
23f. G. Albertin, S. Antoniutti, E. Bordignon, G. Pelizzi, and F. Vitali, *J. Organomet. Chem.* **353,** 229 (1988).
23g. C. F. Barrientos-Penna, C. F. Campana, F. W B. Einstein, T. Jones, D. Sutton, and S. S. Tracey, *Inorg. Chem.* **23,** 363 (1984).
24. K. R. Laing, S. D. Robinson, and M. F. Uttley, *J. Chem. Soc., Dalton Trans.*, 2713 (1973).
25. D. Sellmann, *J. Organomet. Chem.* **44,** C46 (1972); D. Sellmann, A. Brandl, and R. Endell, *Angew. Chem.* **85,** 1121 (1973); *Angew. Chem., Int. Ed. Engl.* **12,** 1019 (1973); D. Sellmann, A. Brandl, and R. Endell, *J. Organomet. Chem.* **49,** C22 (1973); See also: R. A. Back, *Rev. Chem. Intermed.* **5,** 293 (1984).
26. M. R. Smith III, R. L. Keys, G. L. Hillhouse, and A. L. Rheingold, *J. Am. Chem. Soc. 111,* 8312 (1989).
27. R. P. Bennett, *Inorg. Chem.* **9,** 2184 (1970).
28. H. Kisch, *J. Organomet. Chem.* **38,** C19 (1972).
29. A. Albini and H. Kisch, *Angew. Chem.* **87,** 206 (1975); *Angew. Chem., Int. Ed. Engl.* **14,** 182 (1975).
30. G. R. Knox, private communication (1974).
31a. M. M. Bagga, W. T. Flannigan, G. R. Knox, ad P. L. Pauson, *J. Chem. Soc. C*, 1534 (1969).
31b. M. Dekker, G. R. Knox, *J. Chem. Soc., Chem. Commun.*, 1243 (1967).
32. A. Albini and H. Kisch, unpublished results; H. Kisch, P. Reisser, and F. Knoch, *Chem. Ber.* **124,** 1143 (1991).
33. A. Nakamura, M. Aotake, and S. Otsuka, *J. Am. Chem. Soc.* **96,** 3456 (1974).
34. C. Krüger, *Chem. Ber.* **106,** 3230 (1973).
35. F. A. Cotton and B. E. Hanson, *Isr. J. Chem.* **15,** 165 (1976/1977).
36. F. A. Cotton and J. M. Troup, *J. Am. Chem. Soc.* **96,** 3438 (1974).
37. E. H. Schubert and R. K. Sheline, *Inorg. Chem.* **5,** 1071 (1966).
38. M. Herberhold, K. Leonhard, U. Wagner, F. E. Wagner, and J. Danon, *J. Chem. Soc., Dalton Trans.*, 654 (1979), and references therein.
39. H. Inoue, T. Nakagome, T. Kuroiwa, T. Shirai, and E. Fluck, *Z. Naturforsch. B. Chem. Sci.* **42b,** 573 (1987).
40. W. E. Caroll, F. A. Deeney, J. A. Delaney, and F. J. Lalor, *J. Chem. Soc. Dalton Trans.*, 718 (1973).
41. H. Kisch, C. Krüger, H. E. Marcolin, and A. Trautwein, *Z. Naturforsch. B. Chem. Sci.* **42b,** 1435 (1987).
42. H. Kisch, unpublished results.
43. M. Herberhold and K. Leonhard, *J. Organomet. Chem.* **78,** 253 (1974).
44. H. Alper, *J. Organomet. Chem.* **50,** 209 (1973).
45. F. A. Cotton, B. E. Hanson, J. D. Jameson, and B. R. Stults, *J. Am. Chem. Soc.* **99,** 3293 (1977).

46. M. N. Ackermann, L. J. Kou, J. M. Richter, and R. M. Willet, *Inorg. Chem.* **16,** 1928 (1977).
47. R. Battaglia, P. Mastropasqua, and H. Kisch, *Z. Naturforsch. B: Anorg. Chem. Org. Chem.* **35b,** 401 (1980).
48. C. B. Argo and J. T. Sharp, *Tetrahedron Lett.* **22,** 353 (1981); C. B. Argo and J. T. Sharp, *J. Chem. Soc., Perkin Trans. 1*, 1581 (1984).
49. R. G. Little and R. J. Doedens, *Inorg. Chem.* **11,** 1392 (1972).
50. R. J. Doedens, *Inorg. Chem.* **9,** 429 (1970).
51. R. J. Doedens and J. A. Ibers, *Inorg. Chem.* **8,** 2709 (1969).
52. L. G. Kuz'mina, N. G. Bokii, Y. T. Struchkov, A. V. Arutyunyan, L. V. Rybin, and M. I. Rybinskaya, *Zh. Strukt. Khim.* **12,** 875 (1971).
53. A. J. Carty, *Organomet. Chem. Rev. A* **7,** 191 (1971); M. Kilner, *Adv. Organomet. Chem.* **10,** 115 (1972).
54. B. K. Teo, M. B. Hall, R. F. Fenske, and L. F. Dahl, *Inorg. Chem.* **14,** 3103 (1975).
55. R. J. Doedens, *Inorg. Chem.* **7,** 2323 (1968).
56. L. F. Dahl, W. R. Costello, and R. B. King, *J. Am. Chem. Soc.* **90,** 5422 (1968).
57. H. Kisch, C. Krüger, and A. Trautwein, *Z. Naturforsch. B: Anorg. Chem. Org. Chem.* **36b,** 205 (1981).
58. H. A. Patel, A. J. Carty, M. Mathew, and G. J. Palenik, *J. Chem. Soc., Chem. Commun.*, 810 (1972).
59. V. V. Trusov, A. I. Nekhaev, Y. V. Maksimov, and V. D. Tyurin, *Izv. Akad. Nauk SSSR* **8,** 1903 (1985).
60. D. P. Madden, A. J. Carty, and T. Birchall, *Inorg. Chem.* **11,** 1453 (1972).
61. P. Mastropasqua, P. Lahuerta, K. Hildenbrand, and H. Kisch, *J. Organomet. Chem.* **172,** 57 (1979).
62. F. A. Cotton, B. E. Hanson, and J. D. Jamerson, *J. Am. Chem. Soc.* **99,** 6588 (1977).
63. H. Bantel, B. Hansert, A. K. Powell, M. Tasi, and H. Vahrenkamp, *Angew. Chem.* **101,** 1084 (1989); *Angew. Chem., Int. Ed. Engl.* **28,** 1059 (1989).
64. E. J. Wucherer and H. Vahrenkamp, *Angew. Chem.* **99,** 353 (1987); *Angew. Chem., Int. Ed. Engl.* **26,** 355 (1987).
65. P. Lahuerta, J. Latorre, M. Sanau, and H. Kisch, *J. Organomet. Chem.* **286,** C27 (1985).
66. R. J. Haines and A. L. duPreez, *J. Organomet. Chem.* **21,** 181 (1970); R. J. Haines and A. L. du Preez, *Inorg. Chem.* **11,** 330 (1972).
67. M. Herberhold and W. Golla, Chem. Ber. **107,** 3199 (1974).
68. G. Huttner, H. G. Schmid, H. Willenberg, and T. Stark, *J. Organomet. Chem.* **94,** C3 (1975).
69. M. Kooti and J. F. Nixon, *J. Organomet. Chem.* **105,** 217 (1976).
70. M. Herberhold and K. Leonhard, *Angew. Chem.* **88,** 227 (1976); *Angew. Chem., Int. Ed. Engl.* **15,** 230 (1976).
71a. M. N. Ackermann, D. B. Shewitz, and C. R. Barton, *J. Organomet. Chem.* **125,** C33 (1977); M. N. Ackermann, R. M. Willett, M. H. Englert, C. R. Barton, and D. B. Shewitz, *J. Organomet. Chem.* **175,** 205 (1979), and references therein.
71b. M. N. Ackermann, D. J. Dobmeyer, and L. C. Hardy, *J. Organomet. Chem.* **182,** 561 (1979).
71c. M. N. Ackermann, L. C. Hardy, Y. Z. Xiao, D. J. Dobmeyer, J. A. Dunal, K. Felz, A. Sedman, and K. F. Alperovitz, *Organometallics* **5,** 966 (1986).
72a. M. Herberhold and W. Golla, *J. Organomet. Chem.* **26,** C27 (1971).
72b. M. Herberhold, K. Leonhard, and C. G. Kreiter, *Chem. Ber.* **107,** 3222 (1974).
72c. M. Herberhold, K. Leonhard, and A. Geier, *Chem. Ber.* **110,** 3279 (1977).
72d. F. W. B Einstein, D. Sutton, and K. G. Tyers, *Inorg. Chem.* **26,** 111 (1987).

73. M. R. Smith and G. L. Hillhouse, *J. Am. Chem. Soc.* **110,** 4066 (1988).
74. K. R. Dixon, D. J. Eadie, and S. R. Stobart, *Inorg. Chem.* **21,** 4318 (1982).
75. S. K. Kang, T. A. Albright, and C. Mealli, *Inorg. Chem.* **26,** 3158 (1987).
76. C. C. Frazier III and H. Kisch, *Inorg. Chem.* **17,** 2736 (1978) and references therein.
77. M. A. M. Meester, R. C. J. Vriends, D. J. Stufkens, and K. Vrieze, *Inorg. Chim. Acta* **19,** 95 (1976).
78. J. M. Kelly, C. Long, J. G. Vos, J. G. Haasnoot, and G. Vos, *Organomet. Chem.* **221,** 165 (1981), and references therein.
79. C. P. Marabella, J. H. Enemark, W. E. Newton, and J. W. McDonald, *Inorg. Chem.* **21,** 623 (1982), and references therein.
80. R. Battaglia, H. Kisch, C. Krüger, and L. K. Liu, *Z. Naturforsch. B: Anorg. Chem. Org. Chem.* **35b,** 719 (1980).
81. M. Herberhold and M. Süss-Fink, *Z. Naturforsch. B: Anorg. Chem. Org. Chem.* **31b,** 1489 (1976).
82. B. Bovio, *J. Organomet. Chem.* **244,** 257 (1983).
83. H. Kisch, *in* "Chemistry of Diazirines" M. T.H. Liu, Vol. 2, p. 101. CRC Press, Boca Raton, Florida, 1987.
84. W. P. Fehlhammer and A. Liu, unpublished results 1975, cited in Ref. 20b.
85. J. E. Wollrab, L. H. Sharpen, and D. P. Ames, *J. Chem. Phys.* **49,** 2405 (1968).
86. B. Bovio, *J. Organomet. Chem.* **241,** 363 (1983).
87. J. F. Nixon and M. Kooti, *J. Organomet. Chem.* **149,** 71 (1978).
88. H. Kisch and O. E. Polansky, *Tetrahedron Lett.* **10,** 805 (1969).
89. P. Holzmeier and H. Kisch, *Chem. Ber.* (1992), in press.
90. B. Bovio, *J. Organomet. Chem.* **252,** 71 (1983).
91. M. S. Wrighton, *Top. Curr. Chem.* **65,** 37 (1976); M. S. Wrighton, H. B. Abrahamson, and D. L. Morse, *J. Am. Chem. Soc.* **98,** 4105 (1976); R. M. Dahlgren and J. I. Zink, *Inorg. Chem.* **16,** 3154 (1977).
92. A. Albini and E. Fasani, *J. Organomet. Chem.* **273,** C26 (1984).
93. P. Holzmeier, H. Görner, F. Knoch, and H. Kisch, *Chem. Ber.* **122,** 1457 (1989).
94. P. Holzmeier, J. K. Kochi, and H. Kisch, *J. Organomet. Chem.* **382,** 192 (1990).
95a. L. S. Hegedus and A. Kramer, *Organometallics* **3,** 1263 (1984).
95b. H. F. Sleiman and L. McElwee-White, *J. Am. Chem. Soc.* **110,** 8700 (1988).
96. F. A. Cotton, S. A. Duraj, and W. J. Roth, *J. Am. Chem. Soc.* **106,** 4749 (1984).
97. T. Joh, N. Hagihara, and S. Murahashi, *Bull. Chem. Soc. Jpn.* **40,** 661 (1967).
98a. M. I. Bruce, M. Z. Iqbal, and F. G. A. Stone, *J. Organomet. Chem.* **31,** 275 (1971); M. I. Bruce, M. Z. Iqbal, and F. G. A. Stone, *J. Chem. Soc. A*, 3204 (1970).
98b. M. I. Bruce, *J. Organomet. Chem.* **315,** C51 (1986).
98c. A. Smieja, J. E. Gozun, and W. L. Gladfelter, *Organometallics* **6,** 1311 (1987).
99. W. J. Evans, D. K. Drummond, S. G. Bott, and J. L. Atwood, *Organometallics* **5,** 2389 (1986); W. J. Evans, D. K. Drummond, *J. Am. Chem. Soc.* **108,** 7440 (1986); W. J. Evans, D. K. Drummond, L. R. Chamberlain, R. J. Doedens, S. G. Bott, H. Zhang, and J. L. Atwood, *Organometallics* **110,** 4983 (1988).
100. J. E. McMurray, *Acc. Chem. Res.* **7,** 281 (1974); J. E. McMurray, M. P. Fleming, K. L. Kees, and L. R. Krepski, *J. Org. Chem.* **43,** 3255 (1978); J. C. Huffman, K. G. Moloy, J. A. Marsella, and K. G. Caulton, *J. Am. Chem. Soc.* **102,** 3009 (1980); R. S. P. Coutts, P. C. Wailes, and R. L. Martin, *J. Organomet. Chem.* **50,** 145 (1973).
101. G. Fachinetti, G. Fochi, and C. Floriani, *J. Organomet. Chem.* **57,** C51 (1973); G. Fochi, C. Floriani, J. C. J. Bart, and C. J. Giunchi, *J. Chem. Soc., Dalton Trans.*, 1515 (1983).

102. S. Gambarotta, C. Floriani, A. Chiesi-Villa, and C. Guastini, *J. Am. Chem. Soc.* **105,** 7295 (1983).
103. G. Avar, W. Rüsseler, and H. Kisch, *Z. Naturforsch. B: Chem. Sci.* **42b,** 50 (1987).
104. W. Beck, H. Werner, H. Engelmann, and H. S. Smedal, *Chem. Ber.* **101,** 2143 (1968); A. T. McPhail, G. R. Knox, C. G. Robertson, and G. A. Sim, *J. Chem. Soc. A*, 205 (1971).
105. P. Mastropasqua, A. Riemer, H. Kisch, and C. Krüger, *J. Organomet. Chem.* **148,** C40 (1978).
106. R. J. Doedens, *Inorg. Chem.* **8,** 570 (1969).
107. W. Clegg, G. M. Sheldrick, D. Stalke, S. Bhaduri, and H. K. Khwaja, *Acta Crystallogr. Sect. C: Cryst. Struct. Commun.* **40,** 2045 (1984).
108. J. Ashley-Smith, M. Green, N. Mayne, and F. G. A. Stone, *J. Chem. Soc., Chem. Commun.*, 409 (1969).
109a. N. Wiberg, H. W. Häring, and U. Schubert, *Z. Naturforsch. B: Anorg. Chem. Org. Chem.* **33b,** 1365 (1978).
109b. N. Wiberg and M. Veith, *Chem. Ber.* **104,** 3191 (1971).
110. C. Krüger and H. Kisch, *J. Chem. Soc., Chem. Commun.*, 65 (1975).
111. A. Albini and H. Kisch, *J. Organomet. Chem.* **101,** 231 (1975).
112. B. Ulbrich and H. Kisch, *Angew. Chem.* **90,** 388 (1978); *Angew. Chem., Int. Ed. Engl.* **17,** 369 (1978).
113. R. Millini, H. Kisch, and C. Krüger, *Z. Naturforsch, B: Anorg. Chem. Org. Chem.* **40b,** 187 (1985).
114. A. Albini and H. Kisch, *J. Am. Chem. Soc.* **98,** 3869 (1976).
115. R. Battaglia, C. C. Frazier III, H. Kisch, and C. Krüger, *Z. Naturforsch. B: Anorg. Chem. Org. Chem.* **38b,** 648 (1983).
116. A. Gieren, P. Narayanan, K. Burger, and W. Thenn, *Angew. Chem.* **86,** 482 (1974); *Angew. Chem., Int. Ed. Engl.* **13,** 475 (1974).
117. F. R. Estevan, P. Lahuerta, and J. Latorre, *Inorg. Chim. Acta* **116,** L33 (1986).

ADVANCES IN ORGANOMETALLIC CHEMISTRY, VOL. 34

Carbon–Oxygen Bond Acti... by Transition Metal Complexes

AKIO YAMAMOTO

Department of Applied Chemistry
School of Science and Engineering
Waseda University
Tokyo 169, Japan

I

INTRODUCTION

Recently the activation of C–H bonds by transition metal complexes has attracted much attention (*1*). In contrast, the activation of C–O bonds by transition metal complexes has attracted less attention, despite the fact that selective C–O bond cleavage by transition metal complexes, combined with fundamental processes of organotransition metal complexes, can have considerable impact on organic synthesis. An exceptionally well-studied and well-utilized process is the C–O bond cleavage of allylic carboxylates (*2–10*). The objective of this article is to provide an overview of C–O bond cleavage reactions promoted by transition metal complexes, focusing attention on the conceptual and mechanistic aspects of C–O bond activation by transition metal complexes. No effort has been made to make this article a comprehensive review nor to include every reaction involving C–O bond cleavage by transition metal complexes, because there are so many synthetic reactions utilizing allylic oxygen bond cleavage by transition metal complexes (*2–10*).

Carbon–oxygen single bonds as well as carbon–oxygen double bonds may be cleaved by transition metal complexes. The former process constitutes the majority of reactions but the latter is attracting increasing attention. We deal first with the cleavage of the carbon–oxygen single bonds in organic compounds and treat the carbon–oxygen double-bond cleavage later.

Oxygen-containing organic compounds whose C–O bond may be cleaved include the following types: (1) alcohols (including phenols), (2) ethers (including cyclic ethers), (3) esters, (4) anhydrides, and (5) ketones and aldehydes. Of these various processes, cleavage of C–O bonds in organic compounds containing allylic oxygen groups has been studied most extensively. Thus it is appropriate to treat the cleavage of the allylic oxygen bond first.

II

CLEAVAGE OF THE ALLYLIC OXYGEN BOND BY TRANSITION METAL COMPLEXES

Two types of processes have been recognized regarding allylic oxygen bond cleavage. One is the process giving products resulting from the net oxidative addition of allylic compounds, Eq. (1), and the other is the consecutive-type process initiated by transition metal complexes having ligands such as hydride, alkyl, or alkoxide, which may add to the allylic double bond to cause reactions as depicted in Eqs. (2a) and (2b).

X + ML_n ⇌ X, ML_n ⇌ M, X, L ⇌ (L) [M, L, L] X (1)

X = Cl, -OAc, -OC(=O)OR, -OP(=O)(OR)$_2$ etc

OX + YML_n → Y, OX, M, L_n → Y + L_nM-OX

(2a)

OX + YML_n ⇌ (a) Y, M, OX, L_n → Y + L_nM-OX

Y=H, alkyl, aryl, OR etc.

(2b)

Many reactions are known to proceed through a concerted process involving π-allyl–transition metal complexes as shown in Eq. (1), whereas rather limited examples are known to proceed through the consecutive processes as shown in Eq. (2). Of the latter type of process, one is the S_N2' type, Eq. (2a), and the other is the insertion–β-elimination type, Eq. (2b). The process shown in Eq. (2b) consists of two steps, involving initially the insertion of the double bond in the allylic group into the metal–Y bond (Y being an alkyl, hydride, or alkoxide group) followed by the elimination of the OX group attached at the β-carbon. As will be discussed later, the difference in the processes in Eqs. (2a) and (2b) is subtle, sometimes trivial, but in some cases the preequilibrium (a) in Eq. (2b) between the initial system and the allyl-inserted intermediate can be identified.

A. *Allylic Oxygen Bond Cleavage to Give π-Allyl–Transition Metal Complexes*

It was noted in early studies (*11*) that nickel and palladium(0) complexes undergo oxidative addition with allylic compounds to form π-allyl nickel and π-allylpalladium complexes, and that the complexes thus formed are susceptible to attack by nucleophiles on the π-allyl ligand, with formation of carbon–nucleophile bonding. Trost, Tsuji, and other workers developed the concept, which has become one of the most useful metal-mediated synthetic methods known (*2–9*). The applications in organic synthesis have continued to increase at a tremendous pace ever since.

The catalytic process is composed of the elementary steps shown in Scheme 1. As shown, various substrates may be used. These include allylic esters, such as carboxylates, carbonates, phosphates, and carbamates. Other allylic compounds having X groups, such as amino, nitro, and sulfuryl groups, as well as allylic halides, also can be conveniently used for the catalytic reactions. In most cases organic bases are employed to neutralize HX and to drive the catalytic cycle.

Divalent palladium compounds such as palladium acetate, acetylacetonate, and halides are used as the catalyst precursors, often in combination with tertiary phosphines. The latter serve as stabilizing groups as well as ligands controlling the reaction pathway. The divalent palladium compounds are considered to be reduced in the reaction systems to Pd(0) species that carry the catalytic reactions. The exact course of the formation of the Pd(0) species has not been well investigated. Preformed Pd(0) compounds are also employed in organic synthesis. As Pd(0) compounds used in organic synthesis, $Pd(dba)_2$ (dba = dibenzylideneacetone) or $Pd_2(dba)_3CHCl_3$ will be the first choice for organic chemists because of

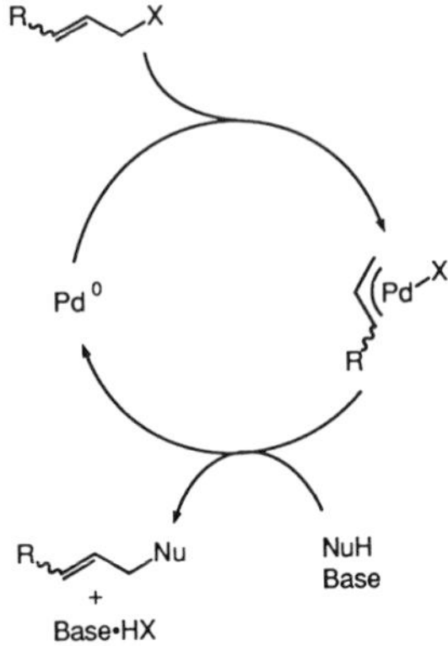

SCHEME 1. Mechanism of nucleophilic substitution of allylic compounds catalyzed by palladium complexes. (Ligand is omitted for simplicity.)

their ease of handling and the freedom of choice of auxiliary ligands, such as tertiary phosphines. Tetrakis(triphenylphosphine)palladium(0), $Pd(PPh_3)_4$, is also frequently employed, because the triphenylphosphine ligand can be readily dissociated in solution to afford a coordination site(s) for the substrate employed in the reaction.

Because of the ready availability of these catalysts and catalyst precursors, these palladium catalyst systems have been extensively used in organic synthesis, but the course of the allylic oxygen bond cleavage has received relatively less attention.

1. *Cleavage of C–O Bonds in Allylic Carboxylates*

On the basis of stereochemical studies of the cleavage of the allylic oxygen bonds with Pd(0) complexes it has been established that the cleavage reactions proceed with anti elimination (backside displacement) of the carboxylate group, such as acetate, from the allylic group. When a soft nucleophile (a resonance-stabilized nucleophile such as malonate) is used, the nucleophilic attack of the nucleophile on the π-allyl ligand coordinated to Pd(II) takes place on the opposite side of palladium (anti attack). The stereochemical course of the catalytic allylic substitution thus proceeds through double inversion and net retention of the configuration at the allylic group as shown in Scheme 2 (*12–19*). In some cases, however, when the anti elimination is hindered, the C–O bond cleavage in the allylic carboxylate or allylic ether can be forced to undergo syn elimination (*20*).

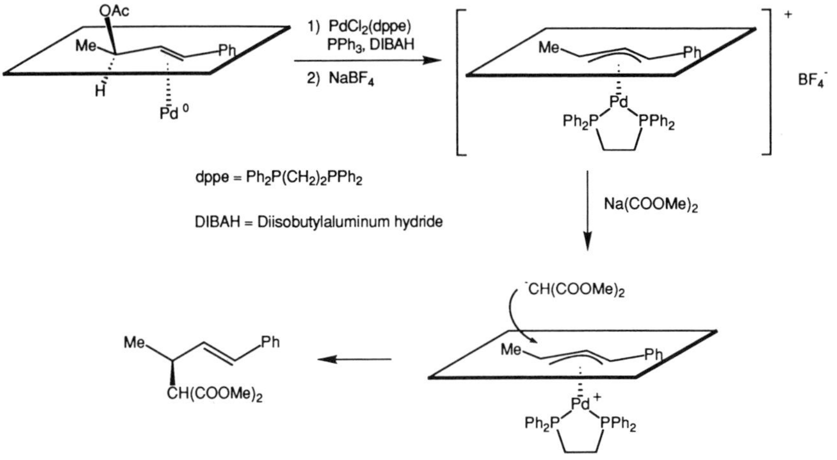

dppe = $Ph_2P(CH_2)_2PPh_2$

DIBAH = Diisobutylaluminum hydride

SCHEME 2.

In contrast, when hard nucleophiles, such as alkylmagnesium compounds, are used in combination with allylic compounds, the alkyl group attacks the metal center first, forming a metal alkyl that reductively eliminates with the allylic entity to produce a coupling product of the allyl and alkyl groups. In this case reductive elimination proceeds with retention of the configuration at the allyl group that is attacked internally (syn attack). Thus, the alkylation of the allylic group proceeds with overall inversion of configuration (*13,15*).

An oxidative addition product of allyl acetate with a Pd(0) complex has been isolated by using tricyclohexylphosphine as ligand (*21–23*). The reaction is accompanied by formation of propenylphosphonium salt (*21–23*).

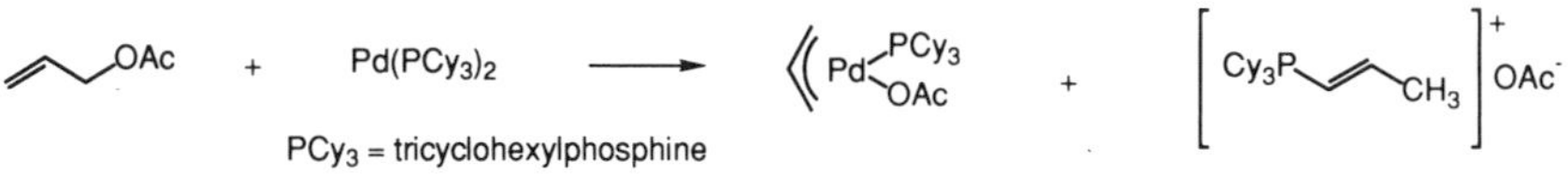

PCy_3 = tricyclohexylphosphine

(3)

On interaction with a Pd(0) complex, the π-allylpalladium complex further forms a dinuclear palladium complex bridged with the allyl and the acetato ligands.

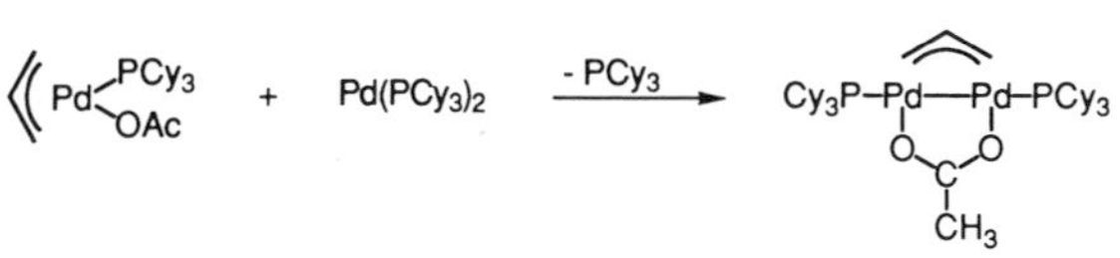

(4)

It is intriguing to note that treatment of allyl acetate with $Pd(PPh_3)_4$, the frequently utilized catalyst for the allylic substitution reaction, gives no apparent sign of allylic oxygen cleavage. However, employment of deuterated allyl acetate, $CH_2{=}CHCD_2OAc$, in the reaction with $Pd(PPh_3)_4$ revealed a clean pairwise 1,3-shift of the allyl acetate to afford a mixture of $CH_2{=}CHCD_2OAc$ and $CD_2{=}CHCH_2OAc$. This result indicates that cleavage of the allylic oxygen bond is a reversible process and C–O bond cleavage and formation of the equilibrium (5) lies far on the side of allylic oxygen bond formation.

$$H_2C{=}CH{-}CH_2{-}OAc + Pd(PPh_3)_4 \rightleftharpoons (\eta^3\text{-}C_3H_5)Pd(PPh_3)(OAc) + 3\,PPh_3 \quad (5)$$

In catalytic processes using allylic acetates, employment of sodium halides is sometimes useful for promoting the reaction. Although the precise reason for this effect has not been clarified, it may be associated with the above reaction. If Eq. (5) is shifted far to the left, the overall catalytic process will be hindered, whereas replacement of the acetate anion by halide may prevent the reverse C–O bond formation process from occurring, and assist the forward process involving nucleophilic attack on the allyl ligand.

The C–O bond in allyl acetate can also be readily cleaved by zero-valent nickel compounds (*24,25*). Treatment of $Ni(cod)_2$ (cod = 1,5-cyclooctadiene) with allyl acetate gives bis(allyl)nickel(II) and nickel(II) acetate. The formation of these compounds can be explained by the initial formation of allylnickel acetate, which subsequently disproportionates to bis(allyl)nickel and nickel acetate. The supposed intermediate, (η^3-allyl)-(acetato)nickel(II), can be trapped as (η^3-allyl)(acetato)$NiPR_3$ by adding tertiary phosphine. The C–O bond in vinyl acetate also can be cleaved on treatment with $Ni(cod)_2$ to give ethylene and nickel acetate.

Cleavage of allyl formate to yield propylene and carbon dioxide is catalyzed by the $Ni(cod)_2$–PPh_3 system. The reaction is considered to proceed via oxidative addition of allyl formate to give a π-allylnickel formate, which liberates propylene by reductive elimination of an allylnickel hydride, formed by decarboxylation of the formate ligand (*25*).

$$HCO_2CH_2CH{=}CH_2 \xrightarrow{Ni(cod)_2\text{-}PPh_3} CH_2{=}CHCH_3 + CO_2$$

$$HCO_2CH_2CH{=}CH_2 + L_nNi(0) \longrightarrow [(\eta^3\text{-}C_3H_5)Ni(O_2CH)L_n] \xrightarrow{-CO_2} [(\eta^3\text{-}C_3H_5)Ni(H)L_n] \longrightarrow C_3H_6 \quad (6)$$

Reductive cleavage of allylic formate is catalyzed by palladium complexes, probably with involvement of the π-allylpalladium formate intermediate. In fact, tertiary phosphine-coordinated π-allylpalladium formate complexes have been prepared by oxidative addition of allyl formate to Pd(0) complexes as well as through other routes [Eqs. (7) and (8)]. The isolated π-allylpalladium formate complex was shown to catalyze the decarboxylative reductive cleavage of allylic formate to olefins, Eq. (9) (*26*).

Pd(styrene)L_2 + R^1CH=C(R^2)CH$_2$OOCH ⟶ [(R^1,R^2-allyl)PdL_2] OCH(=O) (7)

L = PMe_3, $PMePh_2$

[(R-allyl)Pd(μ-Cl)]$_2$ —1) L; 2) HCOOAg⟶ (R-allyl)Pd(L)(OCHO) (8)

L= $PMePh_2$, P(o-tol)$_3$

CH$_3$CH=CHCH$_2$OCHO —(allyl)Pd(L)(OCHO), −CO_2, 10°C, toluene⟶ 1-butene (major) + 2-butene (minor) (9)

L = P(o-tol)$_3$

The C–O bond cleavage of allyl carboxylate to afford π-allylpalladium carboxylate can be combined with the exchange of the carboxylate anion with the formate anion. The resulting π-allylpalladium formate is readily decarboxylated to π-allylpalladium hydride, and reductive elimination of the hydride and the allyl ligand gives the olefin. Thus, by using formic acid, reductive cleavage of allylic carboxylate can be accomplished (*27,28*). The catalytic process has a considerable synthetic utility (*29*).

Cleavage of C–O bonds in allylic carboxylates by transition metal complexes other than Group 10 metals have been explored less extensively. Cleavage of the C–O bond in allylic acetate by a Mo(0) complex has been reported to proceed with stereochemical retention to form a π-allylmolybdenum complex (*30*). The reaction of $MoH_4(dppe)_2$ [dppe = 1,2-bis-(diphenylphosphino)ethane] with allyl carboxylate under light illumination gives olefin, H_2, and MoH(OOCR)$(dppe)_2$ as the products of C–O bond cleavage. The reaction pathway was explained in terms of oxidative addition to give an allylmolybdenum hydride that reductively eliminates propylene (*31*).

2. *Cleavage of C–O Bonds in Allylic Carbonates*

As with allylic carboxylates, the C–O bond in allylic carbonates, phosphates, and carbamates can be readily cleaved by Group 10 transition metal complexes. We deal here with the cleavage of the allylic carbonates as representative examples. Cleavage of allylic carbonates with palladium complexes have been exploited by Tsuji and co-workers (*8*). An advantage of using the allylic carbonates is that the alkyl carbonate entity tends to lose carbon dioxide, generating an alkoxide as a powerful base. The liberated alkoxide anion may abstract hydrogen from active hydrogen compounds, liberating carbon nucleophiles. Thus, by using the allylic carbonates employment of a strong base can be avoided, and the reactions can be advantageously performed under neutral conditions without employing a strong base that may cause unwanted side reactions.

Recently, the product of cleavage of allyl alkyl carbonates by Pd(0) complex has been isolated and characterized. For generating the coordinatively unsaturated Pd(0) complexes, diethylpalladium complexes having two tertiary phosphine ligands can serve as convenient precursors, because they can be readily thermolyzed in the presence of styrene to form a styrene-coordinated Pd(0) complex, Pd(styrene)L_2. The coordinated styrene can be readily displaced from palladium to produce a coordinatively unsaturated, very reactive Pd(0) species that reacts with substrates.

$$\text{trans-PdEt}_2\text{L}_2 + \text{styrene} \longrightarrow \text{Pd(styrene)L}_2 + \text{C}_2\text{H}_4 + \text{C}_2\text{H}_6 \qquad (10)$$

L = tertiary phosphine

The complexes isolated from reaction mixtures of Pd(styrene)L_2 with allylic carbonates have two tertiary phosphine ligands coordinated to the π-allylpalladium entity as a cation that is associated with alkyl carbonate as an anion, Eq. (11) (*32,33*).

$$\text{CH}_2\text{=C(R}^1\text{)CH}_2\text{OCO}_2\text{R}^2 + \text{Pd(styrene)L}_2 \longrightarrow \left[\text{R}^1\text{-}(\eta^3\text{-allyl})\text{PdL}_2\right]^+ {}^-\text{OCO}_2\text{R}^2 \qquad (11)$$

L = PMe_3, $PMePh_2$

The alkyl carbonate moiety readily deprotonates active hydrogen compounds to liberate nucleophiles, which attack the π-allyl ligand to give the allylation product of the nucleophile. Thus, the alkyl carbonate anion can be regarded as the equivalent of an alkoxide base whose potent nucleophilicity is masked by the acidic CO_2 entity. The cleavage of the C–O bond in allylic carbonates on interaction with Pt(0) complexes takes a similar course (*33*).

When the alkoxide liberated from allyl alkyl carbonate combines with a transition metal complex to form a metal alkoxide, the alkoxide may pick up a proton to form an alcohol. When the alkoxide group has a β-hydrogen, the β-hydrogen atom may be abstracted to produce a ketone. The course of the reaction may depend on the metal complex employed, or on the presence or absence of tertiary phosphines (*34*).

Synthetic processes utilizing the cleavage of the C–O bond in related compounds, such as allyl carbamates (*35,36*) and phosphates (*37,38*), followed by nucleophilic substitution, have been developed. In addition to palladium complexes, molybdenum hydrides have been reported to undergo C–O bond cleavage of the alkenyl carbonate under light to give an ethyl carbonate complex of molybdenum hydride (*39*).

3. *Cleavage of C–O Bonds in Allylic Ethers*

Allylic ethers react with zero-valent Group 10 metal complexes in a manner similar to allylic carboxylate reactions to give π-allyl alkoxide (aryloxide) complexes (*25,40,41*).

$Ni(cod)_2$ + $CH_2{=}CHCH_2OPh$ + PPh_3 ⟶ $(\eta^3\text{-allyl})Ni(PPh_3)(OPh)$ (12)

When bis(tricyclohexylphosphine)palladium(0) is used, the reaction is accompanied by formation of a quaternary phosphonium salt that contains the propenyl group derived by isomerization of the allyl moiety.

PdL_2 + $CH_2{=}CHCH_2O{-}C_6H_4{-}Y$ —room temp.→ $(\eta^3\text{-allyl})Pd(PCy_3)(OC_6H_4Y)$ + $[Cy_3P{-}CH{=}CH{-}CH_3]^+[OC_6H_4Y]^-$

$L = PCy_3$ $Y = H, CN$

(13)

Similarly to the reaction given in Eq. (13), reactions of allyl phenyl chalcogenides with PdL_2 ($L =$ tricyclohexylphosphine or *tert*-butylphosphine) or with $Pd(styrene)(PMe_3)_2$ proceed readily with cleavage of the carbon–sulfur bond or carbon–selenium bond. The reaction products have dimeric structures bridged with the allyl and PhS or PhSe ligands (*41,42*).

PdL_2 + $CH_2{=}CHCH_2ZPh$ ⟶ $L{-}Pd(\mu\text{-allyl})(\mu\text{-}ZPh)Pd{-}L$ (14)

$L = PCy_3, P^tBu_3, PMe_3$

$Z = S, Se$

Catalytic processes utilizing the cleavage of allylic phenyl ethers have been developed (*43*).

R–CH=CH–CH$_2$–OX / CH$_2$=CH–CH(R)–OX + PhMgBr $\xrightarrow{\text{[Ni] or [Pd]}}$ R–CH=CH–CH$_2$–Ph + CH$_2$=CH–CH(R)–Ph (15)

X=Ph, SiEt$_3$, SiMe$_3$,

The reaction of diallyl ether with Ni(cod)$_2$ in the presence of tertiary phosphine at room temperature gives a bis-η^2-coordinated diallyl ether–nickel complex, which, on heating, affords the C–O bond cleavage product. Propylene is liberated, presumably through an intermediate allylnickel allyloxide complex (*25*).

Diallyl ether + Ni(cod)$_2$ + nL ⟶ Ni(η^2, η^2-diallyl ether)L

L=PPh$_3$, PCy$_3$

Δ ⟶ [(η^3-allyl)Ni(OCH$_2$CH=CH$_2$)L$_n$] ⟶ CH$_3$CH=CH$_2$ (16)

Among allylic ethers, alkenyloxiranes having epoxide ring have found considerable utility in palladium-catalyzed organic synthesis (*44–47*). Stereospecific reduction of the alkenyloxiranes can be accomplished with palladium catalysts in combination with formic acid and trialkylamine to give homoallyl alcohols.

Ph(H)-epoxide–CH=CH–CO$_2$Et $\xrightarrow[\text{HCO}_2\text{H, Et}_3\text{N, dioxane}]{\text{Pd}_2\text{(dba)}_3\text{CHCl}_3\text{, PR}_3}$ PhCH(OH)–CH(H)–CH=CH–CO$_2$Et (17)

The exact mechanism of the catalytic reaction has not as yet been established. It is assumed that the C–O bond of the epoxide ring adjacent to the allylic entity may be cleaved on interaction with a Pd(0) species to form a π-allylpalladium complex. Interaction of the π-allyl complex thus formed with formic acid, followed by decarboxylation and reductive elimination, gives the homoallylic alcohols, as shown in Scheme 3. Some other examples of the palladium-catalyzed conversion of alkenyloxiranes include amination, alkylation, and alkylstannylation reactions. In these systems, formation of π-allylpalladium intermediates formed by cleavage of the epoxide ring is assumed (*48, 49*).

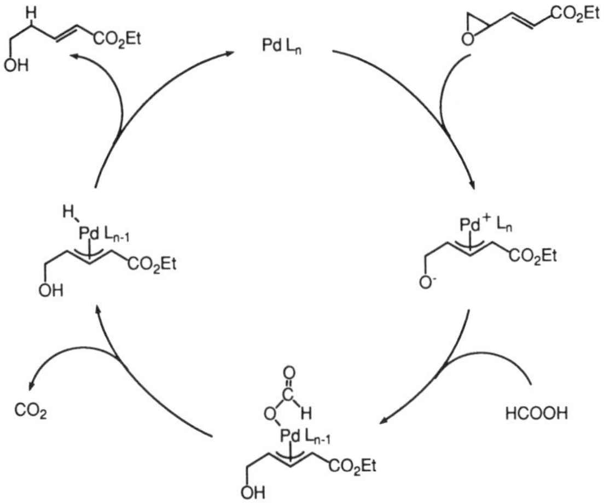

SCHEME 3. Reaction mechanism proposed for stereospecific reduction of alkenyloxiranes with formic acid catalyzed by a Pd(0) complex.

The C–O bond cleavage of alkenyloxiranes to π-allyl complexes can be accomplished with transition metal complexes other than palladium. For example, on treatment of alkenyloxiranes with iron, cobalt, or manganese carbonyls, various η^3-allyloxycarbonyl complexes have been obtained (*50*).

4. *Cleavage of C–O Bonds in Allylic Alcohols*

In the above-mentioned allylic oxygen C–O bond cleavage of allylic compounds, such as allylic carboxylates, carbonates, and alkenyloxiranes, the allylic compounds are usually prepared starting from allylic alcohols. Therefore, if one can cleave the C–O bond in allylic alcohols directly, more efficient processes may be developed. However, examples of direct C–O bond cleavage of allylic alcohols are limited.

An early example includes conversion of allyl alcohol to olefins by cross-coupling with alkylmagnesium bromides, a process catalyzed by $NiCl_2(PPh_3)_4$ (*51,52*). The reaction proceeds however, through magnesium alkoxides formed by reaction of allylic alcohols with alkylmagnesium bromides, and not through the direct reaction of allylic alochols, although this is another example of a C–O bond cleavage reaction.

The reaction of allylic alcohol with $Ni(cod)_2$ and PPh_3 causes dismutation of the allylic alcohol to 1:1:1 mixture of propylene, $Ni(CH_2{=}CHCHO)(PPh_3)_2$, and water (*25*).

$$2CH_2{=}CHCH_2OH + Ni(cod)_2 + 2PPh_3 \xrightarrow{30°C} CH_2{=}CHCH_3 + Ni(CH_2{=}CHCHO)(PPh_3)_2 + H_2O \quad (18)$$

The catalytic dismutation of allylic alcohol to propylene, acrylaldehyde, and water promoted by $RuCl_3H_2O$ was reported previously (*53*).

Interaction of Pd(0) complexes with allylic alcohols leads to cleavage of C–O bonds but the reaction pathway varies depending on the allylic alcohols. Diallylic ether, α,β-unsaturated aldehyde, and their palladium adducts are generated together with olefins and water (*41*). The overall reaction can be accounted for by the following sequence of steps.

$$PdL_2 + R^1CH{=}CR^2CHR^3OH \longrightarrow$$

(19)

Although a clean catalytic process utilizing the C–O bond cleavage of allylic alcohols has not yet been realized, nucleophilic allyl substitution catalyzed by a palladium complex has been achieved indirectly by combining the process with formation of arsenic or boric esters of allylic alcohols (*54,55*). Allylic arsenite can be readily prepared from allylic alcohol and arsenic oxide. The allylic arsenite thus formed oxidatively adds to Pd(0) species to form a π-allylpalladium complex having an anionic arsenate anion. The anion deprotonates NuH, liberating Nu^-, which attacks the π-allyl ligand to give an allylated product of the nucleophile.

(20)

$$\text{CH}_2{=}\text{CHCH}_2\text{OH} + As_2O_3 \longrightarrow \text{CH}_2{=}\text{CHCH}_2\text{OAs}< + H_2O$$

$$\text{CH}_2{=}\text{CHCH}_2\text{OAs}< + PdL_n \longrightarrow [(\pi\text{-allyl})Pd^+<]\ ^-\text{OAs}< \tag{21}$$

Scheme 4 illustrates the proposed mechanism of palladium-catalyzed substitution of allylic alcohol utilizing the ready reactivity of allylic alcohol with arsonic ester.

Another interesting process of industrial significance is the amination of allylic alcohols catalyzed by palladium or platinum complexes (*56*).

$$\text{CH}_2{=}\text{CHCH}_2\text{OH} + NH_4OH\ (\text{or } NH_3) \xrightarrow[\text{dppb, 100°C}]{[Pd]\text{ or }[Ni]} \text{CH}_2{=}\text{CHCH}_2\text{NH}_2,\ (\text{CH}_2{=}\text{CHCH}_2)_2\text{NH},\ (\text{CH}_2{=}\text{CHCH}_2)_3\text{N} \tag{22}$$

dppb: $Ph_2P(CH_2)_4PPh_2$

Before concluding this section, it should be mentioned that most of the synthetic processes involving allylic oxygen cleavage utilizes the electrophilic nature of the π-allyl complexes formed, but the character of the π-allyl complex can be modified from electrophilic to nucleophilic by various means, including treatment with SmI_2 or $SnCl_2$ or by electroreduction. By changing the polarity of the allyl ligand, various processes not dealt with here may be accomplished. However, detailed discussion is beyond the scope of the present article.

Pd (0)
Nu
OAs<
H2O
Pd OAs⁻
HOAs<
NuH
H2O
>AsOAs<
OH

SCHEME 4.

B. *Allylic Oxygen Bond Cleavage without Participation of π-Allyl–Transition Metal Complexes*

1. *C–O Bond Cleavage of Allylic Carbonates*

As discussed in the previous section, formation of stable π-allyl–transition metal complexes contributes toward promoting allylic oxygen bond cleavage. In some cases, however, other types of processes, as shown in Eq. (2a) or (2b), may operate to cause allylic oxygen bond cleavage.

The interaction of allyl carbonates with hydride complexes of cobalt, rhodium, and ruthenium coordinated with tertiary phosphines leads to ready cleavage of the allylic oxygen bond in allyl phenyl carbonates, affording propylene, CO_2, and phenoxides of these metals (*57*).

$$CoH(N_2)L_3 + CH_2{=}CHCH_2OC(O)OPh \xrightarrow{-N_2} CH_2{=}CHCH_3 + CO_2 + Co(OPh)L_3 \qquad (23)$$

$L = PPh_3$

$$RhHL_4 + 2\ CH_2{=}CHCH_2OC(O)OPh \xrightarrow{-L} CH_2{=}CHCH_3 + 2CO_2 + CH_2{=}CHCH_2OPh + Rh(OPh)L_3 \qquad (24)$$

$L = PPh_3$

$$RuH_2L_4 + CH_2{=}CHCH_2OC(O)OPh \xrightarrow{-L} CH_2{=}CHCH_3 + CO_2 + RuH(OPh)L_3 \qquad (25)$$

$L = PPh_3$

In the reactions of allyl phenyl carbonate with these hydride complexes, the phenyl carbonate entity, generated by the C–O bond cleavage, is very susceptible to loss of CO_2, and isolation of an intermediate having the phenyl carbonato ligand was not feasible. Employment of allyl ethyl carbonate, however, allowed the isolation of the intermediate ethylcarbonato complex, which can be further thermolyzed with evolution of CO_2.

$$RuH_2L_4 + CH_2{=}CHCH_2OC(O)OEt \longrightarrow CH_2{=}CHCH_3 + RuH(L)_3(\kappa^2\text{-}O_2COEt)$$

$$RuH(L)_3(\kappa^2\text{-}O_2COEt) \xrightarrow{\Delta} CO_2 + CH_4 + RuH_2(CO)L_3 \qquad (26)$$

The reaction pathway for the allylic oxygen bond cleavage has been studied by using substituted allylic carbonates. Cleavage of the C–O bond in 2-butenyl phenyl carbonate on interaction with a rhodium hydride complex released 1-butene exclusively. This result suggests the involvement of the S_N2' type reaction [cf. Eq. (2a)] accompanying the double-bond shift, as shown below.

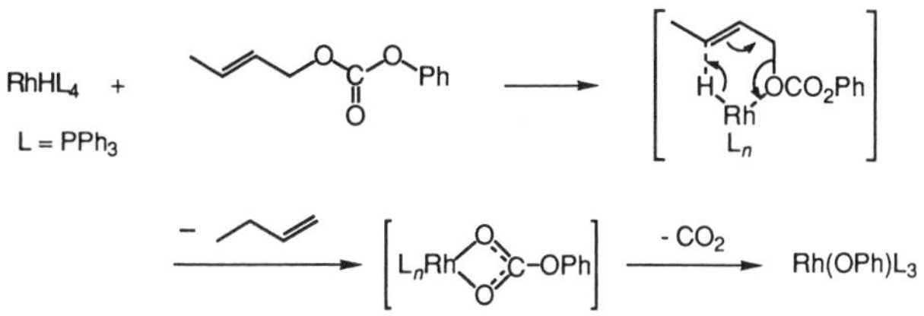

RhHL$_4$ + (2-butenyl phenyl carbonate) → [H···Rh L$_n$ ··· OCO$_2$Ph] (27)
L = PPh$_3$

− (1-butene) → [L$_n$Rh(O$_2$C–OPh)] —(− CO$_2$)→ Rh(OPh)L$_3$

The rhodium hydride complex, RhHL$_4$ (L = triphenylphosphine), and Rh(OPh)L$_3$, which is derived by the reaction of the hydride with allyl phenyl carbonate, catalyze the decarboxylative cleavage of the allylic oxygen bond to give allyl phenyl ether. Interestingly, the decarboxylation of the unbranched 2-butenyl phenyl carbonate gives branched allylic phenyl ether, whereas the decarboxylation of the branched allylic phenyl carbonate, 1-methyl-2-propenyl phenyl carbonate, gives the unbranched 2-butenyl phenyl ether (Scheme 5).

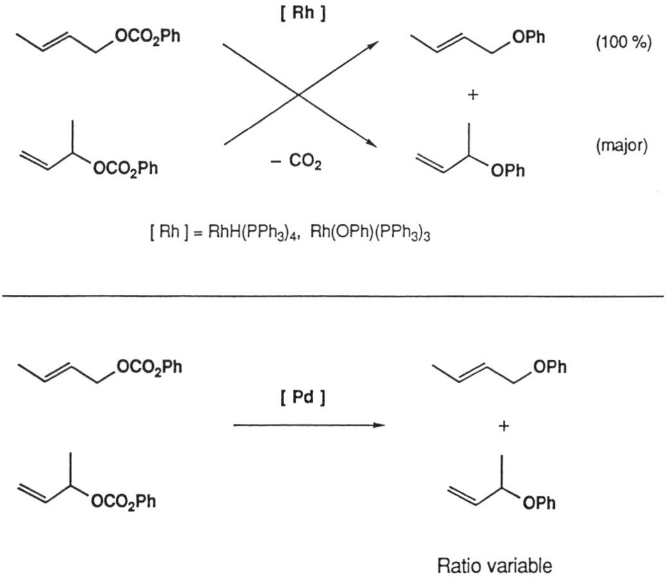

OCO$_2$Ph (2-butenyl) —[Rh], − CO$_2$→ OPh (unbranched) (100 %)
OCO$_2$Ph (branched) → + OPh (branched) (major)

[Rh] = RhH(PPh$_3$)$_4$, Rh(OPh)(PPh$_3$)$_3$

OCO$_2$Ph (2-butenyl) —[Pd]→ OPh (unbranched)
OCO$_2$Ph (branched) + OPh (branched)

Ratio variable

SCHEME 5.

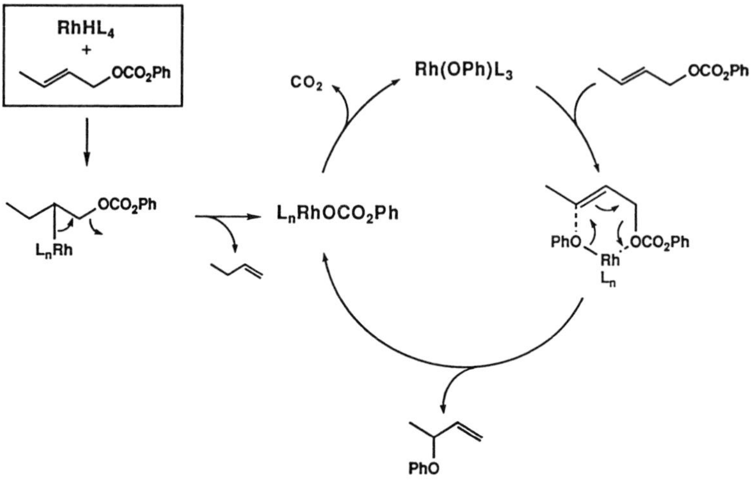

SCHEME 6.

In contract, the corresponding decarboxylative reaction of allylic phenyl carbonate catalyzed by palladium complexes gives mixtures of branched and unbranched allylic ethers. The conspicuous difference between the catalytic behavior of rhodium and of palladium complexes suggests that the rhodium-catalyzed reaction proceeds with the S_N2' type mechanism depicted in Scheme 6, whereas the direct oxidative addition-type mechanism to give π-allylpalladium complexes is operative for the reactions catalyzed by palladium complexes, as in Scheme 7 (*58*).

2. *C–O Bond Cleavage of Allylic Ethers*

Similar to the cleavage of some allylic carbonates, the C–O bonds in allylic ethers can be cleaved by an insertion–elimination mechanism. The reaction of allyl phenyl ether with tertiary phosphine-coordinated cobalt hydride under UV irradiation gives propylene and phenoxycobalt complexes as products of the C–O bond cleavage (*59*). A similar insertion–alkoxide elimination mechanism may be operative in the cleavage of diallyl ether with $CoH(N_2)(PPh_3)_3$ with formation of propylene, allyl alcohol, and a π-allylcobalt complex (*60*).

An early example of the insertion of the double bond in dially ether with subsequent C–O bond cleavage is the reaction of diallyl ether with platinum hydrides. Although zero-valent Group 10 transition metal complexes

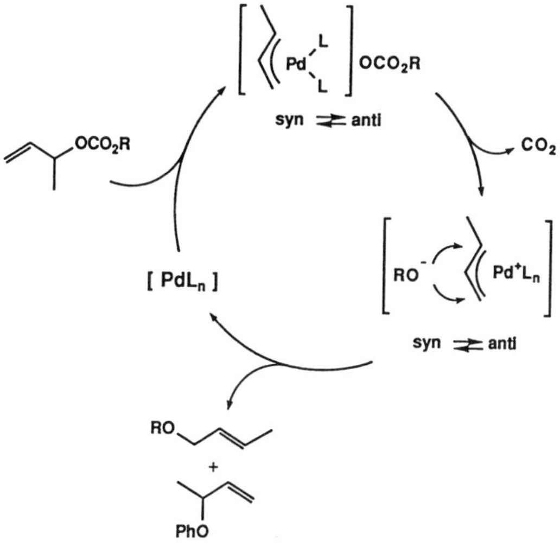

SCHEME 7.

undergo oxidative addition-type reactions with allylic compounds, Pt(II) hydrides react with allylic ethers to cause C–O bond cleavage through an insertion–elimination process. The reaction was shown to proceed through the formation of an intermediate platinum alkyl complex, produced by insertion of the double bond in alkenyl allyl ether, isomerized from diallyl ether, into the Pt–H bond. On cleavage of the C–O bond, with liberation of propionaldehyde, a π-allylplatinum complex is formed (*61*).

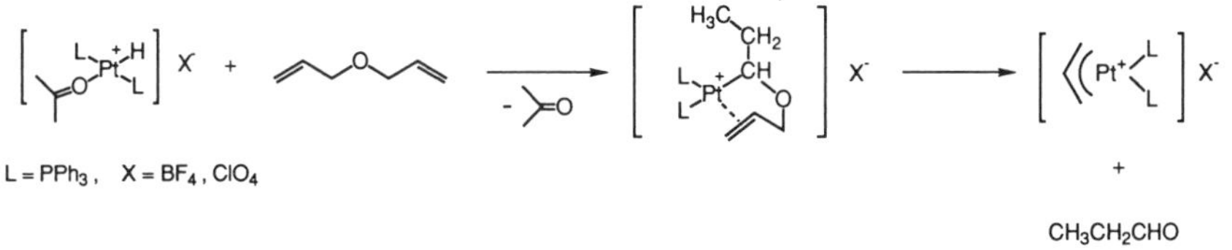

L = PPh_3, X = BF_4, ClO_4

CH_3CH_2CHO

(28)

The C–O bond in allylic silyl ethers can be cleaved by transition metal hydrides via a process without the intermediacy of π-allyl–transition metal complexes (*62*). Cobalt, rhodium, and ruthenium hydrides showed different reactivities toward allylic and vinylic silyl ethers. The complex $CoH(PPh_3)_4$ causes C–O bond cleavage in 2-butenyl trimethylsilyl ethers at room temperature to liberate butenes and $Co(OSiMe_3)L_3$.

$$CoH(N_2)L_3 + \text{〰}\!/\!\!\!\diagup\!\!\!\!\diagdown OSiMe_3 \xrightarrow{-N_2} \text{〰}\!/\!\!\diagdown + \diagdown\!\!\diagup\!\!\!\!= + Co(OSiMe_3)L_3$$

$$CoH(N_2)L_3 + {=}\!\diagdown OSiMe_3 \xrightarrow{-N_2} C_2H_4 + Co(OSiMe_3)L_3 \qquad (29)$$

$L = PPh_3$

In contrast, the rhodium hydride complex, $RhH(PPh_3)_4$, is much less active for C–O bond activation and promotes cleavage of the Si–O bond in a butenyl trimethylsilyl ether to give propylene and $RhH(CO)(PPh_3)_n$, which are considered to be formed by decarbonylation of the butenoxide entity. The ruthenium hydride complex, $RuH_2(PPh_3)_4$, shows an intermediate behavior and promotes both C–O and Si–O bond cleavage. The mechanism of C–O bond cleavage can be accounted for by insertion of the olefinic double bond into the M–H bond, followed by β-elimination of the trimethylsiloxy group by the transition metal complex.

Differentiation of the insertion–elimination mechanism from the S_N2'-type mechanism is often not straightforward. In one case, however, where allyl sulfide bond cleavage was observed on interaction with $RhHL_4$ [Eq. (30)], operation of the insertion–elimination mechanism was established by using the deuterated complex $RhD(P(C_6D_5)_3)_4$ (*63*).

$$R^3(R^4)C{=}C(R^2)\text{–}C(R^1)(H)\text{–}SAr + RhHL_4 \longrightarrow R^3(R^4)(H)C\text{–}C(R^2){=}C(R^1)H + 1/2\ L_2Rh(\mu\text{-}SAr)_2RhL_2 \qquad (30)$$

In the reaction of the deuterated rhodium complex with 3-(phenylthio)propene, the propylene produced by allylic sulfur bond cleavage contained deuterated isomers. It was further found that the allyl phenyl ether recovered from the reaction mixture after the end of the reaction contained deuterium that originated in the rhodium deuteride complex. These results suggest that olefin insertion into the Rh–H bond to give the alkylrhodium complex, and its reverse process, β-hydrogen elimination, are taking place *prior to* the rate-determining C–S bond cleavage involving the abstraction of the ArS group attached at the β-position (Scheme 8).

The regio- and stereospecificity of the butenes obtained in the reaction of 3-phenylthio-1-butene with the rhodium complex also support the insertion–β-thiolato elimination mechanism.

3. *C–O Bond Cleavage of Alkenyl Carboxylates*

Carbon–oxygen bond cleavage involving olefinic insertion into a metal hydride bond, followed by abstraction of the carboxylate group at the

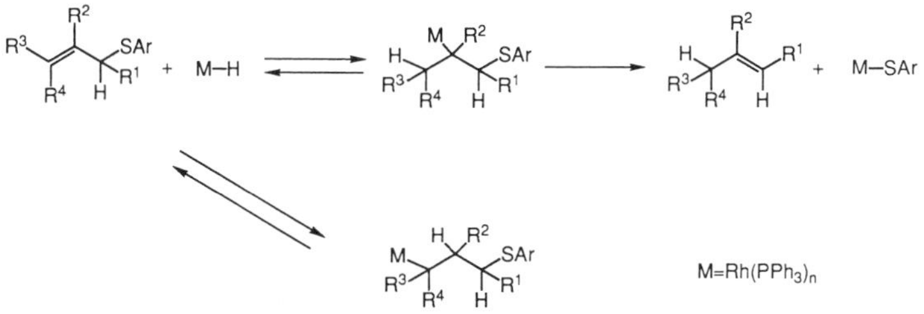

M=Rh(PPh3)n

SCHEME 8.

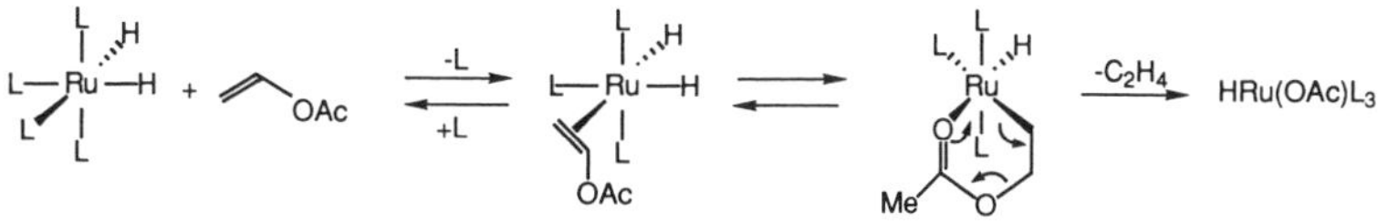

$-C_2H_4$ → $HRu(OAc)L_3$

SCHEME 9.

β-position, was first observed with a ruthenium hydride, $RuH_2(PPh_3)_4$ (*64*).

$$RuH_2L_4 + CH_2{=}CHOAc \longrightarrow C_2H_4 + RuH(OAc)L_3 \qquad (31)$$

$L = PPh_3$

When ruthenium deuteride was used, the ethylene produced contained deuteriums. This result supports the insertion–elimination mechanism shown in Scheme 9. On the other hand, direct cleavage of the C–O bond in vinyl acetate on interaction with a Ru(0) complex to produce vinylruthenium acetate has been recently observed (*65*).

III

CLEAVAGE OF THE CARBON–OXYGEN SINGLE BOND

A. *Cleavage of C–O Bonds by Concerted Processes*

The C–O bond in esters, ethers, and anhydrides can be cleaved by transition metal complexes in a concerted manner. Because the cleavage reaction may give reactive organometallic species, there are potential uses of the reaction in combination with other fundamental organometallic processes, such as olefin insertion and reductive elemination processes.

1. *C–O Bond Cleavage of Esters*

Various esters, including carboxylates, phosphates, and sulfonates, may add to low-valent transition metal complexes in a manner similar to the well-known oxidative addition reactions of organic halides and acid halides, although the established examples are still limited.

a. Carboxylic esters. Some carboxylates are known to add oxidatively to Group 10 metal complexes. An early example was oxidative addition of aryl carboxylates to Ni(0) complexes (*66, 67*).

$$Ni(cod)_2 + 2L + RC(=O)\text{–}OAr \longrightarrow \left[L_nNi\begin{smallmatrix} C(=O)\text{–}R \\ OAr \end{smallmatrix} \right] \tag{32}$$

The primary oxidative addition products of aryl carboxylates, having the acyl and aryoxide ligands, have not been isolated. Further decarbonylations usually take place as shown below.

$$Ni(cod)_2 + EtCOOAr + 3PPh_3 \longrightarrow \left[L_nNi\begin{smallmatrix} C(=O)\text{-}Et \\ OAr \end{smallmatrix} \right] \xrightarrow{-CO} \left[L_nNi\begin{smallmatrix} CH_2CH_3 \\ OAr \end{smallmatrix} \right] \xrightarrow{-C_2H_4} \left[L_nNi\begin{smallmatrix} H \\ OAr \end{smallmatrix} \right] \xrightarrow{+CO} Ni(CO)L_n + ArOH \tag{33}$$

$$Ni(cod)_2 + bipy + CH_3COOPh \longrightarrow \left[(bipy)Ni\begin{smallmatrix} COCH_3 \\ OAr \end{smallmatrix} \right] \xrightarrow{-CO} (bipy)Ni\begin{smallmatrix} CH_3 \\ OPh \end{smallmatrix} \tag{34}$$

The reaction of $Ni(cod)_2$ with triphenylphosphine and phenyl propionate gives ethylene, phenol, and $Ni(CO)(PPh_3)_3$ as the products of the C–O bond cleavage. The course of the reaction can be accounted for by the sequence of processes as shown in Eq. (33). Employment of 2,2′-bipyridine (bipy) as the ligand, and phenyl acetate as the substrate, gives a methylnickel phenoxide complex as the product of oxidative addition followed by decarbonylation [Eq. (34)].

Introduction of one equivalent of carbon monoxide into the alkylnickel bond in an alkylnickel aryloxide complex gives an acylnickel compound as the CO insertion product. Further treatment of the acylnickel aryloxide complex with a π-acid, such as CO, and electron-deficient olefins promotes the reductive elimination of the carboxylate ester.

$$\underset{R = CH_3}{(bipy)Ni\langle^{R}_{OC_6H_4CN}} + CO \longrightarrow (bipy)Ni\langle^{C(=O)-R}_{OC_6H_4CN} \xrightarrow{\pi\text{-acid}} (bipy)Ni(\pi\text{-acid})_n + RCOOC_6H_4CN \quad (35)$$

π–acid : CO, CH_2=CHCOOEt, maleic anhydride

These results indicate that the oxidative addition of carboxylate esters involving acyl oxygen bond cleavage and the reductive elimination of the acyl and aryloxide groups are reversible (*68*).

The rate of oxidative addition of para-substituted phenyl propionate to a Ni(0) complex is first order in the concentration of the nickel complex (*66*). The rate increases when a tertiary phosphine having more electron-releasing substituents is used and when the phenyl propionate has the more electron-withdrawing substituent at the para position. Thus the activation of the acyl oxygen bond may be assumed to be preceded by attack of the electron-rich transition metal complex on the carbonyl group in the aryl carboxylate (Scheme 10).

Although, in general, C–O bond cleavage in aryl esters involves fission of the acyl oxygen bond, alkyl oxygen bond cleavage also takes place when an alkyl carboxylate is treated with a very electron-rich transition metal complex that is capable of cleavage of C–H bonds (*69*).

$$Fe(H)(C_{10}H_7)(dmpe)_2 + PhCOOMe \xrightarrow{-C_{10}H_8} [CH_3OOC\text{–}C_6H_4\text{–}Fe(dmpe)] + Fe(Me)(OC(=O)Ph)(dmpe)_2 \quad (36)$$

dmpe : $Me_2PCH_2CH_2PMe_2$

Transesterification catalyzed by transition metal complexes is another type of process wherein C–O bond cleavage and formation are involved. Although very few mechanistic studies have been made, the chemistry of transition metal alkoxides (*70*) suggests that acyl oxygen bond cleavage is involved in transesterification of carboxylates with alcohols (*71*).

$$R\text{–}\overset{O^{\delta-}}{\overset{\|}{C}}{}^{\delta+}\text{–}OAr \;(\uparrow L_nNi) \longrightarrow L_nNi\langle^{C(=O)-R}_{OAr}$$

SCHEME 10.

$$\text{MeCOOAr} + \text{HOR} \xrightarrow{[\text{Pd}]} \text{MeCOOR} + \text{ArOH} \quad (37)$$

Catalytic exchange of the alkoxide group promoted by a palladium complex is believed to consist of the following two exchange processes, which have been confirmed in a study of the behavior of late transition metal alkoxide complexes (*71*).

$$\text{M-OR} + \text{MeC(=O)-OAr} \longrightarrow \text{M-OAr} + \text{MeC(=O)OR} \quad (38)$$

$$\text{M-OAr} + \text{HOR} \rightleftharpoons \text{M-OR} + \text{HOAr} \quad (39)$$

b. Lactones. Lactones oxidatively add to an Ir(I) complex to afford an iridacycloester complex (*72*).

$$\text{(cyclooctene)IrL}_3\text{Cl} + \beta\text{-lactone} \xrightarrow{\text{-cyclooctene}} \text{ClIrL}_3[\text{OC(O)CH}_2\text{CH}_2] \quad (40)$$

L = PMe_3

β-Lactone and γ-lactone are susceptible to C–O bond cleavage. The reaction is first order in the concentration of the iridium complex and the lactone, and the reactivity decreases with a decrease in the electron density at the metal center.

In contrast, the reaction of Ni(0) complexes with β-lactone catalytically liberates ethylene and CO_2, suggesting that a reaction similar to Eq. (40) occurred, followed by decarboxylation (*73*).

c. Sulfonates. In contrast to the oxidative addition of aryl carboxylates, aryl sulfonates add to low-valent transition metal complexes with cleavage of the aryl oxygen bond. Thus, aryl transition metal complexes can be generated that may be led to react further to give useful organic products (*74*). Oxidative addition of the aryl sulfonates to Pd(0) complexes has been utilized for hydrogenation or coupling reactions in combination with formate or with alkylating reagents (*75a–c*).

$$\text{ArOH} \xrightarrow{(\text{RSO}_2)_2\text{O}} \text{ArOSO}_2\text{R} \qquad \text{R} = \text{CF}_3, p\text{-F-Ph}, p\text{-tolyl, Me, F}$$

$$\text{ArOSO}_2\text{R} \xrightarrow{\text{L}_n\text{M}} [\text{L}_n\text{MAr}]^+\text{OTf}^- \xrightarrow[\text{Pd(OAc)}_2]{\text{HCOOH}} \text{Ar-H}; \quad [\text{L}_n\text{MAr}]^+\text{OTf}^- \xrightarrow{\text{RM}} \text{Ar-R} \quad (41)$$

Enol triflates can be prepared from corresponding ketones, and the C–O bond in the triflates can be readily cleaved on interaction with low-valent transition metal complexes. The enol triflates, therefore, can be regarded as equivalent of vinyl halides, which can likewise oxidatively add to transition metal complexes.

A few examples of the use of the concept of C–O bond cleavage of enol triflates in palladium-catalyzed organic syntheses are shown below.

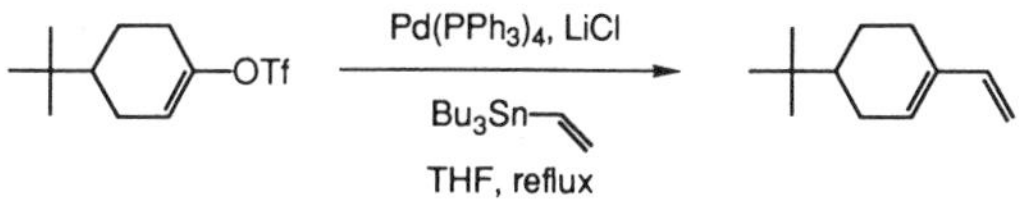

$$\text{4-}t\text{-Bu-cyclohexenyl-OTf} \xrightarrow[\text{Bu}_3\text{SnCH=CH}_2,\ \text{THF, reflux}]{\text{Pd(PPh}_3)_4,\ \text{LiCl}} \text{4-}t\text{-Bu-cyclohexenyl-CH=CH}_2 \tag{42}$$

$$\text{4-}t\text{-Bu-cyclohexenyl-OTf} \xrightarrow[\text{Bu}_3\text{SnH}]{[\text{Pd}]} \text{4-}t\text{-Bu-cyclohexenyl-H} \tag{43}$$

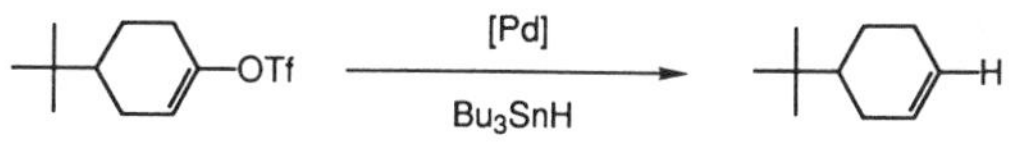

A mechanism shown in Scheme 11 has been proposed for the cross-coupling of enol triflates with a vinyltin compound (*76,77*). Oxidative addition of vinyl triflates to a Pt(0) complex has been studied to establish that the reaction in fact affords a vinylplatinum complex having a triflate anion (*78*).

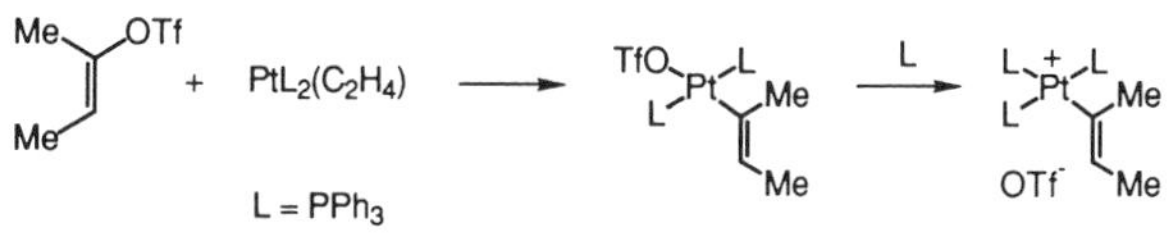

$$\text{Me}_2\text{C=C(Me)OTf} + \text{PtL}_2(\text{C}_2\text{H}_4) \longrightarrow \text{TfO(L)}_2\text{Pt–C(Me)=CMe}_2 \xrightarrow{\text{L}} [\text{L}_3\text{Pt–C(Me)=CMe}_2]^+\,\text{OTf}^- \qquad \text{L} = \text{PPh}_3 \tag{44}$$

The methodology utilizing the C–O bond cleavage of triflates is attracting increasing attention of synthetic organic chemists. One of the recent achievements is the highly stereospecific catalytic asymmetric arylation of

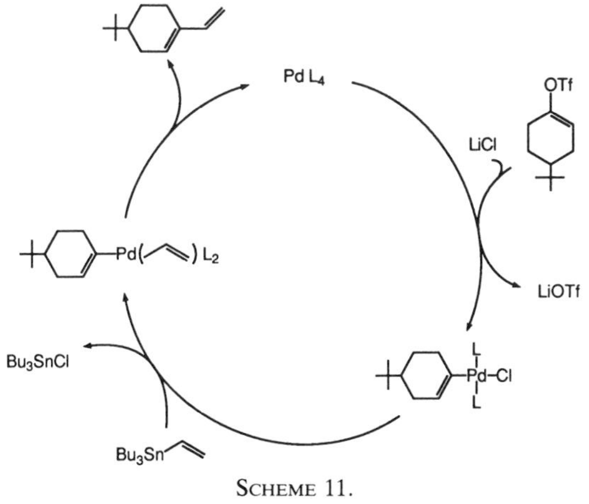

Pd L$_4$; LiCl; OTf (4-*t*-Bu-cyclohexenyl triflate); LiOTf; 4-*t*-Bu-cyclohexenyl–PdL$_2$–Cl; Bu$_3$Sn–CH=CH$_2$; Bu$_3$SnCl; 4-*t*-Bu-cyclohexenyl–Pd(CH=CH$_2$)L$_2$; 4-*t*-Bu-cyclohexenyl–CH=CH$_2$

SCHEME 11.

olefins using the Heck-type arylation catalyzed by a palladium complex coordinated with chiral tertiary phosphine ligand (*79*).

$$\text{2,3-dihydrofuran} + \text{ArOTf} \xrightarrow[{}^{i}Pr_2NEt,\ C_6H_6]{Pd(OAc)_2\text{ - }(R)\text{-BINAP}} \text{2-aryl-2,3-dihydrofuran} \quad (>90\%ee) \tag{45}$$

Another application utilizing C–O bond cleavage of sulfonate is carbonylation of sulfonates to carboxylates (*75c,80–82*).

$$R^1OSO_2R^2 + CO + R^3OH \xrightarrow{[Co]\text{ or }[Pd]} R^1COOR^3 \tag{46}$$

$$R^1C_6H_4\text{–OTf} + R^2SnR^3_3 + CO \xrightarrow[LiCl]{[Pd(0)]} R^1C_6H_4\text{–C(=O)–}R^2 \tag{47}$$

2. *C–O Bond Cleavage of Acid Anhydrides*

Similar to the case for carboxylic esters, the C–O bond in acid anhydride can be readily cleaved on interaction with low-valent transition metal complexes to give acylcarboxylato-type complexes (*83–86*).

$$L_nM + RCOOCOR \longrightarrow RCO\text{–}M(L)_2\text{–}O_2CR \tag{48}$$

$M = Ir(CO)Cl,\ Ni(0) \qquad L = PPh_3(Ir),\ PEt_3(Ni)$

The acylcarboxylato complex can also be prepared by treatment of an alkylcarboxylatonickel complex with one equivalent of CO. Further treatment of the acylcarboxylatonickel complex with CO causes reductive elimination of carboxylic anhydride. Thus, oxidative addition involving cleavage of the acyl oxygen bond acid anhydride and the reductive elimination of acyl and carboxylato ligands in C–O bond formation to give acid anhydride are reversible processes.

$$R'COO\text{–}Ni(L)_2\text{–}Me \xrightarrow{CO} R'COO\text{–}Ni(L)_2\text{–}COMe \xrightarrow{CO} \left[Ni(CO)(L)_2(COMe)(R'COO) \right] \longrightarrow R'COOCOMe + Ni(CO)_nL_{4-n} \tag{49}$$

Cyclic carboxylic anhydrides also oxidatively add to zero-valent Group 9 and 10 metal complexes to afford metallacyclic complexes (*87–90*). Platinum complexes can be obtained as isolable metallacyc[illegible] whereas decarbonylation readily takes place for nickel ana[illegible] metallacyclic esters.

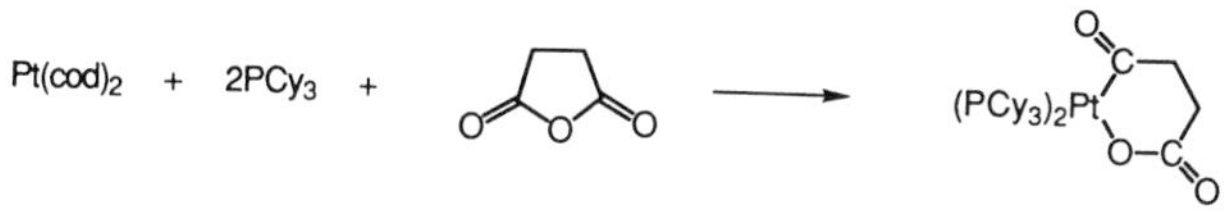

$$\text{Pt(cod)}_2 + 2\text{PCy}_3 + \text{(succinic anhydride)} \longrightarrow (\text{PCy}_3)_2\text{Pt(metallacyclic ester)} \quad (50)$$

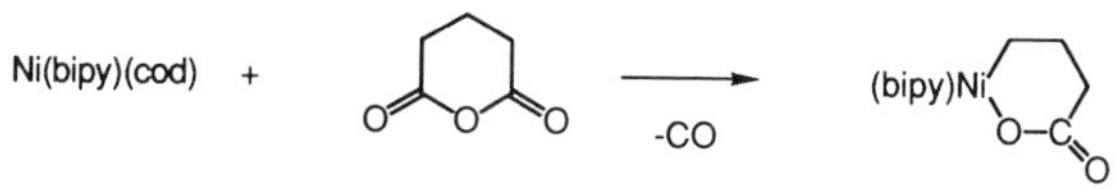

$$\text{Ni(bipy)(cod)} + \text{(glutaric anhydride)} \xrightarrow[-\text{CO}]{} (\text{bipy})\text{Ni(metallacyclic ester)} \quad (51)$$

The metallacyclic ester can be obtained via another route through cyclization of β,γ-unsaturated acid with zero-valent nickel and palladium complexes (*91–93*).

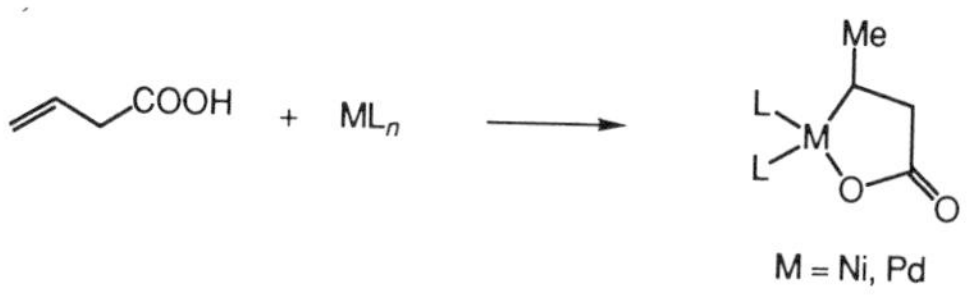

$$\text{CH}_2{=}\text{CHCH}_2\text{COOH} + \text{ML}_n \longrightarrow \text{L}_2\text{M(metallacycle, Me)} \quad (52)$$

M = Ni, Pd

Similar to the behavior of carboxylic esters and acyclic carboxylic anhydrides, the reversibility of the oxidative addition involving the C–O bond cleavage and the reductive elimination involving the C–O bond formation can be observed for the cyclic carboxylic anhydrides as well.

An interesting property of the metallacyclic ester complexes is the skeletal isomerization of the ring compounds. Ring contraction from the six-membered metallacycle to a methyl-substituted five-membered metallacycle takes place through β-hydrogen elimination and reinsertion processes. Employment of a chiral ditertiary phosphine induces formation of the chiral five-membered metallacycle (*87,94*).

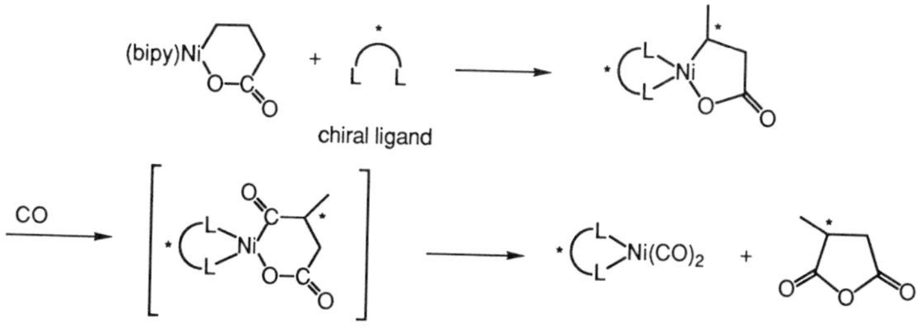

$$\text{(bipy)Ni(metallacycle)} + \text{L}\frown\text{L}^* \text{ (chiral ligand)} \longrightarrow {}^*(\text{L}_2)\text{Ni(5-membered metallacycle)} \xrightarrow{\text{CO}} \left[{}^*(\text{L}_2)\text{Ni(acyl ester)}\right] \longrightarrow {}^*(\text{L}_2)\text{Ni(CO)}_2 + \text{(methylsuccinic anhydride)}^* \quad (53)$$

The result suggests the possibility of catalytic conversion of a six-membered cyclic anhydride such as glutaraldehyde into a chiral methyl-substituted five-membered cyclic anhydride, such as methylsuccinic anhydride, by combining the C–O bond cleavage with other fundamental organometallic processes such as decarbonylation, β-hydrogen elimination, olefin insertion into the M–H bond (followed by CO insertion), and reductive elimination.

3. *C–O Bond Cleavage of Cyclic Ethers*

In Section II,A,3 we reviewed the cleavage of alkenyl oxiranes, which proceeds through π-allyl complex formation. We are concerned here with the cleavage of simple cyclic ethers.

Epoxides oxidatively add to low-valent transition metal complexes to form metallacyclic ethers (*95a–c*). For tricyano ethylene oxide, the C–O bond cleavage takes place at the carbon substituted with two CN groups [Eq. (54)] (*95b*).

PtL4 + (NC)2C–C(CN)(H) epoxide ⟶ L2Pt–C(CN)2–C(H)(CN)–O (54)

L = PPh3, P(p-tol)3, AsPh3

The epoxide C–O bond can be also cleaved on interaction with methylplatinum complexes (*96*). In the case of a nitrogen base-coordinated dimethylplatinum complex, epoxide cleavage is accompanied by CO_2 insertion into the Pt–O bond with formation of a platinum-containing cyclic carbonate (*97,98*).

PtMe2L2 + epoxide(R) ⟶ LnM–O–C–R metallacycle ⟶(CO2) O=C(–O–)–O–Pt(N–N)(Me)(Me)–CH2–C(R)(H) (55)

L = bipyridine, phenanthroline

For the reaction of an iridium complex with epoxides, the C–O bond cleavage of the epoxide ring was accompanied by β-hydrogen elimination to give hydrido and acylalkyl complexes (*99,100*).

(cyclooctene)--IrClL3 + epoxide(R) ⟶(-L) [R–CH(O⁻)–CH2–Ir⁺L3Cl] ⟶ Ir(H)(L)3(Cl)(CH2C(=O)R) (56)

L = PMe3 R = H, Me, Ph

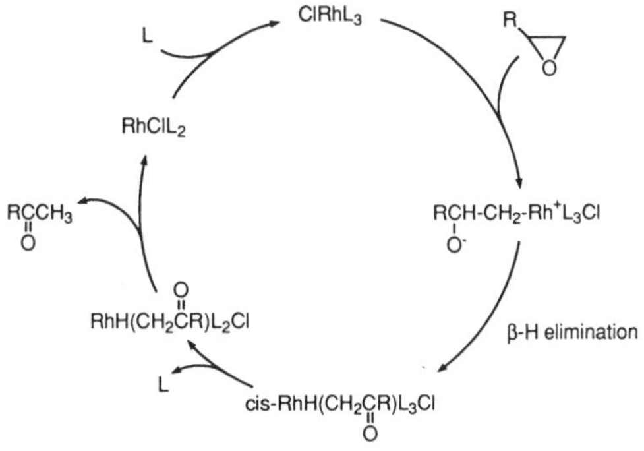

SCHEME 12.

In contrast with the corresponding rhodium complex, further reductive elimination of the hydrido and acylalkyl ligands takes place, liberating ketones or aldehyde. The Rh(I) complex produced in the reaction can react further with the epoxides. Thus a catalytic cycle of converting epoxides to ketones or aldehyde can be constituted (*101*).

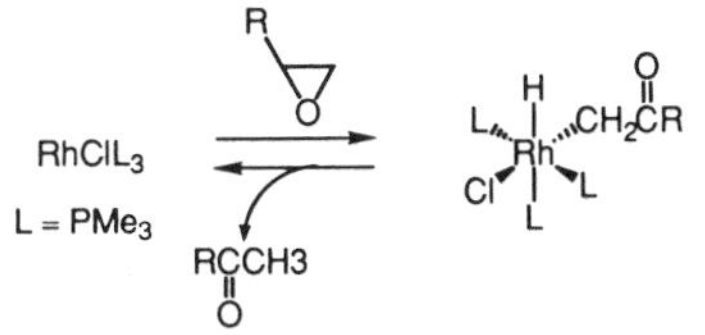

(57)

For the catalytic reaction the pathway in Scheme 12 has been proposed. In this catalytic cycle, dissociation of one of the phosphine ligands from the *cis*-hydrido–acylmethylrhodium complex is rate determining to give the ketone or aldehyde. This is in agreement with the isolation of the *cis*-hydrido–acylalkyl intermediate in the catalytic cycle.

The opening of the epoxide ring can also be induced by cuprate (*102*) and an organotitanium(III) compound (*103–106*); this has been utilized for organic synthesis. A radical mechanism has been proposed (*103–109*).

The C–O bond in four-membered and five-membered cyclic ethers can also be cleaved on interaction with low-valent transition metal complexes (*107–109*). Combination of the C–O bond cleavage in cyclic ethers with CO insertion and ring closure has been applied to lactone synthesis. The CO insertion was observed to occur on the unsubstituted side of the oxetane ring (*107,110*).

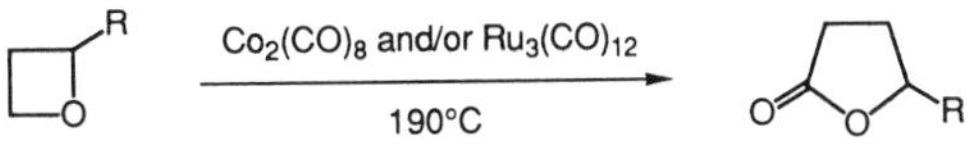

$$\xrightarrow[190^\circ C]{Co_2(CO)_8 \text{ and/or } Ru_3(CO)_{12}} \qquad (58)$$

$$\text{Ph/R-epoxide} + CO \xrightarrow[\text{NR}_4\text{Br, NaOH, C}_6\text{H}_6]{\text{Co}_2(\text{CO})_8\text{, MeI}} \text{(Ph, R, H, HO-substituted furanone)} \quad (59)$$

A different type of C–O bond cleavage involving the cleavage of both C–O bonds in the epoxides has been observed in the reactions with tungsten compexes (*111,112*). The rate of deoxygenation decreases with an increase in the number of methyl substituents and the deoxygenation is regiospecific (i.e., *cis*-epoxide gives *cis*-olefin). The following result using deuterated ethylene oxide strongly suggests a direct transfer of oxygen in ethylene oxide to tungsten and makes the mechanism involving oxametallabutane intermediate unlikely.

$$WCl_2(CH_2{=}CH_2)L_2 + D_2C\overset{O}{—}CD_2 \longrightarrow W(O)Cl_2L_2(CH_2{=}CH_2) + D_2C{=}CD_2 + H_2C{=}CH_2 \quad (60)$$

L = $PMePh_2$

Similar to the behavior of oxiranes, the nitrogen atom in aziridines is abstracted by the tungsten complex. The results have implications in relation to the mechanism of the reverse process, i.e., epoxidation of olefins.

The C–O bond in acetals can be readily cleaved by acidic reagents, but few examples of C–O bond cleavage by transition metal complexes are known. One example is the cleavage of cyclic acetals by acidic alkyltitanium compounds (*113*).

Very few examples of C–O bond cleavage have been reported for noncyclic ethers. However, cleavage of the methyl oxygen bond in methyl phenyl ether has been reported (*69*) when very electron-rich complexes were used.

4. *C–O Bond Cleavage of Alcohols*

Except for allylic alcohols there are very few reports on the cleavage of C–O bonds in alcohols. Abstraction of oxygen from simple alcohols such as methanol or ethanol takes place on interaction with a tungsten complex, liberating alkane and affording both an oxotungsten complex and a bis(alkoxide) complex (*114,115*).

$$WCl_2L_4 + ROH \xrightarrow{-RH} W(O)Cl_2L_3 + W(OR)_2Cl_2L_2 + H_2 \quad (61)$$

L = $PMePh_2$

In contrast to the slow reactions of the tungsten complex with saturated alcohols, allylic alcohols react very rapidly to give the oxo complex and olefin. In this case, cleavage of the C–O bond with formation of an oxo–π-allyl complex has been proposed. The reaction is believed to be driven by formation of a very strong tungsten–oxygen multiple bond.

B. *Cleavage of C–O Bonds by Two-Step Mechanisms*

The C–O single bond in carboxylic esters can be cleaved in reactions with transition metal hydrides or alkyls in a manner different from that discussed in Section III,A,1. The reaction liberates ketone or aldehyde, giving metal alkoxide or aryloxide (*116,117a*).

$$L_nM\text{-}R + R'C(=O)\text{-}OR'' \longrightarrow RC(=O)R' + M(OR'')L_n \tag{62}$$

R = alkyl, H

The reaction products can be accounted for by either of two mechanisms. One involves the direct oxidative addition of carboxylic ester with cleavage of the acyl oxygen bond as mentioned in the previous section. Reductive elimination of the acyl ligand with the hydride or alkyl ligand gives aldehyde or ketone [Eq. (63a)]. The second mechanism [Eq. (63b)] is somewhat reminiscent of the reaction of an ester with an alkylmagnesium compound. Insertion of the carbonyl group of the ester into the M–H or M–alkyl bond followed by elimination of the alkoxide (or aryloxide) group at the β-position liberates aldehyde or ketone with formation of the metal alkoxide (aryl oxide).

$$L_nMR + R'C(=O)\text{-}OR'' \longrightarrow \left[L_nM(R)(OR'')(C(=O)R') \right] \longrightarrow L_nM(OR'') + RC(=O)R' \tag{63a}$$

$$L_nMR + R'C(=O)\text{-}OR'' \longrightarrow \left[L_nM\text{-}O\text{-}C(R')(R)(OR'') \right] \longrightarrow L_nM(OR'') + RC(=O)R' \tag{63b}$$

Although differentiation of the two mechanisms is not easy, at least in one case formation of an alkoxide complex corresponding to the intermediate assumed in Eq. (63b) has been confirmed (*117b*). This suggests the operation of the insertion–elimination mechanism at least in some of the C–O bond cleavages of esters.

$$CoH(N_2)L_3 + CF_3COOEt \xrightarrow{-N_2} L_3Co\text{-}O\text{-}CH(CF_3)(OEt)$$
$$CoH(N_2)L_3 + CF_3CH(OH)OEt \xrightarrow{-N_2,\,-H_2} L_3Co\text{-}O\text{-}CH(CF_3)(OEt) \quad (64)$$

Iridium polyhydride complexes react with trifluoroacetate to cleave the acyl oxygen bond to give iridium alkoxide complexes. On the basis of kinetic studies, a rate-determining insertion of the ester carbonyl group into the Ir–H bond to give a β-alkoxyalkoxide complex, followed by fast alkoxy group abstraction and trifluoroacetaldehyde formation, has been proposed (*118*). The polyhydride complex has been reported to be an excellent transfer hydrogenation catalyst.

IV

CLEAVAGE OF C═O BONDS IN KETONES, ALDEHYDES, OR ESTERS

In the transition metal-catalyzed metathesis of olefins, cleavage of C═C bonds promoted by transition metal complexes has been established (*119*). The corresponding reaction involving cleavage of C═O bonds has been recently found.

The Wittig-type reaction converting carbonyl compounds into olefins has been extensively utilized in organic synthesis. Alkylidene complexes of early transition metals react with carbonyl compounds to give olefins, possibly through an oxametallacyclobutane intermediate (*120*).

$$L_nM{=}CR^1R^2 + R^3R^4C{=}O \longrightarrow \left[\begin{array}{l} L_nM-C(R^1)-R^2 \\ \;\;|\qquad\;\;| \\ \;\;O-C(R^4)-R^3 \end{array} \right] \longrightarrow R^1R^2C{=}CR^3R^4 \quad (65)$$

Tebbe's reagent, formed by the reaction of Cp_2TiCl_2 with trimethylaluminum, can be regarded as a precursor of the alkylidene complex and serves as an excellent olefination reagent (*121–123*). In the synthesis using Tebbe's reagent, a titanium methylidene complex undergoes exchange with organic carbonyl compounds to afford olefins and an oxotitanium compled, probably through an oxametallacyclobutane (*121–123*).

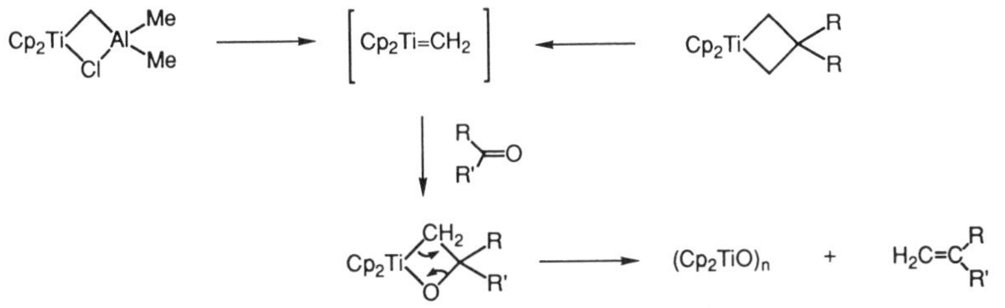

(66)

In fact, the titanaoxacyclobutane could be prepared by reaction of the methylidene complex with ketenes (*124*).

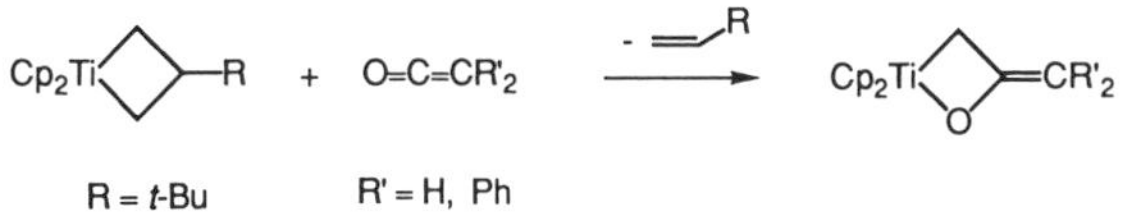

(67)

R = *t*-Bu R' = H, Ph

Carbon–oxygen bond cleavage in a ruthenaoxacyclobutane has been also observed (*125*). The methylidene complex can also be formed from titanacyclobutane by expulsion of olefin (*126*). The advantage of the process using Tebbe's reagent or Grubbs' titanacyclobutane reagents is that they can be applied not only to methylenation of aldehydes and ketones but also to esters and lactones to afford enol ethers. A disadvantage of the process is the sensitivity of the reagents toward air and moisture, and use of Cp_2TiMe_2 has been proposed to overcome this problem (*127*).

An alternative to Tebbe's reagent or Grubbs' titanacycle is to use a mixture of CH_2X_2 ($X = Br, I$), zinc, and $TiCl_4$ ($TiCl_4$ can be replaced with $AlMe_3$) (*128, 129*).

Ketones sometimes from η^2 complexes with transition metals. The oxophilic tungsten complex gives alkylidene oxo–tungsten complexes (*130–132*).

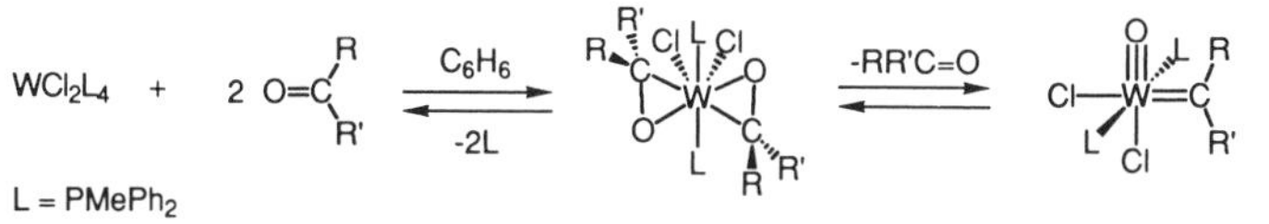

(68)

L = $PMePh_2$

The process is a *four*-electron oxidative addition and is remarkable in that very strong carbon–oxygen double bonds can be cleaved so easily. The reaction is apparently driven by formation of a very strong oxo–tungsten multiple bond (*133*). Insertion of tungsten into the carbon–nitrogen double bond with formation of a tungsten–nitrogen triple bond has been also observed. For aromatic ketones, the C–O bond cleavage reaction takes a different route, liberating olefin and the oxo–tungsten complex.

$$WCl_2L_4 + O{=}C(Ph)(R) \xrightarrow{C_6D_6} Cl{-}W({\equiv}O)(L)_2(Cl){-}L + Ph(R)C{=}C(R)(Ph) \qquad (69)$$

A free radical mechanism has been proposed for the process. The tungsten(II) complex can cause the cleavage of heterocumulenes; for example, CO_2 can be cleaved to form the oxo–carbonyl tungsten complex.

$$WCl_2L_4 + X{=}C{=}Y \longrightarrow Cl{-}W({\equiv}X)(L)_2(Cl){-}C{\equiv}Y + 2L \qquad (70)$$

L = $PMePh_2$ X=C=Y = O=C=O, S=C=O, RN=C=O, RN=C=NR, S=C=NR

Cleavage of the double bond has been explained via formation of an η^2-bond metallacycle (A) that is subsequently cleaved to form the oxo–alkylidene complex, Eq. (71).

$$[A] \longrightarrow [B] \qquad (71)$$

[A] [B]

A related C=O bond cleavage of ketones, observed in reactions with dinuclear tungsten alkoxide complexes (*134,135*), afforded μ-methylene oxo–ditungsten compounds. The μ-alkylidene oxo–ditungsten complexes react further with ketones at room temperature with formation of olefins as the reductive coupling products of the ketones.

$$W_2(OR)_6py_2 + Me_2C{=}O \longrightarrow (RO)_4W(\mu{-}CMe_2)(\mu{-}OR)W(O)(OR)py \xrightarrow{Me_2C=O} Me_2C{=}CMe_2 \qquad (72)$$

R = *i*-Pr, CH_2CMe_3

In contrast, arylketones react with the ditungsten hexaalkoxides somewhat differently (*136*). The reaction of $W_2(OR)_6py_2$ with benzophenone gave a mixture of products, depending on the alkoxide and reaction conditions.

$W_2(OR)_6Py_2$ + Ph_2CO ⟶ { $(RO)_4W[OC(Ph)_2C(Ph)_2O]$ + $[W{=}CPh_2]$; $(RO)_4W(Py)(\eta^2\text{-}OCPh_2)$ + $Ph_2C{=}CPh_2$ + $Ph_2CHCHPh_2$ } (73)

R = CH_2CMe_3, *i*-Pr, *o*-C_6H_{11}

The benzophenone-coordinated tungsten alkoxide complex reacted further with a second equivalent of benzophenone at ambient temperature to form the pinacolate (tetraphenylethylene glycolate) complex. However, no olefin was formed from the isolated pinacolate complex.

Deoxygenative coupling of ketones to olefins promoted by low-valent early transition metal complexes provides a convenient organic synthesis. The most utilized systems are mixtures of titanium chlorides combined with zinc, zinc–copper, or magnesium (*137–139*), but a mixture of tungsten hexachloride with alkyllithium is also effective (*140*). The relationship between the chemistry of tungsten complexes causing the deoxygenation of ketones and that of olefin formation from ketones promoted by the low-valent titanium species is of interest. Formation of the pinacolate intermediate was detected in the titanium case, and the isolated tungsten pinacolate complex did not afford olefin; these results were taken to suggest that the tungsten and titanium systems are different (*136*).

Acknowledgments

The author acknowledges cooperation of his co-workers, whose work has been cited here. Dr. Masato Oshima's assistance in manuscript preparation is also gratefully acknowledged. The author's work on C–O bond cleavage by transition metal complexes described here has been supported by the Ministry of Education, Science, and Culture and Toray Science Foundation.

References

1. R. H. Crabtree and D. G. Hamilton, *Adv. Organomet. Chem.* **28,** 299 (1988); R. G. Bergman, *J. Organomet. Chem.* **400,** 273 (1990).
2. B. M. Trost and T. R. Verhoven, *in* "Comprehensive Organometallic Chemistry (G. Wilkinson, F. G. A. Stone, and E. A. Abel, eds.), Vol. 8, p. 799. Pergamon, Oxford, 1982.
3. B. M. Trost, *Tetrahedron* **33,** 2615 (1977).
4. B. M. Trost, *Acc. Chem. Res.* **13,** 385 (1980).
5. B. M. Trost, *Pure Appl. Chem.* **53,** 2357 (1981).
6. B. M. Trost, *Science* **219,** 245 (1983).
7. J. Tsuji, *Tetrahedron* **42,** 4361 (1986).
8. J. Tsuji, *Acc. Chem. Res.* **2,** 144 (1969).

9. J. Tsuji, "Organic Synthesis by Means of Tansition Metal Complexes." Springer-Verlag, Berlin, 1975.
10. J. Tsuji, "Organic Synthesis with Palladium Compounds." Springer-Verlag, Heidelberg, 1980.
11. N. L. Bauld, *Tetrahedron Lett.*, 859 (1962); K. E. Atkins, W. E. Walker, and R. M. Manyik, *Tetrahedron Lett.*, 3821 (1970); F. Dawans and J. C. Marechal, and P. Teyssie, *J. Organomet. Chem.* **21,** 259 (1970); K. Takahashi, A. Miyake, and G. Hata, *Bull. Chem. Soc. Jpn.* **45,** 230 (1972); T. Tsuda, Y. Chujo, S. Nishi, K. Tawara, and T. Saegusa, *J. Am. Chem. Soc.*, **102,** 6381 (1980).
12. T. Hayashi, T. Hagihara, M. Konishi, and M. Kumada, *J. Am. Chem. Soc.* **105,** 7767 (1983).
13. T. Hayashi, M. Konishi, and M. Kumada, *J. Chem. Soc., Chem. Commun.*, 107 (1984).
14. T. Hayashi, A. Yamamoto, and Y. Ito, *Chem. Lett.*, 177 (1987).
15. T. Hayashi, A. Yamamoto, and T. Hagihara, *J. Org. Chem.* **51,** 723 (1986).
16. P. R. Auburn, P. B. Mackenzie, and B. Bosnich, *J. Am. Chem. Soc.* **107,** 2033 (1985).
17. P. B. Mackenzie, S. Whelan, and B. Bosnich, *J. Am. Chem. Soc.* **107,** 2046 (1985).
18. E. Keinan and N. Greenspoon, *Tetrahedron Lett.* **23,** 241 (1982).
19. B. M. Trost and T. R. Verhoeven, *J. Am. Chem. Soc.* **102,** 4730 (1980).
20. I. Stary and P. Kocovsky, *J. Am. Chem. Soc.* **111,** 4981 (1989).
21. T. Yamamoto, O. Saito, and A. Yamamoto, *J. Am. Chem. Soc.* **103,** 5600 (1981).
22. T. Yamamoto, A. Akimoto, O. Saito, and A. Yamamoto, *Organometallics* **5,** 1559 (1986).
23. T. Yamamoto, M. Akimoto, and A. Yamamoto, *Chem. Lett.*, 1725 (1983).
24. J. Ishizu, T. Yamamoto, and A. Yamamoto, *Chem. Lett.*, 1091 (1976).
25. T. Yamamoto, J. Ishizu, and A. Yamamoto, *J. Am. Chem. Soc.* **103,** 6863 (1981).
26. M. Oshima, I. Shimizu, A. Yamamoto, and F. Ozawa, *Organometallics* **10,** 1221 (1991).
27. H. Hay and H. J. Arpe, *Angew. Chem., Int. Ed. Engl.* **12,** 928 (1973).
28. J. Tsuji and T. Yamakawa, *Tetrahedron Lett.* **7,** 613 (1979).
29. J. Tsuji and I. Shimizu, *Synthesis*, 623 (1986).
30. J. W. Faller and D. Linebarrier, *Organometallics* **7,** 1670 (1988).
31. T. Ito, T. Matsubara, and Y. Yamashita, *J. Chem. Soc., Dalton Trans.*, 2407 (1990).
32. A. Yamamoto, F. Ozawa, K. Osakada, L. Huang, T.-I. Son, N. Kawasaki, and M.-K. Doh, *Pure Appl. Chem.* **63,** 687 (1991).
33. F. Ozawa, T.-I. Son, S. Ebina, K. Osakada, and A. Yamamoto, *Organometallics* **11,** 171 (1992).
34. I. Minami, I. Shimizu, and J. Tsuji, *J. Organomet. Chem.* **296,** 269 (1985).
35. F. Guibe, O. Dangles, and G. Balavoine, *Tetrahedron Lett.* **27,** 2368 (1986).
36. G. Dangles, F. Guibe, G. Balavoine, S. Lavielle, and A. Marquet, *J. Org. Chem.* **52,** 4984 (1987).
37. S.-I. Murahashi, Y. Taniguchi, Y. Imada, and Y. Tanigawa, *J. Org. Chem.* **54,** 3292 (1989).
38. S.-I. Murahashi, Y. Tanigawa, Y. Imada, and Y. Taniguchi, *Tetrahedron Lett.* **27,** 227 (1986).
39. T. Ito, K. Hamamoto, S. Kurishima, and K. Osakada, *J. Chem. Soc., Dalton Trans.*, 1645 (1990).
40. J. J. Eisch and K. R. Im, *J. Organomet. Chem.* **139,** C45 (1977).
41. T. Yamamoto, M. Akimoto, O. Saito, and A, Yamamoto, *Organometallics* **5,** 1559 (1986).
42. K. Osakada, Y. Ozawa, and A. Yamamoto, *J. Organomet. Chem.* **399,** 341 (1990).
43. T. Hayashi, M. Konishi, K. Yokoto, and M. Kumada, *J. Chem. Soc., Chem. Commun.*, 313 (1981).

44. I. Shimizu, M. Oshima, M. Nisar, and J. Tsuji, *Chem. Lett.*, 1775 (1986).
45. M. Oshima, H. Yamazaki, I. Shimizu, M. Nisar, and J. Tsuji, *J. Am. Chem. Soc.*, **111,** 6280 (1989).
46. B. M. Trost and S. R. Angle, *J. Am. Chem. Soc.* **107,** 6123 (1985).
47. B. M. Trost and G. A. Molander, *J. Am. Chem. Soc.* **103,** 5969 (1981).
48. B. M. Trost and A. R. Sudhakar, *J. Am. Chem. Soc.* **109,** 3792 (1987).
49. A. M. Echavarren, D. R. Tueting, and J. K. Stille, *J. Am. Chem. Soc.* **110,** 4039 (1988).
50. R. Aumann, H. Ring, C. Krüger, and R. Goddard, *Chem. Ber.* **112,** 3644 (1979).
51. C. Chuit, H. Felkin, C. Frajerman, G. Roussi, and G. Swierczewski, *J. Organomet. Chem.* **127,** 371 (1977).
52. H. Felkin, E. Jampel-Costa, and G. Swierczewski, *J. Organomet. Chem.* **134,** 265 (1977).
53. J. K. Nicholson and B. L. Shaw, *Proc. Chem. Soc.*, 282 (1963).
54. X. Lu, L. Lu, and J. Sun, *J. Mol. Catal.* **41,** 245 (1987).
55. X. Lu, X. Jiang, and X. Tao, *J. Organomet. Chem.* **344,** 109 (1988).
56. Y. Ishimura, N. Nagato, Japanese Patent, Kokai, S63–2958 (1988); Y. Ishimura, T. Oe, Y. Suyama, and N. Nagato, U. S. Patent 4,942,262 (1990).
57. Y. Hayashi, S. Komiya, T. Yamamoto, and A. Yamamoto, *Chem. Let.*, 977 (1984).
58. Y. Hayashi, Ph.D. Thesis, Tokyo Institute of Technology, Tokyo (1985).
59. S. Oishi, N. Kihara, and A. Hosaka, *Chem. Lett.*, 621 (1985).
60. Y. Kushi, M. Kuramoto, Y. Hayashi, T. Yamamoto, A. Yamamoto, and S. Komiya, *J. Chem. Soc., Chem. Commun.*, 1033 (1983).
61. H. C. Clark and H. Kurosawa, *Inorg. Chem.* **12,** 357 (1973).
62. S. Komiya, R. Srivastava, A. Yamamoto, and T. Yamamoto, *Organometallics* **4,** 1504 (1985).
63. K. Osakada, L. Matsumoto, T. Yamamoto, and A. Yamamoto, *Organometallics* **4,** 857 (1985); K. Osakada, K. Matsumoto, T. Yamamoto, and A. Yamamoto, *Chem. Ind. (London)*, 634 (1984).
64. S. Komiya and A. Yamamoto, *J. Organomet. Chem.* **87,** 333 (1975).
65. S. Komiya, J. Suzuki, K. Miki, and N. Kasai, *Chem. Lett.*, 1287 (1987).
66. T. Yamamoto, J. Ishizu, T. Kohara, S. Komiya, and A. Yamamoto, *J. Am. Chem. Soc.* **102,** 3758 (1980).
67. J. Ishizu, T. Yamamoto, and A. Yamamoto, *Chem. Lett.* 1091 (1976).
68. T. Kohara, S. Komiya, T. Yamamoto, and A. Yamamoto, *Chem. Let.* 1513 (1979); S. Komiya, Y. Akai, K. Tanaka, T. Yamamoto, and A. Yamamoto, *Organometallics* **4,** 1130 (1985).
69. C. A. Tolman, S. D. Ittel, A. D. English, and J. P. Jesson, *J. Am. Chem. Soc.* **101,** 1742 (1979).
70. H. E. Bryndza and W. Tam, *Chem. Rev.* **88,** 1163 (1988); S. E. Kegley, C. J. Schaverien, J. H. Freudenberger, R. G. Bergman, S. P. Nolan, and C. D. Hoff, *J. Am. Chem. Soc.* **109,** 6563 (1987); D. Braga, Pl. Sabatino, C. D. Bugno, P. Leoni, and M. Pasquali, *J. Organomet. Chem.* **34,** C46 (1987); K. Osakada, Y.-J. Kim, and A. Yamamoto, *J. Organomet. Chem.* **382,** 303 (1990); K. Osakada, Y.-J. Kim, M. Tanaka, S. Ishiguro, and A. Yamamoto, *Inorg. Chem.* **30,** 197 (1991).
71. Y.-J. Kim, K. Osakada, A. Takenaka, and A. Yamamoto, *J. Am. Chem. Soc.* **112,** 1096 (1990).
72. A. A. Zota, F. Frolow, and D. Milstein, *Organometallics* **9,** 1300 (1990).
73. T. Yamamoto, J. Ishizu, and A. Yamamoto, *Bull. Chem. Soc. Jpn.* **55,** 623 (1982).
74. W. J. Scott and J. E. McMurray, *Acc. Chem. Res.* **21,** 47 (1988).
75a. For a recent work, see W. Cabri, S. De Bernandinis, F. Francalanci, S. Panco, and R. Santi, *J. Org. Chem.* **55,** 350 (1990).

ᴊ. P. Roth and C. E. Fuller, *J. Org. Chem.* **56,** 3493 (1991).
75c. A. M. Echavarren and J. K. Stille, *J. Am. Chem. Soc.* **109,** 5478 (1987).
76. J. K. Stille, *Angew. Chem., Int. Ed. Engl.* **25,** 508 (1986).
77. W. J. Scott and J. K. Stille, *J. Am. Chem. Soc.* **108,** 3033 (1986).
78. M. H. Kowalski and P. J. Stang, *Organometallics* **5,** 2392 (1986); P. J. Stang, M. H. Kowalski, M. D. Schiavelli, and D. Longford, *J. Am. Chem. Soc.* **111,** 3347 (1989).
79. F. Ozawa, A. Kubo, and T. Hayashi, *J. Am. Chem. Soc.* **113,** 1417 (1991).
80. S. Cacchi, P. G. Ciattini, E. Morera, and G. Otar, *Tetrahedron Lett.* **27,** 3921 (1986).
81. R. E. Dolle, S. J. Schmidt, and L. I. Kruse, *J. Chem., Soc., Chem. Commun.*, 904, (1987).
82. H. Urata, D. Goto, and T. Fuchikami, *Tetrahedron Lett.* **32,** 3091 (1991).
83. D. M. Blake, S. Shields, and L. Wyman, *Inorg. Chem.* **13,** 1595 (1974).
84. H. F. Klein and H. H. Karsch, *Chem. Ber.* **109,** 2515 (1976).
85. S. Komiya, A. Yamamoto, and T. Yamamoto, *Chem. Lett.*, 193 (1981).
86. J. P. Collman and S. R. Winter, *J. Am. Chem. Soc.*, **95,** 4089 (1973).
87. T. Yamamoto, K. Sano, and A. Yamamoto, *J. Am. Chem. Soc.* **109,** 1092 (1987).
88. K. Sano, T. Yamamoto, and A. Yamamoto, *Chem. Lett.* 115, (1983).
89. V. E. Uhlig, G. Fahske, and B. Nestler, *Z. Anorg. Allg. Chem.* **465,** 151 (1980).
90. K. Sano, T. Yamamoto, and A. Yamamoto, *Bull. Chem. Soc. Jpn.* **57,** 2741 (1984).
91. T. Yamamoto, K. Igarashi, J. Ishizu, and A. Yamamoto, *J. Chem. Soc., Chem. Commun.*, 554 (1979).
92. T. Yamamoto, K. Sano, K. Osakada, S. Komiya, A. Yamamoto, Y. Kushi, and T. Tada, *Organometallics* **9,** 2396 (1990).
93. K. Osakada, M.-K. Doh, F. Ozawa, and A. Yamamoto, *Organometallics* **9,** 2197 (1990).
94. K. Sano, T. Yamamoto, and A. Yamamoto, *Chem. Lett.*, 941 (1984).
95a. R. Schlodder, J. A. Ibers, M. Lenarda, and M. Graziani, *J. Am. Chem. Soc.* **96,** 6893 (1974).
95b. M. Lenarda, N. B. Pahor, M. Calligaris, M. Graziani, and L. Randaccio, *J. Chem. Soc., Dalton Trans.*, 279 (1978).
95c. A. Miyashita, R. Shimada, A. Sugawara, and H. Nohira, *Chem. Lett.*, 1323 (1986).
96. T. Hase, A. Miyashita, and H. Nohira, *Chem. Lett.*, 219 (1988).
97. K.-T. Aye, G. Fergusson, A. J. Longh, and R. J. Puddephatt, *Angew. Chem., Int. Ed. Engl.* **28,** 767 (1989).
98. K. T. Aye, L. Gelmini, N. C. Payne, J. J. Vittal, and R. C. Puddephatt, *J. Am. Chem. Soc.* **112,** 2464 (1990).
99. D. Milstein and J. C. Calabrese, *J. Am. Chem. Soc.* **104,** 3773 (1982).
100. D. Milstein, *Acc. Chem. Res.* **17,** 221 (1984).
101. D. Milstein, *J. Am. Chem. Soc.* **104,** 5227 (1982).
102. M. Mitani, H. Matsumoto, N. Gouda, and K. Koyama, *J. Am. Chem. Soc.* **112,** 1286 (1990).
103. W. A. Nugent and T. V. R. Babu, *J. Am. Chem. Soc.* **110,** 856 (1988).
104. T. V. R. Babu and W. A. Nugent, *J. Am. Chem. Soc.* **111,** 4525 (1989).
105. J. S. Yadav, T. Shekharam, and V. R. Gadjil, *J. Chem. Soc., Chem. Commun.*, 843 (1990).
106. T. V. R. Babu, W. A. Nugent, and M. S. Beattie, *J. Am. Chem. Soc.* **112,** 6408 (1990).
107. M. D. Wang, S. Calet, and H. Alper, *J. Org. Chem.* **54,** 20 (1989).
108. D. Jeannin and J. Zakrzewski, *J. Chem. Soc., Chem. Commun.*, 813 (1983).
109. T. Tatsumi, H. Tominaga, M. Hidai, and Y. Uchida, *Chem. Lett.*, 37 (1977).
110. H. Alper, H. Arzoumanian, J.-F. Petrignani, and M. Saldana-Maldonado, *J. Chem. Soc., Chem. Commun.*, 340 (1985).

111. J. C. Bryan, S. J. Geib, A. L. Rheingold, and J. M. Mayer, *J. Am. Chem. Soc.* **109,** 2826 (1987).
112. L. M. Atagi, D. E. Over, D. R. McAlister, and J. M. Mayer, *J. Am. Chem. Soc.* **113,** 870 (1991).
113. A. Mori, K. Maruoka, and H. Yamamoto, *Tetrahedron Lett.* **25,** 4421 (1984).
114. S. Jang, L. M. Atagi, and J. M. Mayer, *J. Am. Chem. Soc.* **112,** 6413 (1990).
115. K. W. Chiu, D. Lyons, G. Wilkinson, M. Thornton-Pett, and M. B. Hursthouse, *Polyhedron* **2,** 803 (1983).
116. T. Yamamoto, S. Miyashita, Y. Naito, S. Komiya, T. Ito, and A. Yamamoto, *Organometallics* **1,** 808 (1982).
117a. Y. Hayashi, T. Yamamoto, A. Yamamoto, S. Komiya, and Y. Kushi, *J. Am. Chem. Soc.* **108,** 385 (1986).
117b. Y. Hayashi, S. Komiya, T. Yamamoto, and A. Yamamoto, *Chem. Lett.* 1363 (1984).
118. A. S. Goldman and J. Halpern, *J. Am. Chem. Soc.* **109,** 7537 (1987); A. S. Goldman and J. Halpern, *J. Organomet. Chem.* **382,** 237 (1990).
119. See for example, J. P. Collman, L. S. Hegedus, J. R. Norton, and R. G. Finke, "Principles and Applications of Organotransition Metal Chemistry." University Science Books, Mill Valley, California, 1987; A. Yamamoto, "Organotransition Metal Chemistry—Fundamental Concepts and Applications." Wiley (Interscience), New York, 1986.
120. R. R. Schrock, *J. Am. Chem. Soc.* **98,** 5399 (1976).
121. K. A. Brown-Wensley, S. L. Buchwald, L. Cannizzo, L. Clawson, S. Ho, D. Meinhardt, J. R. Stille, D. Straus, and R. H. Grubbs, *Pure Appl. Chem.* **55,** 1733 (1983), and references cited therein.
122. K. H. Dötz, *Angew. Chem., Int. Ed. Engl.* **23,** 587 (1984).
123. S. H. Pine, R. J. Pettit, G. D. Geib, S. G. Cruz, C. H. Gallego, T. Tijerrina, and R. O. Pine, *J. Org. Chem.* **50,** 1212 (1985).
124. S. C. Ho, S. Hentges, and R. H. Grubbs, *Organometallics* **7,** 780 (1988).
125. J. F. Hartwig, R. G. Bergman, and R. A. Anderson, *J. Am. Chem. Soc.* **112,** 3246 (1990).
126. L. F. Cannizzo and R. H. Grubbs, *J. Org. Chem.* **50,** 2316 (1985).
127. N. A. Petasis and E. I. Bzowej, *J. Am. Chem. Soc.* **112,** 6392 (1990).
128. K. Takai, Y. Hotta, K. Oshima, and H. Nozaki, *Tetrahedron Lett.*, 2417 (1978).
129. K. Takai, Y. Hotta, K. Oshima, and H. Nozaki, *Bull. Chem. Soc. Jpn.* **53,** 1698 (1980).
130. J. C. Bryan and J. M. Mayer, *J. Am. Chem. Soc.* **109,** 7213 (1987).
131. J. C. Bryan and J. M. Mayer, *J. Am. Chem. Soc.* **112,** 2298 (1990).
132. E. Carmona, L. Sanchez, M. L. Poveda, R. A. Jones, and J. G. Hafner, *Polyhedron* **2,** 797 (1983).
133. W. A. Nugent and J. M. Mayer, "Metal–Ligand Multiple Bonds." Wiley (Interscience), New York, 1988.
134. M. H. Chisholm, K. Folting, and J. A. Klang, *Organometallics* **9,** 602 (1990).
135. M. H. Chisholm and J. A. Klang, *J. Am. Chem. Soc.* **111,** 2324 (1989).
136. M. H. Chisholm, K. Folting, and J. A. Klang, *Organometallics* **9,** 607 (1990).
137. J. E. McMurry, *Acc. Chem. Res.* **16,** 405 (1983).
138. T. Mukaiyama, T. Sato, and J. Hanna, *Chem. Lett.*, 1041 (1973).
139. S. Tyrlik and I. Wolochowicz, *Bull. Soc. Chim. Fr.*, 2147 (1973).
140. K. B. Sharpless, M. A. Umbreit, M. T. Nioth, and T. C. Folld, *J. Am. Chem. Soc.* **94,** 6538 (1972).

ADVANCES IN ORGANOMETALLIC CHEMISTRY, VOL. 34

Charge-Transfer Complexes of Organosilicon Compounds

VALERY F. TRAVEN and SERGEI YU. SHAPAKIN

Mendeleev Chemico-Technological Institute
Moscow, Russia

I

INTRODUCTION

Donor–acceptor interactions play an important role in chemistry. Interactions between electron acceptors (Lewis acids) and electron donors (Lewis bases) underly much of the reactivity of covalent compounds, including organosilicon compounds. In this article the evidence will be examined for weak donor–acceptor complexes, of the charge-transfer type, consisting of donor organosilicon compounds and various acceptor molecules.

The reaction between a donor D and an acceptor A may occur in several steps:

$$D + A \leftrightharpoons D,A \leftrightharpoons D^{+}A^{-} \leftrightharpoons \text{products}$$

The reaction can terminate at the stage of charge transfer or it can include the rupture and formation of covalent bonds. The theory of charge-transfer (CT) complexation has been worked out by Mulliken (*1,2*). According to this theory, the ground state of a CT complex can be described by the wave function φ_n:

$$\varphi_n = a\varphi_0(D, A) + b\varphi_1(D^{+}, A^{-}) \tag{1}$$

where $a^2 \gg b^2$. The contribution of the "without-bond" structure (D, A) to this function dominates, whereas the contribution of the donor–acceptor structure (D^{+}, A^{-}), in which the electron passes from the donor to the acceptor, is insignificant.

The excited state of a CT complex is described by the wave function φ_e:

$$\varphi_e = a^*\varphi_1\,(D^{+}, A^{-}) - b^*\varphi_0\,(D, A) \tag{2}$$

where $a \simeq a^*, b \simeq b^*, a^2 \gg b^2$.

The transition from the state φ_n to the state φ_e on absorption of light thus corresponds to electron transfer from the donor to the acceptor. The difference in the energies of these states is equal to the energy of the corresponding CT band of the absorption spectrum of the complex:

$$h\nu_{CT} = IP - EA - W \tag{3}$$

where W is the dissociation energy of the excited state of the complex, which is approximately constant for related donor molecules and the same acceptor; IP is the ionization potential of the donor and EA is the electron affinity of the acceptor.

In the Koopmans approximation, the value of IP is determined by the energy of the occupied molecular orbital (MO) of the donor, and the value of EA is determined by the energy of the unoccupied MO of the acceptor (*3*):

$$\begin{aligned} IP &= -\varepsilon_{D(i)} \\ EA &= -\varepsilon_{A(j)} \end{aligned} \tag{4}$$

where i is an index of the occupied MO and j is an index of the unoccupied MO.

One can see that the CT energies recorded in visible absorption spectra:

$$h\nu_{CT} = \varepsilon_{A(j)} - \varepsilon_{D(i)} - W \tag{5}$$

are only a part of the energy of orbital interaction between donor and acceptor (*4*):

$$E_{orb} \simeq \sum_{i(D)}\sum_{j(A)} \frac{(c_{D(i)}c_{A(j)}\Delta\beta_{DA})^2}{\varepsilon_{D(i)} - \varepsilon_{A(j)}} + \sum_{i(A)}\sum_{j(D)} \frac{(c_{A(i)}c_{D(j)}\Delta\beta_{DA})^2}{\varepsilon_{A(i)} - \varepsilon_{D(j)}} \tag{6}$$

Nevertheless, charge-transfer complexing follows the general rules of orbital mixing (*5*): (1) the orbitals must have the same symmetry, (2) the orbitals must have similar energies, and (3) overlap of the orbitals should be sterically favorable.

These conditions are often completely fulfilled for π–π^* complexing. Consequently, charge-transfer complexing between aromatic hydrocarbons and π-acceptors has been extensively studied. Derivatives of aromatic hydrocarbons containing silyl and silylalkyl substituents, even when overcrowded, also follow the regularities of π–π^* CT complexing.

However, aromatic hydrocarbons and their derivatives are not the only group of donors in CT complexing with π-acceptors; n-donors, e.g., ethers and sulfides with silyl and silylalkyl groups included, have also been studied in detail, because the oxygen and sulfur atoms in those donors have orbitals of π symmetry.

The capacity of σ bonds to donate electrons to π^* orbitals is not so obvious. For example, σ bonds of aromatic hydrocarbons in π complexes can not participate in charge transfer because orthogonal orbitals do not interact at all. Nevertheless, σ bonds in some compounds can serve as CT donors to π-acceptors. In general, symmetry restrictions do not apply for saturated compounds, because they are usually nonplanar. Saturated hydrocarbons, however, are rather poor σ-donors because of high ionization energies of their C–C and C–H bonding electrons; yet CT complexes with alkanes as donors can be observed, and are discussed in Section II,H.

Polysilanes, having rather lower ionization potentials of Si–Si bonds, are much better σ-donors. The unique properties of oligosilanes, noted earlier *(6–10)* in comparative studies on carbon and silicon compounds, manifest themselves in charge-transfer interactions as well.

The position of the longest wavelength absorption band in the spectra of the CT complexes will be determined by the frontier orbitals of the CT complex, the HOMO of the donor and the LUMO of the acceptor:

$$h\nu_{CT} = \varepsilon_{A(LUMO)} - \varepsilon_{D(HOMO)} - W \tag{7}$$

As the value of $\varepsilon_{A(LUMO)}$ is constant for a series of related donors, the dependencies

$$h\nu_{CT(1)} = aI_1 - b \tag{8}$$

are reported most often *(11)*.

The electron configuration of Eq. (7) is not the only one that can be studied by CT spectroscopy. Transitions between other occupied orbitals of the donor and unoccupied orbitals of the acceptor may also be permitted, and as result, several CT bands can often be recorded.

In principle, the interaction between donor and acceptor could be even more complicated. According to Eq. (6), electron transfer from an occupied MO of the acceptor to an unoccupied MO of the donor should not also be excluded.

Bearing these comments in mind, it is apparent that *(4)*, electronic absorption spectroscopy of CT complexes provides a powerful method for the study of intramolecular structural effects in donor molecules, including organosilicon compounds. Generally, charge-transfer complexing follows all the rules of intermolecular interactions, manifesting itself as a first step of many chemical reactions.

II

ORGANOSILICON COMPOUNDS AS DONORS

A. *Allyl- and Vinylsilanes and Their Cyclic Analogs*

Alkenylsilanes are the simplest organosilicons in which the highest occupied molecular orbitals are of π symmetry. Nevertheless, the number of publications describing the CT complexes of these compounds is not great, because alkenylsilanes have relatively high ionization potentials (*12–14*).

Attempts to record absorption spectra of the CT complexes of tetracyanoethylene (TCNE) with some vinylsilanes $CH_2{=}CH{-}SiX_3$ ($X = Cl$, OEt) failed, because the CT band was superimposed on the TCNE absorption (*12*). However, CT absorption bands have been recorded for several allylsilane complexes (Table I). The monotonic shift of the CT band toward longer wavelengths for the following substituents indicates the following order of decreasing electron-withdrawing properties:

$$SiCl_3 > C(CH_3)_3 > SiH_3 > Si(OEt)_3 > SiMe_3$$

These results are consistent with conjugation of the π orbital with the

TABLE I

FREQUENCIES OF CT BANDS OF COMPLEXES OF ALLYLSILANES WITH TCNE IN DICHLOROMETHANE[a]

Compound	ν_{CT} (cm^{-1})
$CH_2{=}CH{-}CH_2{-}SiCl_3$	30,300
$CH_2{=}CH{-}CH_2{-}Si(OEt)_3$	26,600[b]
$CH_2{=}CH{-}CH_2{-}SiH_3$	27,500
$CH_2{=}CH{-}CMe_3$	29,700
$CH_2{=}CH{-}CH_2{-}SiMe_3$	24,100
$CH_2{=}CH{-}CH_2{-}CMe_3$	28,500[b]
$CH_2{=}CH{-}CH_2{-}CH_2SiMe_3$	27,000[b]

[a] From Ref. 12.
[b] Shoulder.

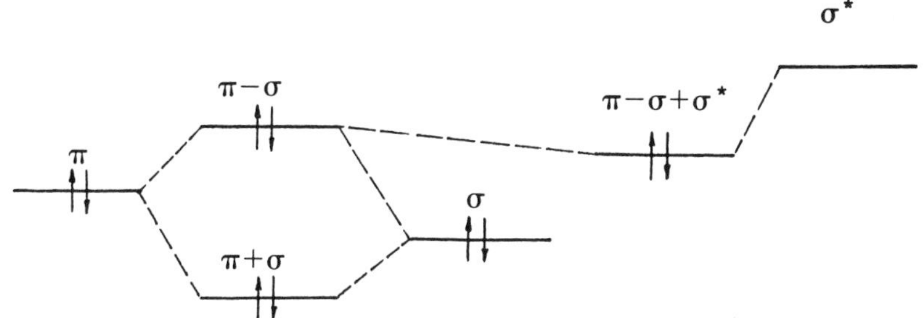

FIG. 1. Before and after σ–π interaction diagram (*12*).

antibonding $\sigma^*_{C\text{-}Si}$ orbital in the molecule of allylsilane, as the energy of the $\sigma^*_{C\text{-}Si}$ orbital decreases with increasing electronegativity of the substituent at the silicon atom; the energy of the HOMO of allylsilane decreases as well (Fig. 1).

Alkenylsilanes also form CT complexes with I_2 (Table II) (*13*). Phenol has also been studied as an acceptor with respect to alkenylsilanes. Egorochkin *et al.* (*13*) established a linear correlation between ν_{CT} and the shift in frequency $\Delta\nu$ of the O–H bond in the infrared spectrum of phenol during its hydrogen bonding with the studied compounds:

$$\Delta\nu = -0.00806\nu_{CT} + 342 \qquad (\rho = 0.980) \tag{9}$$

The value of $\Delta\nu$ characterizes the effect of the substituent on the ground state of the donating molecule, whereas the ν_{CT} value characterizes the excited state. In view of the linear dependence between $\Delta\nu$ and ν_{CT}, it seems that the electronic effects of substituents are similar in both states (*13*).

TABLE II
FREQUENCIES OF CT BANDS AND SHIFTS OF PHENOL OH STRETCHING BANDS[a]

Compound	ν_{CT} (cm^{-1})[b]	$\Delta\nu$ (cm^{-1})[c]
$MeSi(CH_2CH{=}CH_2)_3$	32,700	72
$Me_3SiCH_2CH{=}CH_2$	32,800	80
$Me_2Si(CH_2CH{=}CH_2)_2$	32,800	77
$Me_3Si(CH_2)_2CH{=}CH_2$	34,200	65

[a] From Ref. 13.
[b] Frequencies are those of complexes with I_2.
[c] Shifts of phenol OH stretching bands in IR spectra of complexes with some allylsilanes.

TABLE III

FREQUENCIES OF CT BANDS OF SOME COMPLEXES WITH TCNE IN DICHLOROMETHANE[a]

Compound	Substituent X	$\nu_{CT} \cdot 10^{-3}$ (cm^{-1})	$\nu_{CT}^{X} - \nu_{CT}^{H} \cdot 10^{-3}$ (cm^{-1})
1	H	25.8	0
	Me	24.1	−1.7
	Cl	27.5	1.7
	F	29.1	3.3
2	H	22.4	0
	Me	21.6	−0.8
	Cl	23.7	1.3
	F	25.3	2.9
3	H	23.8	0
	Me	20.1	−3.7
	Cl	28.1	4.3

[a] From Ref. 14.

In order to analyze stereoelectronic effects in the interaction of silyl substituents with $\pi_{C=C}$ bond orbitals, the absorption spectra of CT complexes of some silacyclopentenes (**1** and **2**) and cyclopentenylsilanes (**3**) with TCNE in dichloromethane solution were recorded (*14*). It was shown that with changing electronegativity of X, the shift in $\Delta\nu_{CT}$ is greater for cyclopentenylsilanes (Table III). This observation is explained in terms of conformational differences. In **3**, a conformation is possible in which the

SiX_2 **1** Me SiX_2 **2** SiX_3 **3**

axes of the π and $\sigma_{C\text{-}Si}$ orbitals are nearly parallel, so that effective hyperconjugation can take place. No such favorable conformations are available for **1** or **2**. This correlation between the first ionization potentials and ν_{CT} values has made it possible to use the CT spectra for quantitative estimation of the hyperconjugation effect.

B. *Ethynylsilanes*

Absorption spectra of complexes of some ethynyl derivatives of silicon with iodine have been recorded (*15*). In alkyne–iodine complexes, because the π-donor center is relatively small and the acceptor is bulky, steric effects of substituents can play an important role. The ν_{CT} values should therefore be used with caution for estimation of intramolecular electronic effects in silylacetylenes. Nevertheless, a trustworthy linear dependence of CT band position on the first ionization potential and the sum of σ_p^+ constants for substituents X and Y have been established for compounds of the general structure X—C≡C—Y [where Y = H, Me, Et, *t*-Bu, OMe, OEt, SiH_3; X = SiH_3, $SiMe_3$, $C(SiMe_3)_3$, $SiEt_3$]:

$$IP = 2.20\,\Sigma\,\sigma_p^+ + 10.65 \qquad (\rho = 0.973) \tag{10}$$

$$\nu_{CT} = 3380IP + 2700 \qquad (\rho = 0.950) \tag{11}$$

$$\nu_{CT} = 6200\,\Sigma\,\sigma_p^+ + 37{,}650 \qquad (\rho = 0.951) \tag{12}$$

C. *Aryl- and Hetarylsilanes*

1. *Arylsilanes*

The study of CT complexes of phenylsilanes and benzylsilanes is rather informative for the analysis of intramolecular effects in donor molecules. The CT bands in the absorption spectra of complexes of benzylsilanes with TCNE, for example, are well resolved and their position can be determined to a high degree of accuracy.

A linear dependence ($\rho = 0.961$) of the position of the long-wave band in the CT spectrum on σ_I constants of substituents has been established for 14 benzylsilanes (*16*). A simlar dependence on σ_p constants of substituents has been found for CT complexes of phenylsilanes (*17*). The dependence of CT band energy of complexes with TCNE on the IP value of the donor [Eq. (13)] (*18*)

$$h\nu_{CT} = 0.82IP - 4.28 \tag{13}$$

was used to calculate ionization potentials of some phenyldisilanes and phenyltrisilanes (Table IV) (*19*). Some conclusions about the intramolecular effects in arylpolysilanes were proposed (*20–26*). The study was continued using photoelectron (PE) spectroscopy to measure the ionization potentials of the corresponding compounds.

TABLE IV

CT BANDS OF COMPLEXES OF SOME ARYLSILANES WITH TCNE AND ESTIMATED IONIZATION POTENTIALS[a]

Compound	λ_{CT} (nm)	$h\nu_{CT}$ (eV)	IP (eV)
$PhSiMe_3$	423	2.93	8.80
$PhSi_2Me_5$	493	2.51	8.29
$Ph(SiMe_2)_3Me$	497	2.49	8.27
$PhCH_2SiMe_3$	504	2.46	8.22
$PhCH_2Si_2Me_5$	513	2.42	8.17

[a] From Ref. 19.

Kuznetsov *et al.* (*27*) have discussed the position of the CT band in the absorption spectra of TCNE complexes with some substituted naphthalenes, including 11 siloxy and silyl derivatives. They suggest that the acceptor effect of the Si atom was stronger on the O atom than on the aromatic ring.

The electron-withdrawing properties of the silicon atom in arylsilanes can readily be understood as resulting from the interaction of π orbitals with antibonding orbitals σ^*_{SiX} and $\sigma^*_{CH_2Si}$ (*28*). In order to substantiate this model, the absorption spectra of arylsilane complexes with TCNE were recorded (Table V) (*29*) and orbital energies for the donor molecules were also calculated by the simple Hückel molecular orbital (HMO) method (Tables VI and VII). The SiX_3 group was described on the basis of Bishop's semiempirical model of hyperconjugation (*30*), modified for the

TABLE V

FREQUENCIES OF MAXIMA OF CT BANDS FOR COMPLEXES WITH TCNE IN DICHLOROMETHANE

N	Compound	ν_{CT} (cm^{-1})	ν_{CT} (cm^{-1})[a]
4	$PhSiMe_3$	24,500 ∓ 100	24,650
5	$PhSiH_3$	25,700 ∓ 200	26,000
6	$PhSiCl_3$	29,300 ∓ 300	—
7	$PhSiF_3$	30,600 ∓ 500	—
8	*p*-Me—C_6H_4—SiH_3	23,700 ∓ 200	—
9	*p*-Me—C_6H_4—$SiMe_3$	23,500 ∓ 200	—

[a] From Ref. 29.

TABLE VI

EMPIRICAL PARAMETERS IN THE HYPERCONJUGATION MODEL FOR THE SiX_3 GROUP[a]

X	α_X	β_{SiX}	β_{SiC}	β_{CX}
F	−4.0	−2.0	−0.8	−2.0
Cl	−2.5	−1.7	−0.8	−1.7
H	−1.3	−1.2	−0.8	−1.2
CH_3	−1.0	−1.0	−0.8	−1.0

[a] Values in β units; from Ref. 28.

HMO method. The σ_{SiX} orbitals are transformed into group orbitals, two of which have the π symmetry (Σ_π^{Si} and Σ_π^{X}) and can overlap with the π system of the ring. The linear combination of these orbitals gives an occupied bonding orbital $\Sigma_\pi = \Sigma_\pi^{Si} + \Sigma_\pi^{X}$ and a vacant antibonding orbital $\Sigma_\pi^* = \Sigma_\pi^{Si} - \Sigma_\pi^{X}$.

From these studies Ponec and Chvalovsky (*28*) conclude that the energies of the highest occupied molecular orbitals in arylsilanes are correlated with ν_{CT} values. The tendency of the CT band energy to decrease in the series of substituents $SiF_3 > SiCl_3 > SiH_3 > SiMe_3$ is adequately described by the HMO approach. A similar calculation for the carbon analogs of

TABLE VII

CALCULATED QUANTUM CHEMICAL PARAMETERS FOR ARYLSILANES[a]

N[b]	ε(HOMO)	ε(LUMO)
4	−0.600	0.408
5	−0.614	0.500
6	−0.645	0.619
7	−0.680	0.570
8	−0.594[c]	—
9	−0.597[c]	—

[a] Values in β units; from Ref. 28.

[b] Formulas of compounds are given in Table V.

[c] Values calculated by first-order perturbation theory from the wave function for phenylsilane (**5**) and phenyltrimethylsilane (**4**) using the equation $\Delta\varepsilon_i = c_{i\mu}^2 \Delta\alpha_\mu$; $\Delta\alpha_\mu \simeq 0.1\ \beta$.

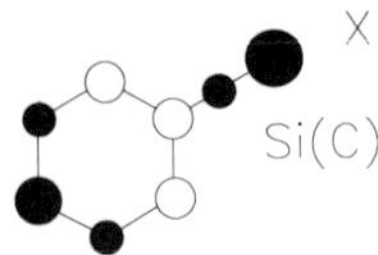

FIG. 2. HOMO of phenylsilanes $PhSiX_3$ and phenylmethanes $PhCX_3$ (*28*).

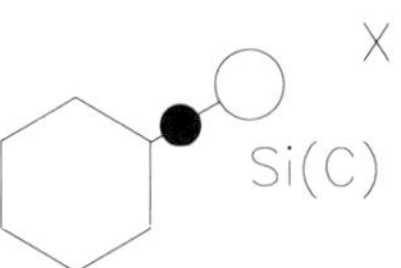

FIG. 3. LUMO of phenylsilanes $PhSiX_3$ and phenylmethanes $PhCX_3$ (*28*).

these compounds using the same standard parameters within the framework of the hyperconjugation model results in a similar sequence of substituents, $CF_3 > CCl_3 > CH_3 > CMe_3$. According to calculations, the HOMO is a $(\pi{-}\Sigma\pi)$ orbital (B_1) (Fig. 2) and the LUMO is, in all cases, a pure Σ_π^* orbital (Fig. 3).

Series of trimethylsilyl-substituted benzenes, naphthalenes, anthracenes, phenanthrenes, and their carbon analogs were analyzed in a similar way. Good correlations between the CT energy and the HOMO energy were established by simple HMO methods. Stability constants were also determined for these complexes (*31*).

Sakurai *et al.* (*32*) have indicated that it is not necessary to use the $d{-}\pi$ conjugation concept to explain electronic effects in phenyl-substituted silanes. The electronic structure of phenyl-substituted silanes was analyzed by the absorption spectra of the CT complexes with TCNE and photoelectron spectra of silanes. The role of the $\sigma{-}\pi$ conjugation between σ_{SiSi}, σ_{SiC}, and π orbitals in the electronic effects of the substituents was demonstrated (*32*). Using the equations derived by Voigt and Reid (*33*),

$$h\nu_{CT} = 0.831\text{IP} - 4.42 \tag{14}$$

$$\Delta E(\text{IP}) = \Delta E(\text{CT})/0.83 \tag{15}$$

Sakurai and Kira (*34*) estimated values of integrals for the interaction between CH_2SiMe_3 or Si_2Me_5 substituents and the aromatic system of naphthalene or benzene. The positions of the CT bands in absorption spectra of the TCNE complexes of the substituted benzene or naphthalene in CH_2Cl_2 were used to calculate the ionization potentials of the donor molecules according to Eqs. (14) and (15) (Tables VIII and IX). In the case of naphthalene, two CT bands were assigned to the electron transition

TABLE VIII

CT BAND MAXIMA OF COMPLEXES OF BENZENE DERIVATIVES (C_6H_5X) WITH TCNE IN CH_2Cl_2[a]

Substituent X	ν_{CT} (cm^{-1})	IP (eV)[b]	IP (eV)[c]	Ref.
H	25,600	9.15	9.245	*35*
Me	23,000	8.76	8.82	*35*
CH_2SiMe_3	20,100	8.33	8.35	*36*
Si_2Me_5	20,400	8.37	8.35	*36*

[a] From Ref. 34.
[b] Values are estimated.
[c] Values obtained by photoelectron spectroscopy.

from the HOMO (ψ_5) and the HOMO-1 (ψ_4) of the donor to the LUMO of the acceptor (TCNE). Using molecular orbital perturbation theory, the interaction between the σ and π orbitals was described schematically by "before–after interaction" diagrams (Fig. 4). The change in energy of interacting orbitals (ΔE) is evaluated using Eq. (16),

$$\Delta E = c_i^2\beta^2/|E_\sigma - E_\pi| \quad (16)$$

TABLE IX

CT BAND MAXIMA OF COMPLEXES OF NAPHTHALENE DERIVATIVES ($C_{10}H_7X$) WITH TCNE IN CH_2Cl_2[a]

	Band 1		Band 2	
Substituent X	ν_{CT} (cm^{-1})	IP (eV)[b]	ν_{CT} (cm^{-1})	IP (eV)[b]
H	18,100	8.03[c]	23,100	8.78[c]
α-Me	17,000	7.86	22,800	8.73
β-Me	17,400	7.92	21,700	8.57
α-CH_2SiMe_3	15,800	7.68	22,700	8.72
β-CH_2SiMe_3	16,400	7.77	20,600	8.40
α-Si_2Me_5	16,900	7.85	22,900	8.75
β-Si_2Me_5	17,200	7.89	20,700	8.42

[a] From Ref. 34.
[b] Values are estimated.
[c] Values of 8.11 and 8.78 eV obtained by photoelectron spectroscopy (*37*).

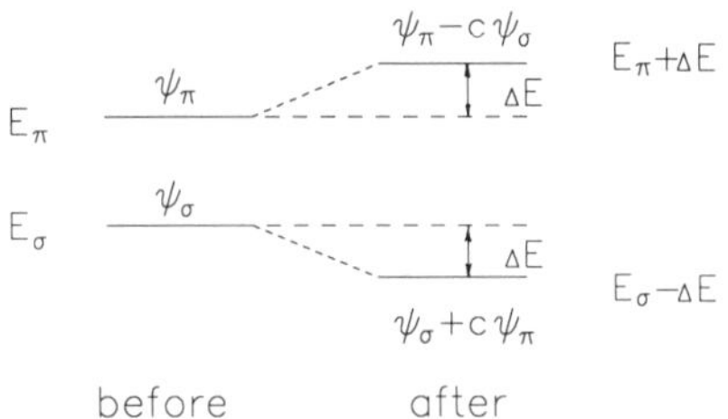

FIG. 4. Before and after σ–π interaction diagram (*34*).

where c_i is the coefficient of the atomic orbital at the site of attachment of the substituent, β is the measure of the extent of the interaction, $E\pi$ is the energy of π orbital, estimated by Eq. (14), and E_σ is the energy of the σ orbitals, determined by appropriate methods (Table X).

The values of c_i coefficients for naphthalene and benzene were determined by HMO calculations. For the determination of β the equation $\Delta E = f(c_i^2/|E_\sigma - E_\pi|)$ was derived for each substituent; the value of β was determined as a square root of the gradient (Table XI) (*38*). It appeared that the conjugation ability of silicon was less than that of carbon or oxygen.

Pitt *et al.* (*39*) used the Mulliken theory (*40*) to derive the dependence of the CT energies of the complexes on IP values of donors. Once the first ionization potential of naphthalene is known [photoelectron spectrum (*41*)] and the position of the first CT band in the absorption spectrum of the naphthalene–TCNE complex is determined, the constant given by $EA + W = 5.88$ eV can be calculated from Eq. (3). In this case, Eq. (3) transforms into Eq. (17):

$$h\nu_{CT(1)} = IP_1 - 5.88 \tag{17}$$

TABLE X

VALUES E_σ FOR SOME σ BONDS[a]

σ Bond	E_σ (eV)	Compound	Ref.
C—H	12.99	CH_4	*35*
C—Si	10.50	Me_4Si	*36*
Si—Si	8.69	Me_6Si_{12}	*36*
O—C	10.85	MeOH	*35*

[a] From Ref. 34.

TABLE XI

EXTENT OF σ–π CONJUGATION[a]

Substituent	β (eV)	Correlatio
Me	1.97	0.996
CH_2SiMe_3	1.69	0.987
Si_2Me_5	0.54	0.999
OMe[b]	2.08	0.968

[a] From Ref. 34, except as noted.
[b] Calculated from data presented by Boc and Alt (*38*).

TABLE XII

CT MAXIMA OF TCNE COMPLEXES AND ESTIMATED IPs OF NAPHTHYLPOLYSILANES[a]

Substituent	λ_{max} (nm)	IP (eV)
H	555	8.12
Me	592	7.98[b]
$SiMe_3$	578	8.03
Si_2Me_5	602	7.95
Si_3Me_7	608	7.93
$Si(SiMe_3)_2Me$	631	7.85
CH_2SiMe_3	638	7.83

[a] From Ref. 39, except as noted.
[b] Photoionization IP, 7.96 eV (*41*).

TABLE XIII

CT MAXIMA OF TCNE COMPLEXES AND IONIZATION POTENTIALS OF p-$MeOC_6H_4$–X[a]

X	λ_{max} (nm)	IP (eV)
H	508[b]	8.18[c]
Me	562[b]	7.91
$SiMe_3$	538	8.03
Si_2Me_5	574	7.85

[a] From Ref. 39, except as noted.
[b] From Ref. 33.
[c] Photoionization IP, 8.20 and 8.22 eV (*41*).

This equation describes the entire series of substituted naphthalenes. It has been used to calculate IP values of silicon derivatives of naphthalene by analysis of the spectra of their complexes with TCNE (Table XII). Using the Eq. (13) (*18*), Pitt *et al.* calculated the ionization potentials for the para-substituted anisoles, p-$MeOC_6H_4$–X (Table XIII). Comparison of the data listed in Tables XII and XIII shows that elongation of the polysilyl chain has a considerably weaker effect on the ionization potentials of naphthalene derivatives than on that of benzene derivatives. This fact can be explained as follows (*39*). The value of the first ionization potential of naphthalene [8.12 eV (*41*)] is lower than that of the σ orbital of polysilanes (Me_6Si_2, 8.69 eV; Me_3Si_8, 8.19 eV). Therefore, the HOMOs of polysilylnaphthalenes have mainly naphthalene character ($c_1 < c_2$, Fig. 5). The

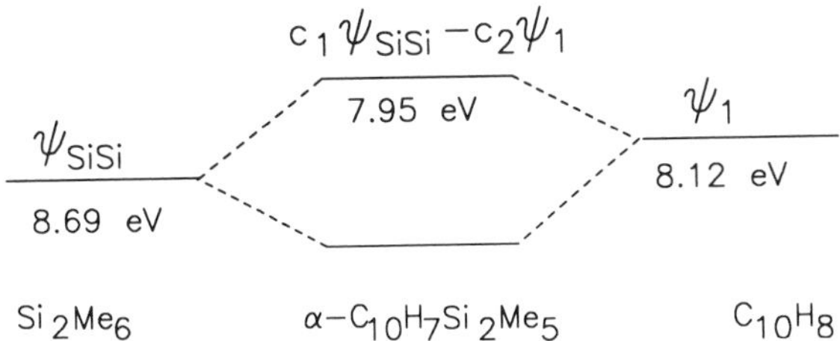

FIG. 5. Molecular orbital diagram of interaction of the Si_2Me_5 group with HOMO of naphthalene (*39*).

ionization potential of benzene is much higher [9.24 eV (*42*)]. The contribution of the σ_{SiSi} orbital to the HOMO of phenylpolysilanes is therefore considerable, and the latter is more responsive to changes in the polysilyl fragment.

Analysis of the absorption spectra of CT complexes suggests that the substituents may be arranged in the following series in the order of their increasing electron-donating power (*39*):

$$H < Me_3Si < Me < Me_5Si_2 < n\text{-}Me_7Si_3 < i\text{-}Me_7Si_3 < Me_3SiCH_2$$

It is important to note that relationships $h\nu_{CT} = f(IP)$ derived for benzene and naphthalene derivatives with simple substituents (such as Cl, Me, or OMe) hold also for the derivatives containing overcrowded Si substituents (such as those mentioned above).

Pyatkina *et al.* (*43*) also used Eq. (14) to estimate the ionization potentials of the para-substituted phenylcyclobutanes containing markedly distorted angles in the silacyclobutane ring.

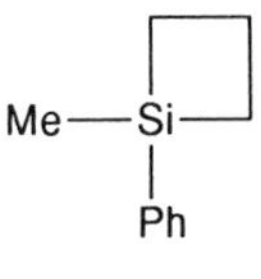

10

Use of similar relationships for disubstituted benzenes and other cyclic compounds of silicon has also been reported. Thus, Traven *et al.* (*44*) used Eq. (18) (*45*)

$$h\nu_{CT} = 0.80\text{IP} - 4.16 \tag{18}$$

to estimate the first and second ionization potentials of some ortho-disubstituted arylsilicon compounds (**11–13**, Tables XIV and XV). The values of IPs estimated by Eq. (18) and the values determined from photoelectron spectra were in adequate agreement.

2. *Multiplicity and Relative Intensities of Charge-Transfer Bands in the Visible Spectra of Arylsilane–TCNE Complexes*

In the study of absorption spectra of arylsilane complexes with TCNE, special attention has been given to analysis of the multiplicity of CT bands and their relative intensities. These problems have been studied in detail (*45,46*).

More than one CT band can be seen in the absorption spectra of most arylpolysilane complexes with TCNE (*29,34,39*). Phenylpolysilane complexes with TCNE have rather overlapping CT bands in the visible region

TABLE XIV

IONIZATION POTENTIALS OF COMPOUNDS OF GENERAL STRUCTURE **11**[a]

11

X	Y	CT complexes IP_1 (eV)	CT complexes IP_2 (eV)	PES IP_1 (eV)	PES IP_2 (eV)
$SiMe_3$	CH_2SiMe_3	8.23	8.88	8.39	8.81
CH_2SiMe_3	CH_2SiMe_3	8.10	8.67	8.07[b]	8.71[b]
Me	CH_2SiMe_3	8.24	8.83	8.31[b]	8.85[b]
Si_2Me_5	CH_2SiMe_3	8.18	8.96	8.21[b]	8.55[b]
Si_2Me_5	Me	8.21	8.80	8.37	8.81

[a] From Ref. 44.
[b] Calculated by the MO LCBO method.

TABLE XV

Ionization Potentials of Silaindanes **12** AND **13**[a]

N	Compound	CT complexes IP_1 (eV)	CT complexes IP_2 (eV)	PES IP_1 (eV)	PES IP_2 (eV)
12	1,1,2,2-tetramethyl-1,2-disilaindane (Me, Me, Si, Si, Me, Me)	8.30	8.95	8.53	8.83
13	2,2-dimethyl-2-silaindane (Si, Me, Me)	8.40	8.86	—	—

[a] From Ref. 44.

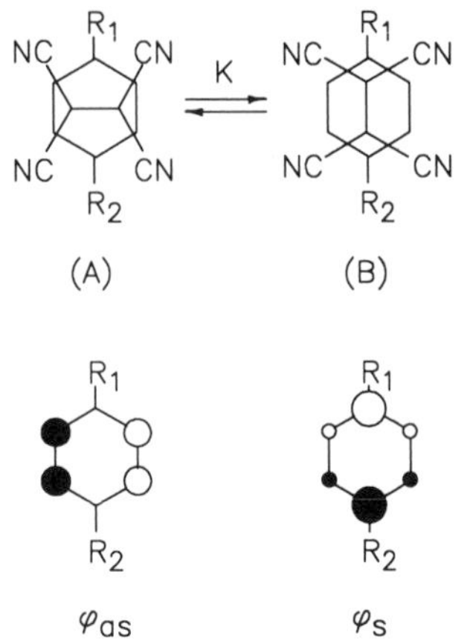

FIG. 6. Two orientations of the TCNE–para-disubstituted benzene CT complex and relative donor MOs (*46*).

of the spectrum. Those of longer wavelengths have lower intensity (*34*). In contrast, the CT bands in the spectra of complexes of benzyltrimethylsilane, anisole, and naphthylpentamethyldisilane are well resolved and have the same intensities. It was suggested that two CT bands in the spectrum may be assigned to the different orientations of the donor and acceptor in CT complexes (*46*). This explanation was suggested first by Orgel for complexes of chloranil with substituted benzenes (*47*). In accordance with Mulliken's maximum overlap principle (*48*), Orgel proposed two orientations of the acceptor (Fig. 6) corresponding to the two higher occupied orbitals φ_{as} and φ_s of the donor. It is believed that the corresponding complexes are in equilibrium.

Holder and Thompson (*49*) studied the complexes of TCNE with benzene, toluene, *p*-xylene, and *t*-butylbenzene and suggested that the equilibrium is shifted toward the isomer (A) with increasing size of the substituent. Thus, the probability of electron transfer from MO φ_s of the donor to the LUMO of the acceptor decreases as the size of substituents increases. As a result, the relative intensity of the longer wavelength CT band in the absorption spectrum of the complex decreases as well. The low intensity of the first CT band in the absorption spectrum of complexes of phenylpentamethylsilane with TCNE could thus be assigned to the steric effect of the substituent. But introduction of another bulky substituent, e.g., Me_3SiCH_2, at the para position does not decrease (as one might expect) the intensity of the first CT band: on the contrary, its relative intensity even increases (Table XVI).

TABLE XVI

BAND PARAMETERS FOR CT SPECTRA OF COMPLEXES OF SOME PARA-SUBSTITUTED PHENYLPENTAMETHYLDISILANES (p-X–Ph–Si_2Me_5) WITH TCNE[a]

X	Band	ν_{CT} (cm^{-1})	Peak height	Area 1/area 2
H	1	25,100	0.57	0.332
	2	20,500	0.26	
Me	1	24,600	0.84	0.591
	2	19,500	0.57	
Me_3SiCH_2	1	24,100	0.24	0.920
	2	17,600	0.23	
MeO	1	25,300	0.60	1.036
	2	17,600	0.70	

[a] From Ref. 46.

An attempt was made to estimate the dependence of relative intensity of CT bands on the electronic effect of the para substituent, as determined by its σ^+ constant (*46*). Figure 7 demonstrates such a linear dependence, but it cannot be regarded as a general rule because the values of σ^+ constants for the p-Me_5Si_2 and p-Me_3SiCH_2 groups are about the same, whereas the ratio of intensities of CT band 1 to CT band 2 of p-$Me_5SiCH_2C_6H_4Si_2Me_5$ is far from being equal to that of p-$Me_5Si_2C_6H_4Si_2Me_5$. It is actually much higher.

In order to explain the effect of the para substituent on the intensity of the longer wavelength CT band in the absorption spectra of substituted phenylpentamethyldisilane complexes with TCNE, Sakurai and Kira (*46*) used perturbation molecular orbital theory. In this analysis, the HOMO of

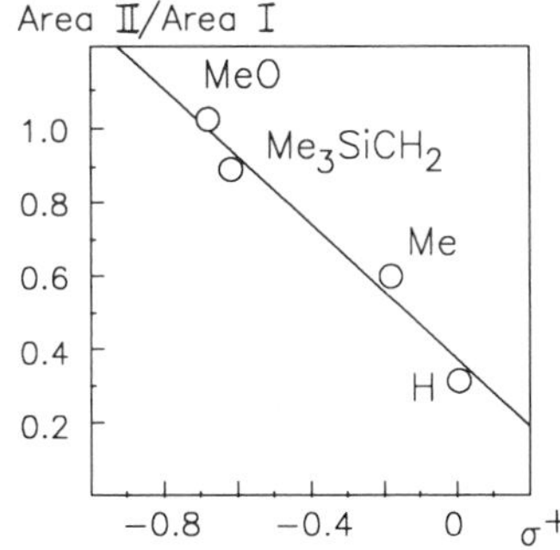

FIG. 7. Plot of relative band area for the absorption spectra of substituted phenylpentamethyldisilane–TCNE complexes versus σ^+ (*46*).

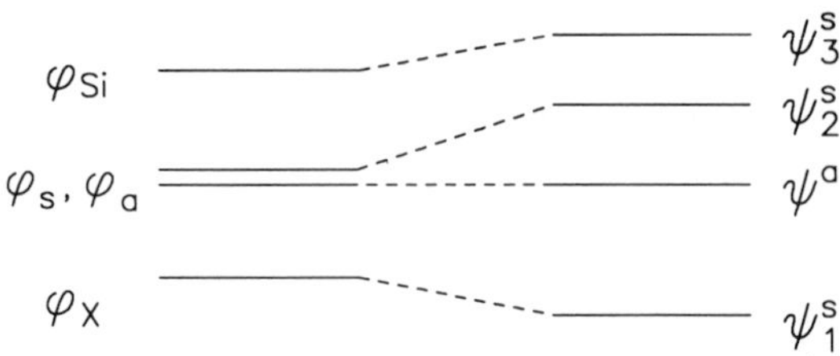

FIG. 8. Qualitative πMO scheme for p-X–$C_6H_4Si_2Me_5$ (*46*).

the substituted phenylpentamethyldisilane p-X–$C_6H_4Si_2Me_5$ was represented by a linear combination of three group orbitals (Fig. 8):

$$\psi^{s}_{HOMO} = a'\varphi_s + b'\varphi_{Si} + c\varphi_X \quad (19)$$

$$a'^2 + b'^2 + c^2 = 1 \quad (20)$$

The energies of group orbitals were estimated, as usual, by the photoelectron spectra of standard compounds (C_6H_6, CH_4, $SiMe_4$, etc.). The results of the analysis are given in Table XVII. The data show that there is good agreement of the ionization potentials of the compound with those calculated from the Voigt–Reid equation (*33*). In addition, there is a good correlation between the relative intensity of the first CT band and the contribution of the σ_{SiSi} orbital to the HOMO of the donor (Fig. 9). It is important to note that the para substituent Si_2Me_5 also obeys this relationship (*46*). It can therefore be stated that the intensity of the longer wavelength CT band in the absorption spectrum of silicon-substituted aromatic compounds decreases with increasing contribution of the σ_{SiSi} orbital to the HOMO of the compound.

TABLE XVII

CALCULATED FIRST IONIZATION POTENTIALS AND WAVE FUNCTIONS OF HOMOS FOR PARA-SUBSTITUTED PHENYLPENTAMETHYLDISILANES (p-X–C_6H_4–Si_2Me_5)[a]

X	IP (eV)[b]	Wave function	σ_{SiSi} fraction
H	8.35 (8.39)	$0.851\varphi_{Si} - 0.526\varphi_s$	0.72
p-Me	8.23 (8.24)	$0.757\varphi_{Si} - 0.631\varphi_s - 0.171\varphi_X$	0.57
p-Me_3SiCH_2	8.02 (7.92)	$0.583\varphi_{Si} - 0.710\varphi_s - 0.395\varphi_X$	0.34
p-MeO	7.93 (7.92)	$0.519\varphi_{Si} - 0.713\varphi_s - 0.471\varphi_X$	0.27

[a] From Ref. 46.

[b] Ionization potentials estimated by the Voigt–Reid equation, $h\nu_{CT} = 0.83\text{IP} - 4.42$ (*33*), are given in parentheses.

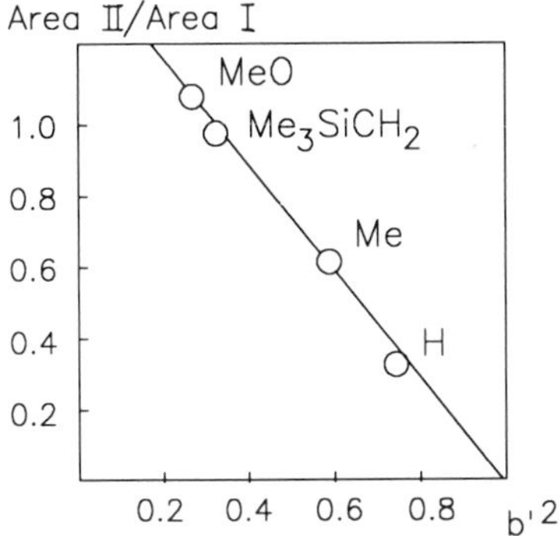

FIG. 9. Plot of relative band area for the absorption spectra of substituted phenylpentamethyldisilane–TCNE complexes versus the σ_{SiSi} fraction for HOMO phenylpentamethyldisilanes (*46*).

Traven *et al.* (*45*) arrived at the same conclusion in their independent studies. They also recorded the absorbtion spectra of TCNE complexes with benzenes containing various silyl substituents (Table XVIII). Most absorption spectra of the complexes contained two well-resolved CT bands. The following linear relationship was thus derived for the first CT

TABLE XVIII

PARAMETERS OF CT BANDS IN ABSORPTION SPECTRA OF TCNE COMPLEXES WITH ARYLPOLYSILANES AND PHOTOIONIZATION POTENTIALS OF DONORS[a]

		CT band 1		CT band 2		PES	
N	Compound	ν_{CT} (cm^{-1})	$h\nu_{CT}$ (eV)	ν_{CT} (cm^{-1})	$h\nu_{CT}$ (eV)	IP_1 (eV)	IP_2 (eV)
14	$C_6H_5CH_2SiMe_3$	20,100	2.49	25,100	3.12	8.35	9.15
15	$C_6H_5CH(SiMe_3)_2$	19,200	2.38	25,000	3.10	—	—
16	$C_6H_5C(SiMe_3)_3$	18,870	2.33	24,700	3.06	8.20	9.08
17	2-$C_6H_5Si_3Me_7$	18,700	2.32	24,500	3.04	8.15	9.20
18	C_6H_6	26,000	3.23	—	—	9.25	—
19	*p*-$Me_5Si_2C_6H_4CH_2SiMe_3$	17,800	2.19	24,100	2.98	7.97	8.82
20	*p*-$Me_5Si_2C_6H_4Si_2Me_5$	18,400	2.28	23,500	2.92	8.06	8.82
21	*m*-$Me_5Si_2C_6H_4Si_2Me_5$	20,300	2.52	25,400	3.16	8.39	9.20
22	$C_6H_5Si_2Me_5$	20,300	2.52	24,700	3.06	8.39	9.11
23	*p*-$MeC_6H_4Si_2Me_5$	19,600	2.43	24,300	3.02	8.26	9.02
24	1-$C_6H_5Si_3Me_7$	19,000	2.36	24,700	3.06	8.12	9.13
25	1-(*p*-MeC_6H_4)Si_3Me_7	18,200	2.25	24,300	3.02	8.01	9.09

[a] From Ref. 45.

bands of the spectra:

$$h\nu_{CT(1)} = 0.81IP_1 - 4.32 \qquad (\rho = 0.996) \tag{21}$$

The parameters of this equation are very close to those of Eq. (14) derived earlier (*33*) for the TCNE complexes with benzene derivatives containing less bulky substituents (Me—, Cl—, Br—, MeO—). Thus, despite the obviously different steric conditions of complexing of TCNE with this series of donors, the energies of CT bands of these complexes obey the same linear dependence.

The appearance of several CT bands in the absorption spectra of complexes of aromatic hydrocarbons with π-acceptors has been reported in several publications. The values for the second CT bands were first included in a correlation of the type $h\nu = f(\mathrm{IP})$ by Traven *et al.* (*45*):

$$h\nu_{CT} = 0.80IP - 4.16 \qquad (\rho = 0.997) \tag{22}$$

An important feature of Eq. (22) is the incorporation of the energies of electron transitions from π-donor orbitals of different symmetries. The second CT bands are assigned to the electronic transition from the ψ_{as} orbitals and the first CT bands are assigned to the transition from the ψ_s orbitals, which are altered to a different extent by the influence of the substituent. Analysis of Eqs. (21) and (22) is especially helpful in the study of the nature of phenylpolysilane HOMOs, which are mixed σ–π orbitals (*37*). It is important to note that phenylpolysilanes obey the same equations as benzene and those of its derivatives, for which there is no doubt about the π character of the HOMOs (benzylsilanes, for instance). The practical value of Eqs. (21) and (22) is that they can be used to estimate the ionization potentials of compounds for which these values are unknown or unmeasurable by other methods (e.g., by photoelectron spectra).

TABLE XIX

Complexation Constants (K) and Extinction Coefficients (ε) of CT Complexes of TCNE with Arylsilanes in Two Solvents[a]

	Chloroform		Cyclohexene	
Compound	K (liter/mol)	ε (liter/mol cm)	K (liter/mol)	ε (liter/mol cm)
26	0.510 ∓ 0.030	2510 ∓ 100	2.230 ∓ 0.010	2130 ∓ 10
27	0.720 ∓ 0.020	320 ∓ 3	—	—
28	—	—	0.164 ∓ 0.002	2190 ∓ 20

[a] From Ref. 45.

The increasing contribution of the group σ_{SiSi} orbitals to the HOMO of the donor molecule is probably not the only reason for the decreasing intensity of CT bands in the absorption spectra of CT complexes with TCNE. Steric hindrance to overlapping of the donor orbitals of arylsilane and the acceptor orbital of TCNE is another cause. Influence of the steric factor on the complexing with TCNE in the series of α-(trimethylsilyl)toluenes $C_6H_4CH_{3-n}(SiMe_3)_n$ (**26**, **27**, and **28**, Table XIX; $n = 1$, 2, and 3, respectively) has been discussed (*45*). It was noted that the intensity of color of the complexes of TCNE with silanes **27** and **28** was markedly lower than that of the complex of silane **26** at the same concentrations of the compounds. The color intensities of complexes **27** and **28** decreased markedly as their solutions were cooled to −80°C. This behavior appears to be typical for weak CT complexes (*50*).

The parameters for the complexes TCNE with compounds **26**, **27**, and **28** were determined by the method proposed by Benesi and Hildebrand (*51*) (Table XIX, Fig. 10). If chloroform was used as a solvent, negative values

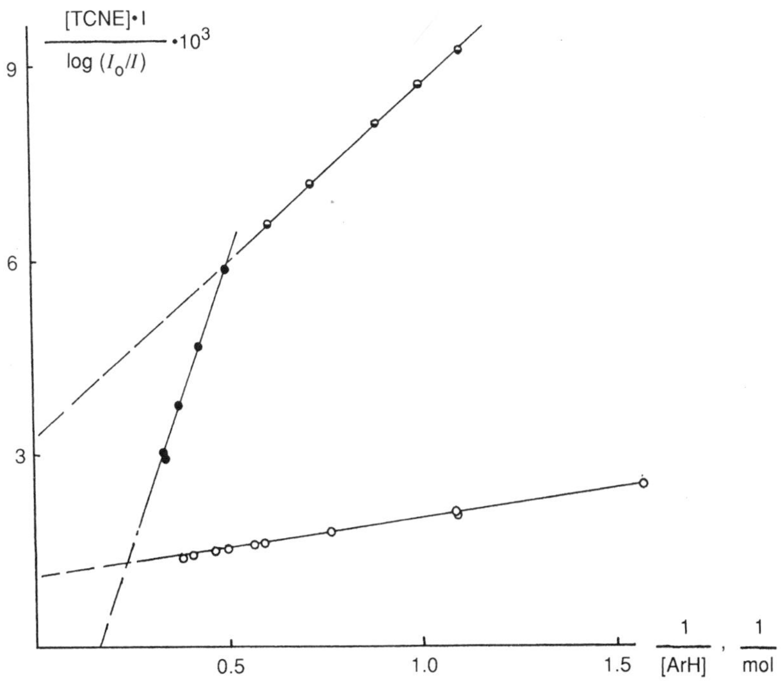

Fig. 10. Dependence of [TCNE]/log(I_0/I) on ArH concentration for solutions of TCNE complexes with arylsilanes in $CHCl_3$: (○), $C_6H_5CH_2SiMe_3$; (○), $C_6H_5CH(SiMe_3)_2$; (●), $C_6H_5C(SiMe_3)_3$ (*45*).

of parameters of complexation of α,α,α-tris(trimethylsilyl)toluene with TCNE were obtained. The result is characteristic for weak CT complexes: solvent competition probably occurs during the formation of complexes with TCNE (*52*). In fact, if a more inert solvent, such as hexane, is used, the values of ε and K are positive. The marked diminution of K values in the transition from compounds **26** and **27** to compound **28** cannot be due to the different electronegativity of CH_2SiMe_3 and $C(SiMe_3)_3$ groups (*53*). It seems to be probable that steric hindrances to overlapping of the donor and acceptor orbitals are responsible for these results.

3. *Hetarylsilanes*

Absorption spectra of CT complexes have been used to estimate electronic effects of substituents in a series of substituted heterocyclic aromatic compounds such as furans (*54–57*) and thiophenes (*55,58–60*), which are π-donors as well.

The absorption spectra of CT complexes of some silicon-substituted furans with TCNE were recorded (*54*). It was found that electron-donating substituents produce a bathochromic shift and electron acceptors cause a hypsochromic shift of the CT band of the spectra. A linear relationship between ν_{CT} and the chemical shift (τ) in proton magnetic resonance spectra of the donors has been established. Similar results were obtained for a series of silicon-substituted thiophenes (*58*): dependence of ν_{CT} on τ has been found as well.

It was established (*55*) by analysis of absorption spectra of CT complexes that the degree of conjugation of the silicon atom with the π system of heterocycles decreases in the series of substituted silanes: 2-furyl > 2-thienyl > 3-furyl. Lopatin *et al.* (*59*) have found a linear dependence between the energies of CT bands in absorption spectra of complexes of TCNE with furans (ν_{CT}^{F}) and those of thiophenes (ν_{CT}^{T}) containing organosilicon substituents:

$$\nu_{CT}^{T} = 0.46\nu_{CT}^{F} + 1100 \qquad (23)$$

There was also a linear correlation between the positions of the CT bands in spectra of complexes of 2-furyl and those of 3-furyl derivatives of silanes (furylSiR_3, where R = H, Cl, Br, Me, etc.) (*56*):

$$\nu_{CT}^{(3)} = 0.386\nu_{CT}^{(2)} + 13{,}300 \qquad (24)$$

The sensitivity of HOMOs of 2-furyl derivatives to the inductive effect of SiR_3 groups is higher than that of 3-furyl derivatives. This is probably due to the different values of coefficients at the 2- and 3-positions of the HOMO of the furan (Fig. 11).

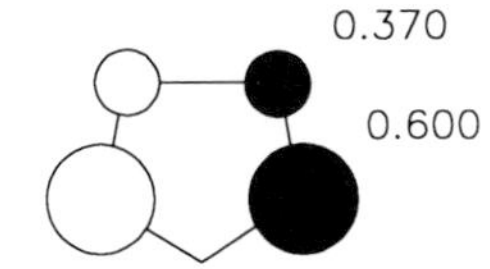

FIG. 11. HOMO of furan (*6*).

Egorochkin *et al.* (*57*) studied the absorption spectra of CT complexes of TCNE with 32 silicon-disubstituted furans of the general structure:

29

(e.g., R = H, Me, Me_3Si; R_1 = CH_2SiMe_3, $SiEt_3$) and derived the following linear relationships:

$$\nu_{CT} = (6570 \mp 540)\mathrm{IP} - (35{,}650 \mp 4630) \qquad (\rho = 0.992) \qquad (25)$$

and

$$\nu_{CT} = (7500 \mp 650)\Sigma\sigma_p^+ + (21{,}450 \mp 200) \qquad (\rho = 0.980) \qquad (26)$$

It should be noted that similar dependencies were established (*33*) for benzene derivatives as well:

$$\nu_{CT} = 6700\mathrm{IP} - 35{,}650 \qquad (27)$$

and

$$\nu_{CT} = 7600\sigma_p^+ + 25{,}800 \qquad (28)$$

The possibility of using σ_p^+ constants (determined earlier for the series of substituted benzenes) in the furan series was discussed. From a comparison of the corresponding relationship for benzenes and furans, Egorochkin *et al.* concluded that the electron-donating power of 2-furyl derivatives is higher.

Relationships were obtained between the values of ν_{CT} for TCNE complexes and shifts ($\Delta\nu$) of the OH stretching vibration in the IR spectra of complexes with phenol, as well as between the values of ν_{CT} and the σ_p^+ constants of substituents, for some silicon-containing thiophene derivatives (*60*).

The absorption spectra of mixtures of silicon derivatives of ferrocene with CBr_4 in heptane and dimethylformamide (DMF) have been studied and interpreted as spectra of CT complexes (*61*) (Table XX). The degree

TABLE XX

SPECTRAL CHARACTERISTICS OF CT COMPLEXES OF SILICON DERIVATIVES OF FERROCENE WITH CBr_4 AND IPs OF DONORS[a]

N	Compound[b]	Solvent	λ_{CT} (nm)	IP (eV)
30	Fc–$SiMe_2$–Fc	Heptane	382	6.10
		DMF	365	
31	[Fc–$SiMe_2$]NH	Heptane	380	6.12
		DMF	367	
32	[Fc–$SiMe_2$]O	Heptane	381	6.11
		DMF	368	
33	Fe(C₅H₄–$SiMe_2$)₂NH	Heptane	378	6.13
		DMF	370	
34	Fe(C₅H₄–$SiMe_2$)₂O	Heptane	378	6.13
		DMF	368	
35	FcH	Heptane	382	6.10
		DMF	368	

[a] From Ref. 61.
[b] Fc, Ferrocenyl.

of complex formation of all compounds listed in Table XX was measured. With heptane as solvent, the values were 0.1–0.13; with DMF these values were 0.06–0.07. It has been shown that equimolecular 1 : 1 complexes are formed in all cases (*62*). The ionization potentials (IP_X) of the donors (X-substituted ferrocenes) were estimated using the following formula (*63*):

$$h\nu_{CT}^{X} - h\nu_{CT}^{Fc} = IP_X - IP_{Fc} \tag{29}$$

It is interesting to note that the ionization potentials of the compounds remain about the same, whereas the degree of complex formation for compounds (**30–32**) is about twice as high as that of ferrocene. Shvekhgeymer *et al.* associate the phenomenon with the presence of two ferrocene groups in these compounds.

Studies have also been carried out by Du *et al.* (*64*) concerning complexing of ferrocene-containing polysilanes with TCNE, TCNQ, chloranil, and iodine as acceptors.

4. *Arylmetals of Group IV*

Absorption spectra of complexes of TCNE with arylmetals of Group IV were studied in order to estimate the electronic effects of substituents

TABLE XXI

CHARACTERISTICS OF CT COMPLEXES OF TCNE WITH GROUP IVA TETRAPHENYLS (Ph_4M) IN DICHLOROMETHANE[a]

M	λ_{CT} (nm)	ε_{max} (liter/mol cm)	K (liter/mol)
Si	395	171 ∓ 17	2.31 ∓ 0.23
Ge	395	329 ∓ 24	1.14 ∓ 0.08
Sn	395	237 ∓ 28	2.36 ∓ 0.28
Pb	395	274 ∓ 27	2.53 ∓ 0.25

[a] At 21°C; from Ref. 68.

containing atoms of other Group IV elements, i.e., Ge, Sn, and Pb (*13,17,32,54–57,59,60,65–67*). Thus the absorption spectra of CT complexes of TCNE with phenyl-substituted germanes and silanes have been compared (*17,32*). A correlation between the σ_p constants of substituents and the energies of CT bands in the visible spectra has been established. It seems quite probable that atoms of those elements have the same ability for conjugation with the π orbitals of the benzene ring (*17*).

Absorption spectra of CT complexes of the tetraphenyl derivatives of Group IV metals, Ph_4Si, Ph_4Ge, Ph_4Sn, and Ph_4Pb, with TCNE have also been reported (*68*) (Table XXI). The complexes of all four compounds absorb in the same region of the spectrum and have about the same complex formation constant values. Presumably these are complexes in which the aromatic rings are the charge donors, and the effects of electronegativity and the size of the central atom on the electronic structure are quite small.

The positions of CT bands in the absorption spectra of the CT complexes of TCNE with Group IV metal aryl compounds are adequately described by one equation for the different elements (C, Si, Ge, and Sn). For substituted furans, the σ_p^+ constants for the Group IV substituents are correlated with ν_{CT}, as well as with the chemical shift for protons on the furan ring (*54,57*).

Effects of substituents containing elements of Group IV on the position of CT bands in the absorption spectra of TCNE complexes of furan and thiophene derivatives were compared (*59*). The electron-withdrawing effect followed the order C < Si > Ge > Sn > Pb. The visible spectra of CT complexes of nonaromatic π systems containing substituents with the Group IVA elements were studied as well (*13,62,67*).

D. *Silyl Ethers and Silylsulfides*

The absorption spectra of CT complexes of TCNE with silicon-containing sulfides were first recorded by Traven *et al.* (*69*) (Table XXII). The addition of a silicon atom to sulfur decreases the energy of the lone pair orbital, as shown by a hypsochromic shift of the CT band in the spectra. The first ionization potentials of silicon-containing sulfides were determined from the equations

$$h\nu_{CT} = 0.45\text{IP} - 1.38 \qquad (30)$$

and

$$\text{IP} = 1.18\Sigma\sigma^*_{R,R'} + 8.63 \qquad (\rho = 0.965) \qquad (31)$$

derived for some sulfides based on the absorption spectra of their CT complexes with TCNE and inductive constants σ^*_S of the corresponding alkyl substituents R (*70*). The ionization potentials of silicon-containing sulfides, as determined from the absorption spectra of their CT complexes, are much higher than predicted using only the inductive properties of the $SiMe_3$ group (Table XXIII), due to back-donation of unshared electrons of sulfur into σ^* orbitals of the silicon substituents.

It is interesting to compare the effects of the $SiMe_3$ and Si_2Me_5 groups on the ionization potential values of sulfides. The bathochromic shift of the CT band of the TCNE complex shows that the ionization potential of butyl(pentamethyldisilyl)sulfide is lower than that of butyl(trimethylsilyl)-sulfide by about 0.5 eV. In addition to the electron-donating inductive effect and the electron-withdrawing conjugative effects of the silicon atom in the sulfides $RSSi_2Me_5$, the σ orbital of the Si–Si bond may interact with the lone pair n orbital of sulfur. This would significantly increase the

TABLE XXII

SPECTRAL PARAMETERS OF CT COMPLEXES OF TCNE WITH SULFIDES RSR′[a]

R	R′	ν_{CT} (cm^{-1})	IP (eV)[b]
n-C_4H_9	n-C_4H_9	18,900	8.27
n-C_4H_9	i-C_3H_7	18,850	8.27
n-C_4H_9	$SiMe_3$	20,750	8.78
$SiMe_3$	$SiMe_3$	24,000	9.65
n-C_4H_9	CH_2SiMe_3	18,200	8.09
CH_2SiMe_3	CH_2SiMe_3	17,300	7.84

[a] From Ref. 69.
[b] Calculated from Eq. (30).

TABLE XXIII

IONIZATION POTENTIALS OF SULFIDES RSR′[a]

R	R′	$\Sigma\sigma^*_{R,R'}$	$h\nu_{CT}$ (eV)	IP (eV)[b]	IP (eV)[c]	ΔIP (eV)
$SiMe_3$	$SiMe_3$	−1.80[d]	2.95	9.63	6.51	3.12
		−0.88[e]			7.59	2.04
CH_2SiMe_3	CH_2SiMe_3	−0.52	2.14	7.83	8.02	−0.19
Bu	$SiMe_3$	−1.03[d]	2.56	8.76	7.41	1.35
		−0.57[e]			7.96	0.80
Bu	CH_2SiMe_3	−0.39	2.26	8.08	8.17	−0.09
Bu	Si_2Me_5	—	2.34	8.28	—	—

[a] From Ref. 70.
[b] Calculated from Eq. (30).
[c] Calculated from Eq. (31).
[d] According to Ref. 71.
[e] According to Ref. 29.

energy of the HOMO. Insertion of a CH_2 group between the sulfur and silicon atoms, as in $BuSCH_2SiMe_3$, ruling out electron acceptance by silicon, decreases the ionization potential of the sulfide markedly. A still lower ionization potential, from the ν_{CT} value for the TCNE complex, is shown for $S(CH_2SiMe_3)_2$. In silyl sulfides containing the fragment S–CH_2–Si, interaction between the occupied n_S and σ_{CSi} orbitals can take place (*70*). The effects of the interaction between the unshared electrons of the sulfur atom and the σ orbitals of the SiSi and CSi bonds were estimated using the perturbation of molecular orbital (PMO) method (*72*): $H_{n\sigma_{SiSi}} = -0.354$ eV, $H_{n\sigma_{CSi}} = -1.000$ eV. The value of $H_{n\sigma_{SiSi}}$ was used to evaluate the ionization potential of phenylthiopentamethyldisilane. In this molecule, the HOMO is formed by mixing the π_{Ph}, n_S, and σ_{SiSi} orbitals. The calculated ionization potential, 8.34 eV, can be compared with the experimental value of 8.28 eV from the absorption spectrum of the CT complex of phenylthiopentamethyldisilane with TCNE, and Eq. (32) for the complexes of TCNE with arylsulfides (*73*). The interaction integral between the n_S and σ_{CSi} orbitals was supported in a similar way. For several arylthiosilanes the ionization potentials obtained from CT spectra of complexes agree well with those measured by photoelectron spectroscopy.

$$\nu_{CT} = 7331\text{IP} - 41{,}830 \qquad (32)$$

Conjugation of the n_S and σ orbitals in silyl sulfides has been treated rather thoroughly by Wojnowski and co-workers (*74*). Their data for the absorption spectra of CT complexes of some silicon-containing sulfides with TCNE are shown in Table XXIV. The CT bands were identified using

TABLE XXIV

VIS ABSORPTION DATA FOR CT COMPLEXES WITH TCNE FOR TRISALKOXYSILYLTHIO DERIVATIVES OF PERMETHYLPOLYSILANES[a]

Compound	ν_{CT} (cm^{-1})[b]		IP (eV)
$(t\text{-BuO})_3SiS(SiMe_2)_2Me$	—		—
	21,240	(σ)	8.23
$(t\text{-BuO})_3SiS(SiMe_2)_2SSi(OBu\text{-}t)_3$	17,360	(n_S)	7.82
	21,160	(σ)	8.22
$(i\text{-PrO})_3SiS(SiMe_2)_4Me$	—		—
	20,620	(σ)	8.14
$(i\text{-PrO})_3SiS(SiMe_2)_3SSi(OPr\text{-}i)_3$	—		—
	20,860	(σ)	8.18
$(t\text{-BuO})_3SiS(SiMe_2)_3SSi(OBu\text{-}t)_3$	—		—
	20,980	(σ)	8.19
$(t\text{-BuO})_3SiS(SiMe_2)_4SSi(OBu\text{-}t)_3$	16,560	(n_S)	7.60
	—		—
$(i\text{-PrO})_3SiSH$	16,460	(n_S)	7.57
	21,000	(σ)	8.19
$(t\text{-BuO})_3SiSH$	16,690	(n_S)	7.64
	22,000	(σ)	8.35
$[(t\text{-BuO})_3SiS]_2$	17,460	(n_S)	7.84
	—		—

[a] From Ref. 74.

[b] Symmetries of donor MOs are given in parentheses.

the Sandorfy C method: the molecular orbital of the sulfide was regarded as a linear combination of group orbitals, n_S and σ_{SiMe_2} orbitals in this particular case. Table XXIV specifies the types of orbitals whose contribution to the corresponding occupied orbitals of the donor molecules is dominant. The first long-wave CT band is assigned to the $n_S \rightarrow \pi^*$ (TCNE) transition, and the second CT band is assigned to the $\sigma \rightarrow \pi^*$ (TCNE) transition. The first ionization potentials IP(n_S) listed in Table XXIV were calculated from the absorption spectra of CT complexes by Eq. (33) (*75*),

$$\nu_{CT} = 3690\text{IP} - 11{,}941 \qquad (s = 42\ \text{cm}^{-1}) \tag{33}$$

and IP(σ) values were calculated by Eq. (34) (*76*),

$$\nu_{CT} = 0.771\text{IP} - 3.716 \qquad (\rho = 0.999) \tag{34}$$

The interaction of the acceptor with alkoxysilanes does not always end at the complexation stage. Iodine is reported to be an active catalyst of the alkoxy–alkoxy exchange reaction, its catalytic activity being explained by the formation of a CT complex with alkylalkoxysilane (*77*). The following

TABLE XXV

WAVELENGTH OF ABSORPTION PEAKS OBSERVED IN SOLUTIONS OF IODINE OR IODINE MONOBROMIDE WITH ALKOXYSILANES (IN CCl_4)[a]

	λ_{CT} (nm)			
	I_2 solution		IBr solution	
Alkoxysilane	UV	Visible	UV	Visible
$Me_2Si(OEt)_2$	228	474	229	428
$Me_2Si(OEt)(OPr\text{-}n)$	243	474	244	428
$Me_2Si(OPr\text{-}n)_2$	233	478	238	428
$Me_2Si(OEt)(OBu\text{-}n)$	228	475	234	428
$Me_2Si(OBu\text{-}n)_2$	225	475	225	430
$Me_2Si(OEt)(OBu\text{-}t)$	230	482	223	430
$Me_2Si(OBu\text{-}t)_2$	237	492	218	436
$Me_2Si(OEt)(OBu\text{-}i)$	238	482	229	429
$Me_2Si(OBu\text{-}i)_2$	240	475	230	435
Et_3SiOEt	238	475	228	425
$Et_3SiOPr\text{-}n$	238	490	226	450
$Et_3SiOBu\text{-}n$	229	495	224	425
$Et_3SiOBu\text{-}t$	233	498	225	445
$Et_3SiOBu\text{-}i$	238	490	233	455

[a] From Ref. 77.

reaction occurs:

$$R_3SiOR' + R''OH \overset{I_2}{\rightleftharpoons} R_3SiOR'' + R'OH$$

During studies on the mechanism of this reaction, absorption spectra of mixtures of some disiloxanes and alkoxysilanes with I_2 and IBr were recorded (Table XXV). CT bands were not found in the absorption spectra of disiloxanes with acceptors, but were observed in the spectra of alkoxysilanes with I_2 (or IBr) in CCl_4. The complex formation constants listed in Table XXVI were determined using the equation given by Ketelaar *et al.* (*78*),

$$\frac{C_I}{A - A_0} = \frac{1}{(\varepsilon_c - \varepsilon_I)K_f C_D} + \frac{1}{\varepsilon_c - \varepsilon_I} \quad (35)$$

where A is the observed absorbance of the sample solution, A_o is the absorbance of the I_2–CCl_4 solution, C_I is the initial concentration of iodine, C_D is the initial concentration of alkoxysilane, ε_c is the molar extinction coefficient of the complex, and ε_I is the molar extinction coefficient of iodine. The values of K_f fall within a range between $K_f = 0.92$

TABLE XXVI

FORMATION CONSTANTS OF CT COMPLEXES OF I_2 WITH ALKYLALCOXYSILANES[a]

Compound	K_f (solvent–CCl_4; acceptor–I_2)
Et_3SiOEt	0.55 ∓ 0.01
$Me_2Si(OEt)_2$	0.61 ∓ 0.02

[a] From Ref. 77.

[I_2–EtOH complex in heptane at 20°C (*79*)] and $K_f = 0.176$ [I_2–benzene complex in CCl_4 at 25°C (*80*)]. Alkoxysilanes behave as *n*-donors toward I_2; their complexes are less stable than the I_2–EtOH complex but are more stable than the I_2–benzene complex. Using these data and the findings of kinetic studies, Ito and Ibaraki (*77*) suggest the reaction mechanism given by Scheme 1.

$$Et_3SiOR + [R'OH\text{---}I]^+I^- \rightleftharpoons \left[Et_3Si \overset{OR}{\frown} H^+ \;\; \underset{\underset{I}{\vdots}}{R'O} \right] I^- \rightleftharpoons$$

$$[Et_3SiOR'\text{---}I]^+I^- + ROH \rightleftharpoons Et_3SiOR' + [ROH\text{---}I]^+I^-$$

or

$$R'OH + [Et_3SiOR\text{---}I]^+I^- \rightleftharpoons \left[H \overset{OR}{\frown} \overset{+}{Si}Et_3 \;\; \underset{\underset{I}{\vdots}}{R'O} \right] I^- \rightleftharpoons$$

$$[ROH\text{---}I]^+I^- + Et_3SiOR' \rightleftharpoons ROH + [Et_3SiOR'\text{---}I]^+I^-$$

SCHEME 1

E. *Arylsilyl Ethers and Aroxymethylsilanes*

In these silicon compounds, the highest occupied molecular orbitals are combinations of π_{Ph}, n_O, and σ_{CSi} (or σ_{SiSi}) orbitals. Thus such compounds might conceivably act both as *n*- and π-donors.

The donor properties of (aroxymethyl)triethylsilanes (p-X–C_6H_4-OCH_2SiEt_3) have been investigated (*81*); data for the absorption spectra

TABLE XXVII

WAVENUMBERS AND IONIZATION POTENTIALS OF SOME (AROXYMETHYL)TRIETHYLSILANES AND CORRESPONDING ANISOLES[a]

	p-X–$C_6H_4OCH_2SiEt_3$		p-X–C_6H_4OMe	
X	ν_{CT} (cm^{-1})	IP (eV)[b]	ν_{CT} (cm^{-1})	IP (eV)
MeO	14,700	7.74	15,700	7.87
Me	16,500	7.97	17,800	8.14
H	18,300	8.21	19,700	8.40
I	17,200	8.06	18,500	8.24
Br	18,000	8.17	19,300	8.35
Cl	18,100	8.18	19,500	8.37
F	18,400	8.22	19,900	8.42

[a] From Ref. 81, except as noted.
[b] Calculated by the equation $IP = 5.769 + 1.335 \cdot 10^{-4}\nu_{CT}$ (*82*).

of their complexes with TCNE are listed in Table XXVII. A linear correlation between the IP values of the corresponding anisoles (IP_{an}) and (aroxymethyl)triethylsilanes (IP_{ar}) was established:

$$IP_{ar} = 0.89IP_{an} + 0.7 \qquad (\rho = 0.990) \tag{36}$$

Based on this correlation, Golovanova *et al.* (*81*) concluded that these compounds act mainly as π-donors toward TCNE.

The absorption spectra of the CT complexes of (aroxymethyl)triethylsilanes and aroxytrimethylsilanes (p-X–$C_6H_4OSiMe_3$, where X = MeO, Me, H, F, Cl, Br, I) with TCNE have been compared (*83*). Aroxytrimethylsilanes are weaker π-donors that the corresponding anisoles, whereas (aroxymethyl)triethylsilanes behave as stronger π-donors.

Complexes of (aroxymethyl)triethylsilanes, (aroxymethyl)triethoxysilanes [p-X–$C_6H_4OCH_2Si(OEt_3)$], and the corresponding anisoles with I_2 have also been studied (*84, 85*) (Table XXVIII). The long-wavelength CT bands for all of these complexes were assigned to a transition of the type $\pi \rightarrow \sigma^*$: π-HOMO of the donor $\rightarrow \sigma^*$ orbital of I_2. A second CT band is also observed and its energy is almost independent of the substituent on the benzene ring. This absorption band was assigned to the $\pi \rightarrow \sigma^*$ transition from the next to highest occupied molecular orbital, HOMO-1. Unlike the HOMO, which has b_1 symmetry, the HOMO-1 orbital has a_2 symmetry and therefore its energy does not depend on the nature of the substituent.

CT complexes of TCNE with the phenyl ethers C_6H_5OR, where R = $SiMe_3$, CH_2SiMe_3, $CH_2Si(OMe)_3$, and Si_2Me_5, have been studied by

TABLE XXVIII

CT BANDS OF COMPLEXES OF STUDIED COMPOUNDS (p-X–$C_6H_4OCH_2Y$) WITH I_2[a]

	λ_{CT} (nm)		
X	Y = $SiEt_3$	Y = $Si(OEt)_3$	Y = H
MeO	405	387	384
Me	377	367	364
H	355	348	345
Cl	358	352	349
Br	357	351	350
F	354	—	—

[a] From Ref. 84.

Knyazhevskaya *et al.* (*86*); data are given in Table XXIX. As expected, the CT wavelengths for the silicon complexes cannot be calculated using a purely inductive model [Eq. (37)]; interaction between the n_O and σ orbitals of the C–Si or Si–Si bonds must also be included.

$$h\nu_{CT} = 2.41 + 0.36\sigma_R^* \qquad (\rho = 0.989) \tag{37}$$

For these compounds, HOMOs can be considered as linear combinations of three group orbitals:

$$\psi = c_1\psi_{\pi_{Ph}} + c_2\psi_{n_O} + c_3\psi_{\sigma_{CSi\ (or\ SiSi)}} \tag{38}$$

TABLE XXIX

ABSORPTION SPECTRA OF CT COMPLEXES WITH TCNE AND IONIZATION POTENTIALS OF ETHERS (C_6H_5OR)[a]

R	σ_R^*	$h\nu_{CT}$ (eV)	$h\nu_{CT}$ (eV)	$\Delta h\nu_{CT}$ (eV)	IP (eV)
$SiMe_3$	−0.90	2.64	2.09	0.55	8.50
$SiMe_3$	−0.44	—	2.25	0.39	—
Si_2Me_5	—	2.53	—	—	8.36
CH_2SiMe_3	−0.26	2.30	2.32	0.02	8.07
$CH_2Si(OMe)_3$	—	2.42	—	—	8.22
$CH_2Si(OEt)_3$	—	2.38	—	—	8.18
$CH_2Si(OPr)_3$	—	2.34	—	—	8.12

[a] From Ref. 86.

[b] Calculated by Eq. (37): $h\nu_{CT} = 2.41 + 0.36\sigma_R^*$ ($\rho = 0.989$).

[c] Calculated by the equation $h\nu_{CT} = 0.80\text{IP} - 4.16$ (*45*).

TABLE XXX
IONIZATION POTENTIALS OF SOME SILYL ETHERS[a]

Compound	$h\nu_{CT}$ (eV)	IP^{exp} (eV)	IP^{calc} (eV)	ΔIP (eV)
p-$Me_3SiOC_6H_4OSiMe_3$	2.20	7.95	7.93	0.02
o-$Me_3SiOC_6H_4OSiMe_3$	2.31	8.09	8.16	0.07
p-$MeOC_6H_4OSiMe_3$	2.18	7.93	7.79	0.14
o-$MeOC_6H_4OSiMe_3$	2.20	7.95	8.03	0.08
p-$Me_5Si_2OC_6H_4OMe$	2.07	7.76	7.76	0.00

[a] From Ref. 86.

The interaction integrals between pairs of orbitals were calculated using the LCBO method, as

$$H_{n_O\sigma_{CSi}} = -0.934 \text{ eV}$$

$$H_{n_O\sigma_{SiSi}} = -0.387 \text{ eV}$$

For aroxymethylsilicon compounds the effect of substitution of an alkoxy group for a methyl group at the silicon atom on the energy of the HOMO was also investigated. Substitution of an OMe group for a Me group increases the IP, consistent with the greater electronegativity of oxygen compared with carbon. This should decrease the energy of the σ_{CSi} orbital, which is mixed with the unshared electron pair of the oxygen atom.

The values of the ionization potentials of some aryl(silyl)ethers were calculated by the MO LCBO method as mentioned above (see Section II,D).

Good agreement between these calculated IP values (IP^{calc}) and the experimentally found IP values (IP^{exp}), determined from the spectra of CT complexes, confirms the mixing of the π_{Ph}, n_O, and σ orbitals (Table XXX). It should also be added that according to photoelectron spectra, substitution of the $SiMe_3$ group for the Me group at the oxygen atom decreases the energy of the unshared electrons of oxygen by an average of 0.58 eV (*87*).

The effects of the interaction of the n-orbital of the oxygen atom with σ_{SiC} and σ_{SiSi} orbitals (*86*) are compared with similar effects in the series of the sulfides (*70*) in Table XXXI. It can be seen that the efficiencies of n–σ mixing in the series of ethers and sulfides are about the same.

The absorption spectra of CT complexes of TCNE were also used to estimate the electron structure of tricyclic donors containing silicon and oxygen (or silicon and sulfur) in the central ring (*88*) (**36–39**).

TABLE XXXI

INTEGRALS OF INTERACTION OF n AND σ ORBITALS[a]

	X = O	X = S
$H_{n_X\sigma_{SiC}}$	−0.934 eV	−1.000 eV
$H_{n_X\sigma_{SiSi}}$	−0.387 eV	−0.354 eV

[a] From Refs. 70 and 86.

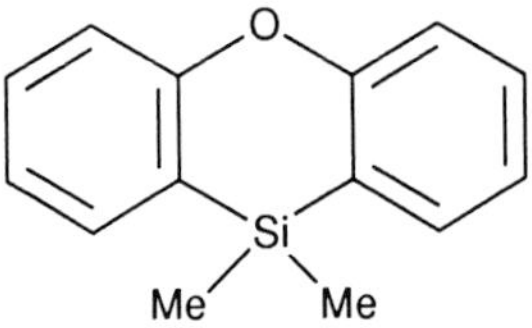

36 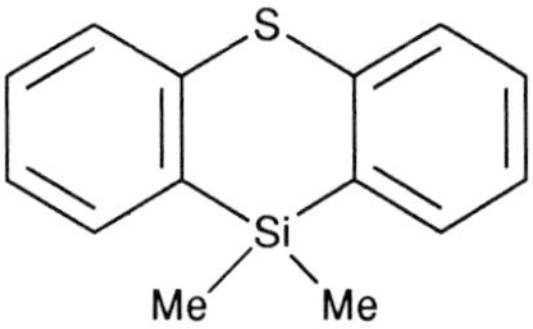 **37**

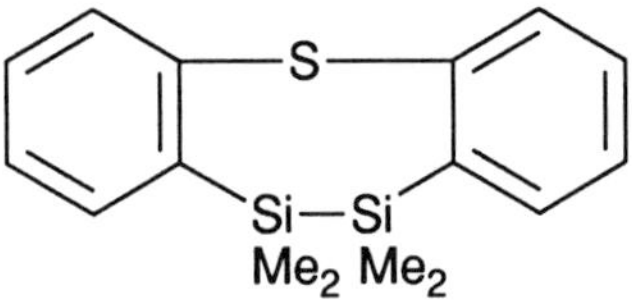

38 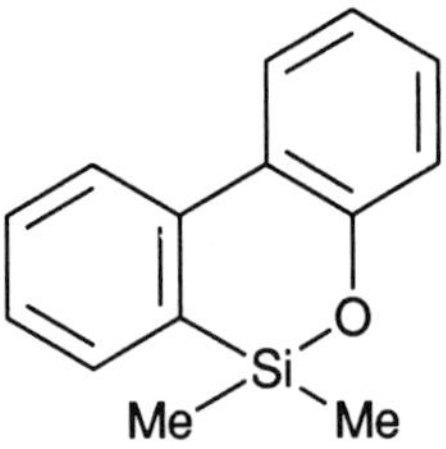 **39**

The reliability of estimations of orbital interactions in heterocyclic compounds using the absorption spectra of CT complexes is demonstrated by the adequate agreement between the IP values calculated from Eqs. (22) and (32) and the data from photoelectron spectra (Table XXXII) (*90, 91*).

Conformational effects in the Si heterocycles were estimated from spectra of their CT complexes and from LCBO MO calculations. Distortion of the molecule geometry by the introduction of an additional $SiMe_2$ group into the heterocycle, as in compounds **37** and **38**, was confirmed by X-ray structural analysis. The angle between the benzene rings diminishes to 98°. Disturbed mixing of the group orbitals results in marked increases in the ionization potentials. These disturbances, however, do not impair estimation of the ionization potentials using the absorption spectra of CT complexes with TCNE. Steric hindrance in the donor molecule (as in other π-donors) decreases the intensity of the CT bands, but has essentially no

TABLE XXXII

ABSORPTION SPECTRA OF CT COMPLEXES WITH TCNE AND IONIZATION POTENTIALS OF SILICON-CONTAINING HETEROCYCLES[a]

	CT complexes				PES		
Compound	λ^1_{CT} (nm)	λ^2_{CT} (nm)	IP_1 (eV)[b]	IP_2 (eV)[b]	IP_1 (eV)	IP_2 (eV)	Ref.
36	573	389	7.90	9.18	7.92	9.06	*90*
37	645	408	7.82	9.05[c]	7.85	9.12	*90*
38	510	410	8.24	8.98	8.13	9.01	*91*
39	574	410	7.90	9.24	—	—	—

[a] From Refs. 88 and 89.
[b] Calculated from Eq. (22).
[c] Calculated from Eq. (32).

effect on the relationships between the positions of the CT bands and the IP values as determined from photoelectron spectra.

The absorption spectra of CT complexes of TCNE with heterocyclic donors containing silicon and oxygen, sulfur, or nitrogen in the ring have been also studied (*65,90,92*).

F. *Peralkylpolysilanes*

As mentioned in the introduction, polysilanes, in which the HOMOs are mainly formed from σ_{SiC} and σ_{SiSi} orbitals, can also serve as donors in CT complexes. Moreover, the CT absorption bands for complexes of such σ-donors follow the same relationship $h\nu_{CT} = f(\text{IP})$ in their spectra as the complexes of benzene derivatives, in which HOMOs are π orbitals only slightly transformed by substitution. However, the intensities of CT bands of complexes diminish markedly with the growth of the contribution of the σ orbitals. The low intensity of the CT bands in the visible region is probably the main reason why the donor properties of peralkysilanes and peralkylpolysilanes toward acceptors were not noticed for such a long time (cf. Ref. *59*).

1. *Peralkylpolysilanes and π-Acceptors*

Photoelectron spectra of permethylpolysilanes show that those molecules have relatively low ionization potentials that decrease with increasing length of the polysilyl chain (*91*). It can be seen from Table XXXIII that their IP values are much lower than those of benzene. Therefore, it should

TABLE XXXIII

IONIZATION POTENTIALS OF PERMETHYLPOLYSILANES[a]

Compound	IP_1 (eV)	IP_2 (eV)	IP_3 (eV)	σ_{SiC} (eV)
$Me(SiMe_2)Me$	—	—	—	10.5
$Me(SiMe_2)_2Me$	8.69	—	—	10.4
$Me(SiMe_2)_3Me$	8.19	9.14	—	10.4
$Me(SiMe_2)_4Me$	7.98	8.76	9.30	10.5
$Si(SiMe_3)_4$	9.24	(9.9)	—	10.4
$(SiMe_2)_5$	7.94	8.91	9.80	10.7
$(SiMe_2)_6$	7.79	8.16	9.12	10.4

[a] From Ref. 91.

not be surprising that electrons in delocalized, molecular σ orbitals of polysilanes can, when excited by light, pass to the vacant orbitals of π-acceptors such as TCNE.

The first reports on CT interaction between permethylpolysilanes and TCNE were published in 1973 (*76, 93*). Solutions of polysilanes with TCNE and similar π-acceptors in inert solvents ($CHCl_3$, CH_2Cl_2, and the like) showed new absorption bands in the visible region. The depth of color and the wavelengths of absorption increase with the number of silicon atoms in the polysilane chain (*76,93*) (Table XXXIV). A detailed study of the absorption spectra has shown that CT complexes are formed in solution, permethylpolysilane being the donor and TCNE the acceptor (Fig. 12).

TABLE XXXIV

PARAMETERS OF CT BANDS IN ABSORPTION SPECTRA OF COMPLEXES OF PERMETHYLPOLYSILANES WITH TCNE[a]

	Band 1		Band 2		Band 3	
Compound	λ_{CT} (nm)	$h\nu_{CT}$ (eV)	λ_{CT} (nm)	$h\nu_{CT}$ (eV)	λ_{CT} (nm)	$h\nu_{CT}$ (eV)
$Me(SiMe_2)_2Me$	417	2.97	—	—	—	—
$Me(SiMe_2)_3Me$	480	2.58	370	3.35	—	—
$Me(SiMe_2)_4Me$	520	2.38	390	3.18	—	—
$Si(SiMe_3)_4$	458	2.71	—	—	—	—
$Me(SiMe_2)_6Me$	540	2.30	435	2.85	—	—
$(SiMe_2)_6$	555	2.23	477	2.60	360	3.44

[a] From Ref. 93.

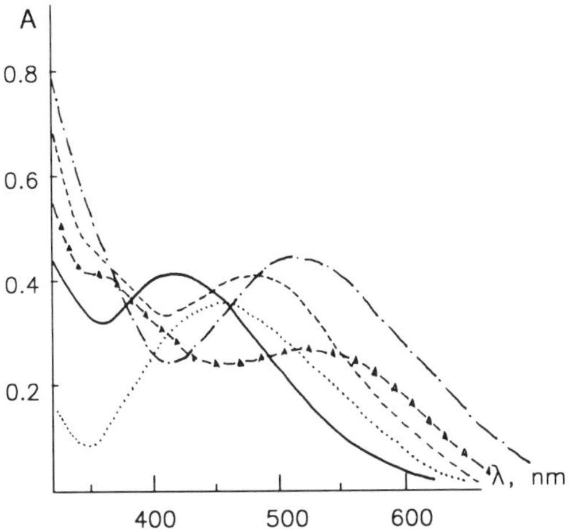

FIG. 12. Spectra of charge-transfer complexes of TCNE (0.015 *M*) in chloroform with permethylpolysilanes: (——), $Me(SiMe_2)_2Me$; (----), $Me(SiMe_2)_3Me$; (-▲-▲-), $Me(SiMe_2)_4Me$; (••••••), $Si(SiMe_3)_4$; (-·-·-), $(SiMe_2)_6$ (*93*).

Good linear correlations between the CT band energies and the ionization potential values of permethylpolysilanes were reported (Fig. 13). This correlation was expressed in an analytical form as well (*76*):

$$h\nu_{CT} = 0.771\text{IP} - 3.710 \qquad (\rho = 0.999) \tag{39}$$

The positions of the CT bands for this equation were initially evaluated from the maxima of experimental curves of the visible absorption spectra (*76*). Some CT bands were, however, overlapped in those spectra. In order to estimate the position of individual bands, Briegleb's relationship has

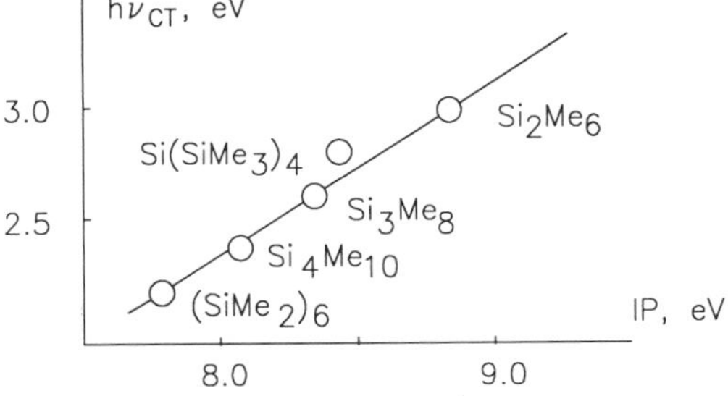

FIG. 13. Plot of charge-transfer energies $h\nu_{CT}$ for TCNE complexes against the first ionization potentials of polysilanes (*93*).

been used (*93,94*):

$$\frac{\nu_h - \nu_l}{2(\nu_{max} - \nu_l)} \simeq 1.1 \qquad (40)$$

(where ν_h and ν_l are wavenumbers at half the maximum intensity on the high- and low-energy sides of the peak located at ν_{max}); the Gauss equation for asymmetric distribution curves (*95*) is also used:

$$D = D_{max} \exp\left[-\left(\frac{\nu_{max}}{d} \cdot \frac{\nu_{max} - \nu}{\nu}\right)^2\right] \qquad (41)$$

The accuracy of Eq. (41) was estimated by comparison of experimental and calculated absorption curve areas. In the absorption spectrum for the complex of Si_2Me_6 with TCNE in $CHCl_3$, the difference in the corresponding areas does not exceed 2%. Good agreement of the calculated and experimental nonoverlapping absorption CT bands has made it possible to use Eq. (41) for the analysis of more complicated absorption spectra of the CT complexes of some permethylpolysilanes. This analysis gave a more accurate dependence of the energies of the first CT bands on the first ionization potentials of permethylpolysilanes:

$$h\nu_{CT(1)} = 0.83IP_1 - 4.20 \qquad (\rho = 0.984) \qquad (42)$$

The similarity of the parameters in Eq. (42) and those in the equations derived earlier for π-donors (*18,33,96*) is of special interest. Equation (42) proves unambiguously the donor function of permethylpolysilanes and the acceptor function of TCNE in the systems $Me(SiMe_2)_nMe$–TCNE. Evidently, the different symmetry of the donor σ orbitals of polysilanes and acceptor π orbital of TCNE does not rule out their mixing, which is followed by electron transfer.

The electron transfer from permethylpolysilanes to TCNE has also been proved by direct observation of the electron spin resonance (ESR) signal of the anion radical of TCNE (*76*). When a solution containing TCNE (0.1 *M*) and permethylpolysilane (0.1 *M*) in dichloroethane was exposed to the light of a mercury lamp, a strong ESR signal of the TCNE anion radical was recorded. As soon as irradiation was discontinued the ESR signal disappeared immediately. A radical or a cation radical of permethylpolysilane was probably also present in solution, but was not identified. The intensity of the ESR signal was directly directly proportional to the length of the polysilane chain. A mixture of tetramethylsilane with TCNE did not produce a signal under the same conditions (*76*).

The longest wavelength absorption band in the visible spectrum of the CT complex is assigned to the electron transition from the HOMO of the donor (its energy is determined by the value of first ionization potential) to the LUMO of the acceptor. However, electron transfer can also take place from other high-lying orbitals of the donor molecule. Several CT bands can therefore be seen in the absorption spectra of the complexes of TCNE with those permethylpolysilanes, which have several ionization potentials in the region below 9 eV.

It should be noted that Eq. (42) is transformed only slightly when the data for all CT bands of the visible spectra are included in the equation of the type $h\nu_{CT} = f(\mathrm{IP})$ (*95*):

$$h\nu_{CT(1,2,3)} = 0.84\mathrm{IP}_{1,2,3} - 4.25 \qquad (\rho = 0.981) \tag{43}$$

Equations (42) and (43) can be used to determine the IP values for higher permethylpolysilanes whose photoelectron spectra are difficult to obtain because of low volatility of the samples.

Strict adherence to the relationships of the $h\nu_{CT} = f(\mathrm{IP})$ type in describing the interaction of TCNE with permethylpolysilanes suggests that this interaction is rather regular, at least from the standpoint of the spatial arrangement of the donor and acceptor molecules.

An attempt was made to measure the thermodynamic parameters of the complexation (*95*). When $CHCl_3$ was used as a solvent and the dependence of CT band absorption on the concentration of the components was treated by the Benesi–Hildebrand method (*51,52*), it was impossible to determine the complex formation constant K and the extinction coefficient ε of the complexes; experimental straight lines in the coordinates $\{[\mathrm{TCNE}]l/\log(I_0/I)\}(1/D)$ intersected the negative portion on the ordinate. Complexes of permethylpolysilanes with TCNE therefore appear to be rather unstable. Substitution of cyclohexane for chloroform gives better results: the straight line intersects the positive region on the ordinate, but the measured values of K and ε are within the limit of the error of the experiment in this case (*95*). The low values of K, as well as the low intensities of CT bands in the absorption spectra, are probably due to steric hindrances imposed on orbital overlap. The methyl groups on silicon doubtless decrease the extent of overlap of LUMO of TCNE with the HOMO of the polysilane, located in the region of the Si–Si bonds.

It has already been mentioned that when a mixture of Si_4Me_{10} and TCNE ($CHCl_3$) was exposed to mercury lamp irradiation, on ESR signal of the anion radical of TCNE was recorded (*76*). Irradiation of permethylpolysilanes **40** can also break the Si–Si bonds and liberate dimethylsilene **41**, which is also able to give up its electron to the acceptor (*97–99*):

$$\underset{\mathbf{40}}{\mathrm{Me(SiMe_2)_{\mathit{n}}Me}} \xrightarrow{h\nu} \mathrm{Me(SiMe_2)_{\mathit{n}-1}Me} + \underset{\mathbf{41}}{:\mathrm{SiMe_2}}$$

However, electron transfer in the TCNE–permethylpolysilane $(SiMe_2)_n$ system was also observed in the absence of irradiation (*95*). An ESR signal characteristic of the TCNE anion radical was recorded in a solution containing a polysilane $(SiMe_2)_n$ in $CHCl_3$. It was shown that the derivatives of silanaphtalenes **40** and **41**, the formation of which might be anticipated in the synthesis of permethylpolysilane, were not donors in those experiments (*100*).

The molecular orbitals formed by linear combinations of σ_{SiSi} orbitals can thus interact effectively with vacant π orbitals of TCNE and other π-acceptors. The IP values of the donor molecules are probably the decisive factor in this interaction.

2. *Peralkylpolysilanes and σ-Acceptors*

The electrons of delocalized HOMOs of permethylpolysilanes can be transferred to σ-acceptors as well as π-acceptors. Charge-transfer absorptions have been observed between tetraalkyl derivatives of Group IV elements and the σ-acceptors WF_6, WCl_6, and MoF_6 (*101*) (Table XXXV). It was later found that in contrast to tetraalkylsilanes (*102*), permethylpolysilanes actively react with σ-acceptors (*103,104*). The reaction is rather exothermic:

$$Me_3Si(SiMe_2)_nSiMe_3 + WCl_6 \rightarrow Me_3SiCl + Cl(SiMe_2)_nSiMe_3 + WCl_{6-k}$$

where $n = 0$–4 and $k = 2$ or 3. It is believed that rupture of the Si–Si bond is preceded by formation of the corresponding CT complexes of polysilane, $Me(SiMe_2)_nMe$, with WCl_6. Comparison of the breakdown rates of the Si–Si bonds in polysilanes with their ionization potentials (which directly characterize their donor properties) reveals a linear correlation

TABLE XXXV

SPECTRAL CHARACTERISTICS OF CT COMPLEXES Me_4Si WITH σ-ACCEPTORS[a]

Acceptor	ν_{CT} (cm^{-1})
NbF_5	43,500; 37,900
MoF_6	27,900
WF_6	40,300
WF_5OMe	40,300

[a] From Ref. 101.

TABLE XXXVI

RELATIVE CLEAVAGE RATES OF REACTION WITH WCl_6 AND FIRST IONIZATION POTENTIALS OF POLYSILANES[a]

$Me(SiMe_2)_2Me$	1	8.69
$Me(SiMe_2)_3Me$	2.05	8.19
$Me(SiMe_2)_4Me$	2.75	7.98
$Si(SiMe_3)_4$	1.08	8.24
$Me(SiMe_2)_6Me$	4.10	7.70

[a] From Ref. 102.

(Table XXXVI) in which the cleavage rate of polysilanes increases with decreasing first ionization potential, consistent with the important role of charge transfer at the stage of complexation. Deviation from linear dependence in case of $Si(SiMe_3)_4$, which reacts with WCl_6 at about the same rate as hexamethyldisilane despite a lower ionization potential, is probably due to steric hindrance imposed on complexation of the pentasilane molecule.

Other donor–acceptor reactions involving Si–Si bonds have also been studied. Oxidation of organodisilanes by perbenzoic acid (*105*) is attended by introduction of the oxygen atom into the Si–Si σ bond:

$$R_3SiSiR_3 + PhCO_3H \rightarrow R_3SiOSiR_3 + PhCO_2H$$

Four organodisilanes were examined kinetically. The visible spectra of their CT complexes with TCNE were also recorded and the thermodynamic characteristics of the transition state were determined (Table XXXVII). As the structure of disilane changes, the value of entropy of activation $\Delta S^{\ddagger}$ of this reaction remains about the same. The enthalpy of activation $\Delta H^{\ddagger}$, on the contrary, depends on the structure of disilane. A good linear dependence between $\Delta H^{\ddagger}$ values and the energies CT bands E_{CT} has been established:

$$\Delta H^{\ddagger} = 0.737E_{CT} - 36.8 \qquad (\rho = 0.991) \tag{44}$$

Equation (44) shows that the enthalpy of activation of the reaction depends mostly on the energy of the HOMO, reflecting the dominant role of the donor–acceptor interaction in the transition state for oxidation.

Organic derivatives of silicon were also found to form CT complexes with Br_2 and I_2 at 25°C (CCl_4, n-C_6H_{14}, and CH_2ClCH_2Cl were used as solvents) (*106,107*). Data obtained are listed in Table XXXVIII. The

TABLE XXXVII

CT ENERGIES AND RATE PARAMETERS FOR PERACID OXIDATION OF DISILANES[a]

Compound	ν_{CT} $(cm^{-1})^b$	E_{CT} (kcal/mol)	$\Delta H^\ddagger$ (kcal/mol)	$\Delta S^\ddagger$ (eu)
$Me_3SiSiMe_3$	24,000 ∓ 75	68.6 ∓ 0.2	14.0 ∓ 0.5	−26.8 ∓ 1.6
42 (cyclic $SiMe_2$–$SiMe_2$, seven-membered ring)	23,500 ∓ 57	67.2 ∓ 0.2	12.3 ∓ 0.1	−28.1 ∓ 0.1
43 (cyclic $SiMe_2$–$SiMe_2$, six-membered ring)	22,500 ∓ 48	64.3 ∓ 0.1	10.7 ∓ 0.7	−28.5 ∓ 2.2
44 (cyclic $SiMe_2$–$SiMe_2$, five-membered ring)	21,600 ∓ 45	61.8 ∓ 0.1	8.75 ∓ 0.2	−29.3 ∓ 0.7

[a] From Ref. 105.
[b] CT band frequencies of complexes with TCNE in dichloromethane at room temperature.

TABLE XXXVIII

SPECTRAL CHARACTERISTICS OF CT COMPLEXES OF HALOGENS WITH PERALKYLSILANES AND DISILANES[a]

Compound	Halogen	Solvent	λ_{CT} (nm)
Me_4Si	I_2	CCl_4	268 ∓ 2
Et_4Si	I_2	CCl_4	273 ∓ 1
Si_2Me_6	I_2	CCl_4	297 ∓ 1
		CH_2ClCH_2Cl	295 ∓ 5
		n-C_6H_{14}	298 ∓ 1
	Br_2	CCl_4	295 ∓ 5
Si_2Et_6	I_2	CCl_4	300 ∓ 2

[a] From Ref. 106.

TABLE XXXIX

KINETIC PARAMETERS AND ARRHENIUS EQUATION PARAMETERS FOR REACTION OF DISILANES WITH HALOGENS[a]

Compound	Halogen	Solvent	k_2 (23.5°C; liter/mol cm)	E_a (kcal/mol)	log(A)
Me_6Si_2	Br_2	CCl_4	2.5	10.2 ∓ 0.5	7.9 ∓ 0.4
		CH_2ClCH_2Cl	8.1	9.2 ∓ 0.5	7.7 ∓ 0.4
	I_2	CCl_4	$3.34 \cdot 10^{-4}$	10.5 ∓ 0.8	—
		CH_2Cl_2	$3.00 \cdot 10^{-3}$	—	—
		CH_2ClCH_2Cl	$3.50 \cdot 10^{-3}$	9.7 ∓ 0.1	4.7 ∓ 0.5
		PhCl	$0.97 \cdot 10^{-3}$	14.2 ∓ 0.5	7.2 ∓ 0.6
Et_6Si_2	Br_2	CCl_4	0.24	10.4 ∓ 0.6	7.0 ∓ 0.4
		CH_2ClCH_2Cl	0.25	8.7 ∓ 0.5	5.9 ∓ 0.3
	I_2	PhCl	$7.00 \cdot 10^{-5}$	15.1 ∓ 0.5	7.2 ∓ 0.6

[a] The Arrhenius equation is $k_2 = A \exp(-E_a/RT)$; data from Ref. 107.

reaction of disilanes R_6Si_2 with halogens is overall a second-order reaction, first order in each component (*107*) (Table XXXIX). Rapid establishment of equilibrium has been suggested:

$$R_6Si_2 + Hal_2 \rightleftharpoons R_6Si_2 \cdot Hal_2$$

The reaction is not retarded in the presence of inhibitors of radical chain reactions, so a radical mechanism is unlikely. An ionic mechanism is also improbable, because the reaction occurs in low-polar, poorly solvating solvents. The interaction of disilanes with halogens therefore seems to be a synchronous process; its most probable intermediate is the CT complex $R_6Si_2 \cdot Hal_2$.

G. *1,1-Dialkyl-1-silacyclobutanes*

Owing to the high ionization potentials of the Si–C bonds, tetraalkylsilanes hardly show any donor properties in their interaction with the π-acceptors. But strong distortion of the valency angle at silicon in a small ring can facilitate ionization; the first ionization potential of 1,1-dimethyl-1-silacyclobutane is only 9.4 eV compared to 10.5 eV for tetramethylsilane (*108*). It is believed that the highest occupied molecular orbitals of 1,1-dialkyl-1-silacyclobutanes have some ethylene-like character. The analogy of silacyclobutanes with unsaturated compounds is also confirmed by their reactivity (*109*). Moreover, the low ionization potentials of these molecules allow them to serve as donors in CT complexing to π-acceptors.

Mixtures of 1,1-dialkyl-1-silacyclobutanes with TCNE (chloroform as solvent) absorb in the near UV and visible regions of the spectrum, although the separate components do not. This absorption is due to charge transfer from the silacyclobutane to TCNE. As the electron-donating power of the alkyl substituent in 1,1-dialkyl-1-silacyclobutane increases, the CT band shifts markedly to longer wavelength.

Analysis of the data listed in Table XL shows that there is a good linear correlation between the position of the CT bands for TCNE complexes of 1,1-dialkyl-1-silacyclobutanes (**45–51**) and the sum of inductive σ^* constants of the substituents at the silicon atom:

$$h\nu_{CT} = 2.38\sigma^*_{R,R'} + 3.56 \qquad (\rho = 0.992) \tag{45}$$

For silacyclobutanes (**52–54**; 1-methyl-, 1-trimethylsilylmethyl-, 1,1-diisopropyl-, and 1,1-dicyclohexyl-1-silacyclobutanes), the energies of the CT bands are much higher than calculated from Eq. (45), probably because steric hindrance prevents effective overlap of the donor HOMO with the accepting LUMO of TCNE. Such steric interactions are especially important because the silacyclobutane ring is not planar [the dihedral angle is 30° (*20*)].

TABLE XL

POSITION OF CT BAND IN SPECTRA OF MIXTURES OF 1,1-DIALKYL-1-SILACYCLOBUTANES WITH TCNE[a]

N	R	R′	$\Sigma\sigma^*_{RR'}$	$\Sigma E^S_{RR'}$ [b]	λ_{CT} (nm)	$h\nu_{CT}$ (eV)[c]	$h\nu_{CT}$ (eV)[d]	IP (eV)[e]
45	Me	Me	0.000	0.00	350	3.54	3.56	9.31
46	Me	Et	−0.100	−0.07	370	3.35	3.32	9.08
47	Me	n-C_3H_7	−0.115	−0.36	375	3.31	3.29	9.03
48	Me	i-C_3H_7	−0.190	−0.47	400	3.10	3.11	8.78
49	Me	n-C_4H_9	−0.130	−0.39	380	3.26	3.25	8.97
50	Me	t-C_4H_9	−0.125	−1.13	380	3.26	3.26	8.97
51	Me	C_6H_{11}	−0.150	−0.79	388	3.19	3.20	8.89
52	Me	CH_2SiMe_3	−0.260	—	382	3.25	2.94	—
53	i-C_3H_7	i-C_3H_7	−0.380	−0.84	395	3.14	2.65	—
54	C_6H_{11}	C_6H_{11}	−0.300	−1.58	390	3.17	2.75	—

[a] From Ref. 110.
[b] From Ref. 111.
[c] Experimental findings.
[d] Calculated using Eq. (44).
[e] Calculated using Eq. (42).

TABLE XLI
ESTIMATION OF IPs OF SILACYCLOBUTANES BY SPECTRA OF CT COMPLEXES WITH TCNE[a]

Compound	λ_{CT} (nm)	IP (eV)[b]	IP (eV)[c]
1,1-dimethyl-1-silacyclobutane ($SiMe_2$ ring) **55**	350	9.31	9.40
Me_2Si/$SiMe_2$ ring **56**	400	8.78	8.80

[a] From Ref. 110.
[b] Calculated using Eq. (42).
[c] By PES (*108*).

Equation (42), describing the interaction of TCNE with permethylpolysilanes, can be used to determine IPs of silacyclobutanes as well (Table XL). The ionization potentials of 1,1-dimethyl-1-silacyclobutane and 1,1,3,3-tetramethyl-1,3-silacyclobutane (**55** and **56**) calculated from Eq. (42) are compared with those measured from photoelectron spectra (*108*) in Table XLI. Because silacyclobutanes obey Eq. (42), the direct effect of their structure on the energy of their HOMOs can be estimated:

$$\mathrm{IP} = 2.77\sigma^*_{\mathrm{R,R'}} + 9.33 \qquad (\rho = 0.993) \tag{46}$$

The sensitivity of the energy of HOMOs of dialkylsilacyclobutanes to the effect of alkyl substituents at the silicon atom is comparable to the HOMO sensitivity to the effects of the same substituents in ethylenes, aliphatic iodides, alcohols, and dialkyl sulfides, compounds in which donor HOMOs are undoubtedly the π or n orbitals (*112,113*):

Ethylenes: $$\mathrm{IP} = 1.98\sigma^*_{\mathrm{R}} + 9.74 \tag{47}$$

Aliphatic iodides: $$\mathrm{IP} = 1.69\sigma^*_{\mathrm{R}} + 9.82 \tag{48}$$

Alcohols: $$\mathrm{IP} = 3.17\sigma^*_{\mathrm{R}} + 10.97 \tag{49}$$

Dialkylsulfides: $$\mathrm{IP} = 1.18\sigma^*_{\mathrm{R}} + 8.63 \tag{50}$$

H. *Peralkyl polymetals of Heavier Group IV Elements as Donors*

Because the ionization potentials of Group IV compounds decrease from Si to Ge, Sn and Pb, compounds of the heavier elements in the group should also form CT complexes with electron acceptors. In fact, CT bands have been recorded in the absorption spectra of permethylpolygermane–TCNE solutions (*76,114*) and a strong ESR signal of the TCNE anion radical was found in studies on the mixture of Ge_2Me_6 with TCNE. As expected, the CT bands in the absorption spectra of permethylpolygermane complexes with TCNE are shifted bathochromically compared with spectra of polysilane complexes. The intensity of CT band of the Ge_2Me_6–TCNE complex does not change for several hours (*95*), but TCNE complexes with higher polygermanes, e.g., Ge_3Me_8, Ge_4Me_{10}, Ge_5Me_{12}, are quite unstable. The visible absorption of a mixture of $(GeMe_2)_6$ with TCNE proves to be most unstable: the CT band is registered only at the moment of mixing of the reagents, and rapidly disappears. However, absorption of the anion radical of TCNE can be observed in the 350- to 500-nm region for 30 minutes. This indicates complete electron transfer from permethylpolygermane to TCNE:

$$\text{TCNE} + (\text{GeMe}_2)_6 \rightarrow \text{TCNE}^{\overline{\cdot}} + (\text{GeMe}_2)_6^{\,+\cdot}$$

An ESR signal of the TCNQ anion radical was recorded in solutions of mixtures of higher linear polygermanes $Me(GeMe_2)_nMe$ ($n > 5$) with TCNQ in CH_2Cl_2.

Saturated compounds of tin give up their electrons to π-acceptors even more easily (Table XLII). CT bands are detected in the absorption spectra during mixing of TCNE even with tetramethylstannane (*114,115*). Tetraalkylstannanes react with TCNE rather rapidly; the PMR signal related to tetraethylstannane decreases by a factor of two in 15 minutes when $SnEt_4$ is

TABLE XLII
CT SPECTRA OF COMPLEXES OF TETRAALKYLTINS AND TCNE[a]

Compound	λ_{CT} (nm)	ν_{CT} (cm^{-1})	IP (eV)
$SnMe_4$	350	28,600	9.70
$SnEt_4$	426	23,500	8.93
$SnBu_4$	417	24,000	8.83

[a] From Ref. 115.

TABLE XLIII

CT FREQUENCIES OF TETRAALKYLLEAD–TCNE COMPLEXES[a]

$Et_{4-n}PbMe_n$ $(0 \leq n \leq 4)$	λ_{CT} (nm)[b]	ν_{CT} (cm^{-1})[b]	k_{TCNE} (liter/mol sec)[c]	IP (eV)[d]
Me_4Pb	412	24,300	0.032	8.90
Me_3PbEt	430	23,300	0.52	8.65
Me_2PbEt_2	454	22,000	3.1	8.45
$MePbEt_3$	491	20,400	12	8.26
$PbEt_4$	—	—	48	8.13

[a] From Refs. 115 and 116.
[b] In 1,2-dichloropropane at −35 to −52°C.
[c] Second-order rate constant for insertion determined spectrophotometrically for first 10% in CH_3CN at 25°C.
[d] IP from He(I) PES.

mixed with TCNE. An addition reaction occurs at the C=C bond (*115*):

$$SnEt_4 + TCNE \rightarrow Et_3Sn(NC)_2CC(CN)_2Et$$

The CT complex of $Me_3SnSnMe_3$ with TCNE is so unstable that its absorption spectrum could not be recorded, although the electronic absorption specuum of the TCNE anion radical was observed (*114*). Complete charge transfer in the mixture $Me_3SnSnMe_3$–TCNE was also demonstrated by the signal assigned to the anion radical of TCNE in the ESR spectrum (*114*).

CT bands are observed in the absorption spectra of mixtures of Alk_4Pb–TCNE only at low temperatures of −35 to −52°C (*115,116*) (Table XLIII). The CT band for the Me_4Pb–TCNE complex is slightly more stable. At −10°C in chloroform, the absorption band disappears ($t_{1/2} = 20$ minutes) and absorption at $\lambda_{max} = 417$ nm of the ion $(NC)_2CC(CN){=}C(CN)_2^-$ increases accordingly (Fig. 14).

Kinetic studies suggest the reaction mechanism given in Scheme 2 occurring through the CT stage (*115*):

$$R_3'PbCH_2R + TCNE \rightarrow R_3'Pb(NC)_2CC(CN)_2CH_2R \xrightarrow{-HCN}$$

$$R_3'Pb(NC)_2C\overset{NC}{\overset{|}{C}}{=}CHR \xrightarrow[-HCN]{TCNE} R_3'\overset{+}{Pb}(NC)_2C{=}\overset{NC}{\overset{|}{C}}\underset{\underset{R}{|}}{\overset{-}{C}}\overset{CN}{\overset{|}{C}}{=}C(CN)_2$$

SCHEME 2

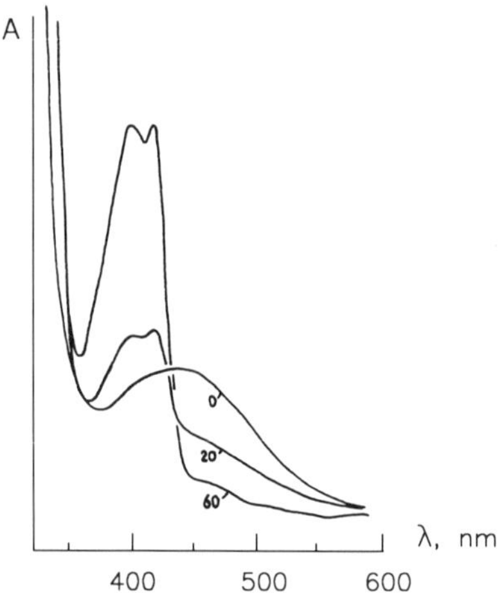

FIG. 14. Changes in the visible absorption spectrum (A) during the reaction of 0.3 *M* Me_4Pb and 0.01 *M* TCNE in chloroform at −10°C: 0, 20, and 60 minutes (*116*).

A linear correlation was found between the IP values of tetraalkylleads and tetraalkylstannanes and the energies of CT bands in the absorption spectra of their complexes with TCNE.

The appearance of CT bands in the absorption spectra of TCNE complexes with peralkyl derivatives of silicon, germanium, tin, and lead, and the dependence of their positions on the values of ionization potentials characterizing the energy of the donor HOMOs, suggest the possibility of charge transfer in systems including carbon compounds as σ-donors. This suggestion seems to be proved correct by observation of CT bands in the absorption spectra of mixtures of some saturated hydrocarbons with TCNE (*117*). CT bands were observed in the absorption spectra of TCNE mixtures with the compounds listed in Table XLIV. As the first ionization potential of the hydrocarbon decreases, a bathochromic shift of the CT bands becomes quite evident.

Charge-transfer phenomena were also observed in the interaction of peralkyl derivatives of the Group IV elements with σ-acceptors. Dilute solutions of NbF_5, MoF_6, and WF_6 containing tetraalkyl compounds of the Group IV elements are colored, and CT bands can be observed in the absorption spectra of these mixtures (Table XLV). CT complexes were also formed on mixing tetraalkyl compounds of germanium and tin with I_2 and Br_2 in neutral solvents (Tables XLVI and XLVII) (*106,118*); complex formation constants for the tin compounds were determined by the Benesi–Hildebrand method (Table XLVIII) (*118*).

TABLE XLIV

SPECTRAL CHARACTERISTICS OF CT COMPLEXES OF TCNE WITH HYDROCARBONS AND CORRESPONDING IONIZATION POTENTIALS[a]

Compound	λ_{CT} (nm)	IP (eV)
Hexamethylethane	310	—
Cyclohexane	320	9.88
Dicyclohexyl	330	9.40
Decalin	330	9.40
Adamantane	355	9.25
1,1-Diadamantyl	380	—

[a] From Ref. 117.

TABLE XLV

SPECTRAL CHARACTERISTICS OF CT COMPLEXES OF TETRAALKYLMETALS OF GROUP IV ELEMENTS WITH TRANSITION METAL FLUORIDES[a]

	ν_{CT} (cm^{-1})		
Compound	NbF_5	MoF_6	WF_6
Me_4Ge	42,900	43,400	34,800
Me_4Sn	36,300	—	29,800
n-Pr_4Sn	—	—	43,700
			34,500
n-Bu_4Sn	—	—	41,300
			31,700

[a] From Ref. 104.

TABLE XLVI

LONG-WAVE ABSORPTION BANDS IN SPECTRA OF CT COMPLEXES Et_4Ge WITH HALOGENS[a]

Acceptor	Solvent	λ_{CT} (nm)
I_2	n-C_6H_{14}	277 ∓ 1
	CCl_4	279 ∓ 1
Br_2	n-C_6H_{14}	277 ∓ 1
	CCl_4	278 ∓ 1

[a] From Ref. 106.

TABLE XLVII

SPECTRAL CHARACTERISTICS OF CT COMPLEXES OF TETRAALKYLTIN WITH HALOGENS[a]

		λ_{CT} (nm)	
R_4Sn	IP (eV)	Br_2	I_2
Me_4Sn	9.70	275	279
Et_4Sn	8.93	310	300
Pr_4Sn	—	312	300
Bu_4Sn	8.83	312	300

[a] From Ref. 118.

TABLE XLVIII

STABILITY CONSTANTS AND MOLAR EXTINCTION COEFFICIENTS OF CT COMPLEXES OF TETRAALKYLTIN WITH TCNE[a]

R_4Sn	K (liter/mol)	ε (liter/mol cm)
Me_4Sn	0.34 ∓ 0.03	4650 ∓ 100
Et_4Sn	0.37 ∓ 0.03	5400 ∓ 150

[a] From Ref. 118.

In contrast to tetraalkyl compounds of germanium and tin, analogous compounds of lead react rapidly with π-acceptors. For example, tetraalkylleads react with hexachloroiridate (IV) ($IrCl_6^{2-}$) at 25°C in acetonitrile or acetic acid (*116*) as follows:

$$Me_4Pb + 2IrCl_6{}^{2-} \xrightarrow{HOAc} Me_3PbOAc + MeCl + IrCl_6{}^{3-} + (IrCl_5{}^{2-})$$

A linear correlation was established between the first ionization potentials of organolead compounds and the logarithms of the reaction rate constants (Table XLIX). It was proposed that the rate-limiting step was electron transfer from the organolead compound to the $IrCl_6{}^{2-}$ anion:

$$R_4Pb + IrCl_6{}^{2-} \xrightarrow{k} R_4Pb^{\dot{+}} + IrCl_6{}^{3-}$$

$$R_4Pb^{\dot{+}} \xrightarrow{fast} R^{\cdot} + R_3Pb^{+}$$

and so on.

TABLE XLIX

CORRELATION OF THE RATES OF OXIDATION OF $Et_{4-n}PbMe_n$ BY $IrCl_6^{2-}$ IN ACETONITRILE WITH THE VERTICAL IP[a]

$PbMe_nEt_{n-4}$	k (liter/mol s)	IP (eV)
$PbEt_4$	26	8.13
$PbEt_3Me$	11	8.26
$PbEt_2Me_2$	3.3	8.45
$PbEtMe_3$	0.57	8.65
$PbMe_4$	0.02	8.90

[a] From Ref. 116.

Peralkyl polymetals react with σ-acceptors more rapidly as the metal–metal chain grows (*112*). In the course of the reaction, the E–E bond breaks:

$$m\mathrm{Me_3E(EMe_2)_nEMe_3} + \mathrm{WCl_6} \rightarrow \mathrm{Me_3ECl} + \mathrm{Cl(EMe_2)_nEMe_3} + \mathrm{WCl_{6-2m}}$$

where E = Ge, Sn; $m \geqslant 1$; $n \geqslant 0$. In the case of organotin compounds, the Sn–C bond breaks as well.

Saturated hydrocarbons can also form CT complexes with σ-acceptors. There is evidence (*114,119*) that these hydrocarbons can form CT complexes with WF_6 and MoF_6 (as with I_2 and Br_2). The resultant complexes are sufficiently stable that complex formation constants can be calculated. Analysis of the CT band position of hydrocarbon complexes with halogens (based on the Mulliken theory) has shown good agreement between the calculated values and experimental findings (*119*).

III

ORGANOSILICON COMPOUNDS AS ELECTRON ACCEPTORS

We have seen many examples of organosilicon compounds acting as donors in CT complex formation. Can silicon compounds also act as acceptors in CT complexes? Halosilanes are known to serve as Lewis acids in formation of acid–base complexes, such as those of Me_3SiCl with pyridine, DMF, quinoline, and Et_3N (*120–123*).

The most likely candidates for CT acceptors are probably the silicon tetrahalides. Studies of the absorption spectra of mixtures of π-donors such as naphthalene and hexamethylbenzene with $SiCl_4$ failed to show evidence for complex formation, even though weak complexes were observed with CCl_4 (*120,121*). However, new absorption bands assigned to CT complexes have been observed in solutions of $SiBr_4$ with aromatic π-donors (Figs. 15, 16, and 17). Molar polarization findings have also been reported in support of this assignment (*121*).

IV

CONCLUSION

According to the orbital mixing rules (*5*), the interaction of σ and π orbitals in organic molecules is prohibited by symmetry. This approximation, suitable for planar molecules formed by the same atoms of the

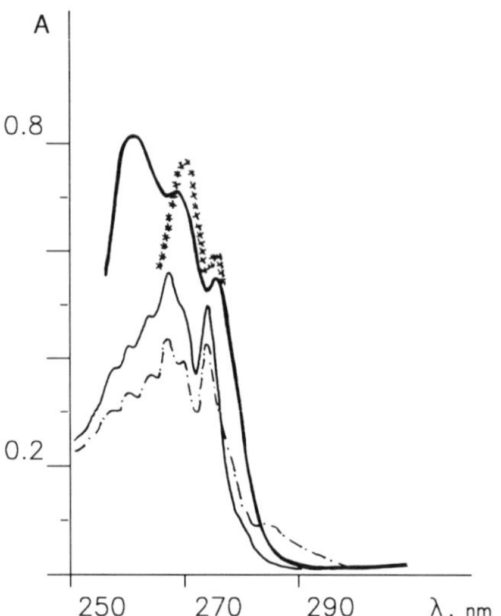

FIG. 15. Absorption spectra (A) of 1,3,5-trimethylbenzene in cyclohexane (——), $SiCl_4$ (-·-·-), $SiBr_4$ (-x-x-x-x-x-x) and CCl_4 (——) (*121*).

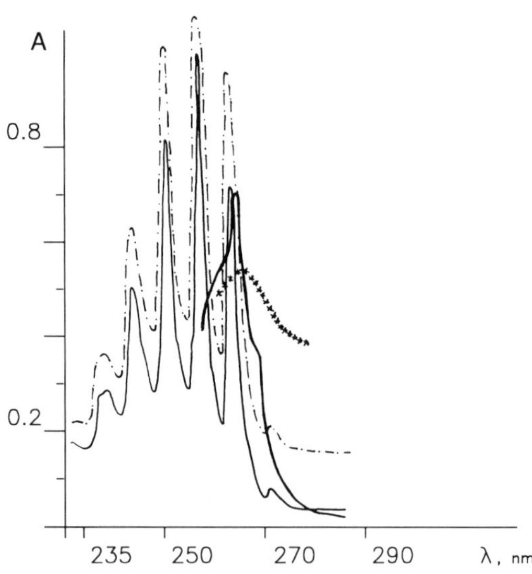

FIG. 16. Absorption spectra (A) of benzene in cyclohexane (——), $SiCl_4$ (-·-·-), $SiBr_4$ (-x-x-x-x-x-x), and CCl_4 (——) (*121*).

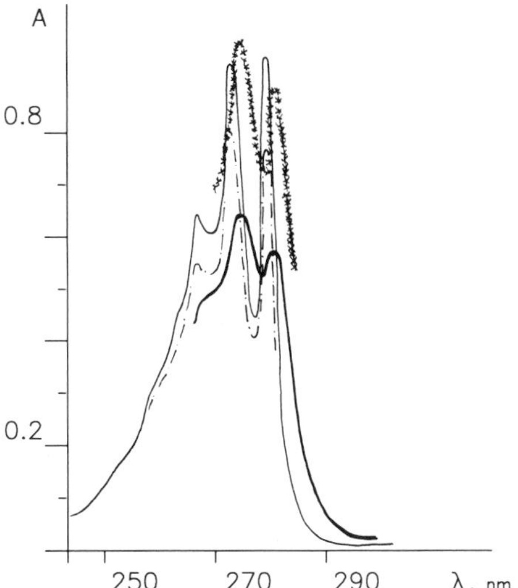

FIG. 17. Absorption spectra (A) of anisole in cyclohexane (——), $SiCl_4$(·····) $SiBr_4$ (-x-x-x-x-x-x), and CCl_4 (——) (*121*).

second-row elements, does not hold strictly for organic compounds of the higher row elements. The σ bond orbitals of compounds having nonplanar structures can mix with the neighboring π orbitals in the same molecules and with those of other molecules in molecular complexes.

Direct measurement of energies of the occupied orbitals (by photoelectron spectra) and vacant electron levels (by electron transmission spectroscopy) shows that organic compounds of silicon can be both effective donors and acceptors (Fig. 18) (*124*).

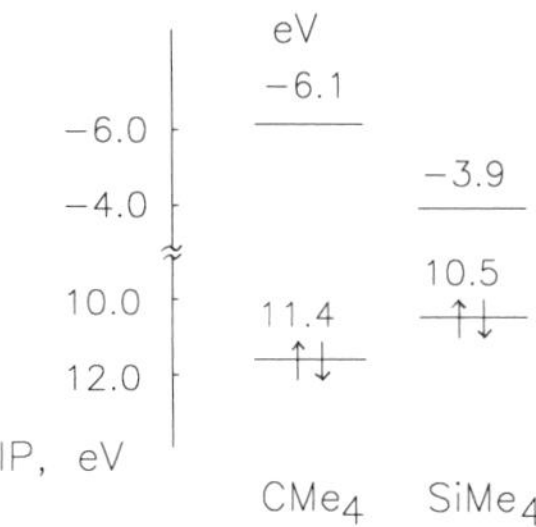

FIG. 18. Values of first ionization potentials (IP) and electron affinity (EA) of neopentane and tetramethylsilane.

It is necessary to emphasize again the unique properties of permethylpolysilanes. The position of CT bands in the absorption spectra of their complexes with TCNE is substantially determined by the same equation that is used to estimate the position of the CT bands in the absorption spectra of complexes of π-donors,

$$h\nu_{CT} = a\mathrm{IP} + b \quad (51)$$

where $a = 0.80$–0.84 and $b = 4.16$–4.24. The low intensities of CT bands in the absorption spectra of permethylpolysilane complexes are probably explained by steric hindrance to complexing.

One can compare the effectiveness of electron delocalization through mixing of σ orbitals in polysilanes and polygermanes with those of π orbitals in π-donors. The values of the corresponding integrals H, are given below (*125*):

Mixing orbitals	H (eV)
σ_{SiSi}	−0.50
σ_{GeGe}	−0.39
$\pi_{C=C}$	−1.23

The "aromatic" properties of cyclic polysilanes such as dodecamethylcyclohexasilane (*8, 126, 127*) are especially noteworthy. This polysilane forms a stable anion radical and can be converted into a stable cation radical as well. As it reacts with π-acceptors in solutions, charge transfer becomes quite obvious. The detachment of the electron from the dodecamethylcyclohexasilane molecule gives a very stable cation radical: the molecular ion $(SiMe_2)_6^{+}$ has a 100% abundance in the mass spectrum of

TABLE L

INTENSITIES OF MOLECULAR IONS IN MASS SPECTRA OF SOME PERMETHYLMETALS OF GROUP IVA[a]

Permethylmetal	Intensity (%)
Si_2Me_6	68.7
Ge_2Me_6	54.7
Sn_2Me_6	32.2
Si_3Me_8	76.3
Ge_3Me_8	58.6
$(SiMe_2)_6$	100.0

[a] From Ref. 128.

this compound (the energy of ionizing electrons was 12 eV) (*128*) (Table L). Dodecamethylcyclohexasilane absorbs strongly in the long-wavelength region of the UV spectrum (*127*). These properties all resemble those of aromatic hydrocarbons.

References

1. R. S. Mulliken, *J. Am. Chem. Soc.* **74,** 811 (1952).
2. R. S. Mulliken, *J. Phys. Chem.* **56,** 801 (1952).
3. T. Koopmans, *Physics* **1,** 104 (1934).
4. G. Klopman, ed., "Chemical Reactivity and Reaction Paths," p. 383 Wiley, New York, 1974.
5. R. Hoffmann, *Acc. Chem. Res.* **4,** 1 (1971).
6. R. Muller and E. G. Rochow, *J. Chem. Educ.* **42,** 41 (1965).
7. C. J. Attrige, *Organomet. Chem. Rev. Sect. A* **5,** 323 (1970).
8. H. Kwart and K. King, "d-Orbitals in the Chemistry of Silicon, Phosphorus and Sulfur." Springer-Verlag, Berlin, 1977.
9. M. G. Voronkov, V. P. Malishkevich,, and Yu. A. Yugilevskiy, "Siloxane Bond." 1976, Nauka, Novossibirsk.
10. R. West and E. Carberry, *Science* **189,** 179 (1975).
11. R. Foster, "Organic Charge-Transfer Complexes," p. 33. Academic Press, London.
12. R. Ponec and V. Chvalovsky, *Collect. Czech., Chem. Commun.* **38,** 3845 (1973).
13. A. N. Egorochkin, V. A. Kuznetsov, M. A. Lopatin, S. E. Skobeleva, V. F. Mironov, V. A. Sheludyakov, and V. I. Zhun, *Dokl. Akad. Nauk SSSR* **250,** 111 (1980).
14. R. Ponec, V. Chvalovsky, E. A. Cernysev, and N. G. Komalenkova, *Collect. Czech., Chem. Commun.* **39,** 1177 (1974).
15. A. N. Egorochkin and M. A. Lopatin, *Metalloorg. Khim.* **1,** 1290 (1988).
16. H. Sakurai, M. Kira, and M. Ochiai, *Chem. Lett.*, 87 (1972).
17. T. A. Zakomoldina, P. G. Sennikov, V. A. Kuznetsov, A. N. Egorochkin, and V. O. Reikhsfel'd *Zh. Obshch. Khim.* **50,** 898 (1980).
18. R. G. Farrell and J. Newton, *J. Phys. Chem.* **69,** 3506 (1965).
19. V. F. Traven and T. V. Pyatkina, *Zh. Obshch. Khim.* **43,** 685 (1973).
20. C. G. Pitt, M. S. Haberson, M. M. Bursey, and P. F. Rogerson, *J. Organomet. Chem.* **15,** 359 (1968).
21. H. Sakurai and M. Kumada, *Bull. Chem. Soc. Jpn.* **37,** 1894 (1964).
22. H. Gilman, W. H. Atwell, and G. L. Schwebke, *J. Organomet. Chem.* **2,** 369 (1964).
23. D. N. Hague and R. H. Prince, *Chem. Ind.* (*London*), 1492 (1964).
24. D. N. Hague and R. H. Prince, *J. Chem. Soc.*, 4690 (1965).
25. C. G. Pitt, L. L. Jones, and B. G. Ramsey, *J. Am. Chem. Soc.* **89,** 5471 (1967).
26. H. Sakurai, H. Yamamori, and M. Kumada, *Chem. Commun.*, 198 (1968).
27. V. A. Kuznetsov, A. N. Egorochkin, S. Yu. Khovryakov, and D. V. Muslin, *Zh. Obshch. Khim.* **44,** 1958 (1974).
28. R. Ponec and V. Chvalovsky, *Collect. Czech., Chem. Commun.* **39,** 1185 (1974).
29. H. Bock and H. Alt, *J. Am. Chem. Soc.* **92,** 1569 (1970).
30. D. M. Bishop and H. Ito, *Theor. Chim. Acta* **16,** 377 (1970).
31. J. Reffy, J. Karger-Kocsis, and J. Nagy, *Period. Polytech. Chem. Eng.* **24,** 7 (1980).
32. H. Sakurai, M. Ichinose, M. Kira, and T. G. Traylor, *Chem. Lett.*, 1383 (1984).
33. E. M. Voigt and C. Reid, *J. Am. Chem. Soc.* **86,** 3930 (1964).
34. H. Sakurai and M. Kira, *J. Am. Chem. Soc.* **96,** 791–4.

35. K. Watanabe, *J. Chem. Phys.* **26,** 542 (1975).
36. C. G. Pitt and H. Bock, *J. Chem. Soc., Chem. Commun.*, 28 (1972).
37. M. J. S. Dewar, E. Haselbach, and S. D. Worsley, *Proc. R. Soc. London* **315,** 431 (1970).
38. H. Bock and H. Alt, *Chem. Ber.* **102,** 1534, (1969).
39. C. G. Pitt, R. N. Carey, and E. C. Toren, *J. Am. Chem. Soc.* **94,** 3806 (1972).
40. R. Foster, "Organic Charge-Transfer Complexes." Academic Press, London, 1969; W. B. Person and R. S. Mulliken, "Molecular Complexes: A Lecture and Reprint Volume." Wiley, New York, 1969.
41. J. L. Franclin, J. G. Pillard, H. M. Rosenstock, *et al.*, "Ionization Potentials, Apparent Potentials, and Heats of Formation of Gaseous Positive Ions," U. S. Department of Commerce, Washington, D. C., 1969.
42. D. W. Turner, C. Baker, A. D. Baker, and C. R. Brundle, "Molecular Photoelectron Spectroscopy." Wiley (Interscience), London, 1970.
43. T. V. Pyatkina, V. F. Traven, and E. D. Babich, *Izv. Akad. Nauk SSSR, Ser. Khim.*, 1998 (1975).
44. V. F. Traven, M. Yu. Eismont, V. V. Redchenko, and B. I. Stepanov, *Zh. Obshch. Khim.* **50,** 2007 (1980).
45. V. F. Traven, R. West, V. F. Donyagina, and B. I. Stepanov, *Zh. Obshch. Khim.* **45,** 824 (1975).
46. H. Sakurai and M. Kira, *J. Am. Chem. Soc.* **97,** 4879 (1975).
47. L. E. Orgel, *J. Chem. Phys.* **23,** 1352 (1955).
48. R. S. Mulliken, *J. Chim. Phys. Phys. Chim. Biol.* **51,** 341 (1954).
49. D. D. Holder and C. C. Thompson, *J. Chem. Soc., Chem. Commun.*, 277 (1972).
50. P. R. Hammond and L. A. Burkhard, *Chem. Commun.* 986 (1968).
51. H. A. Benesi and I. H. Hildebrand, *J. Am. Chem. Soc.* **71,** 2703 (1949).
52. S. Carter, J. N. Murrel, and E. J. Rosch, *J. Chem. Soc.*, 2048 (1965).
53. A. R. Bassindale, C. Eaborn, D. R. M. Walton, and D. J. Young, *J. Organomet. Chem.* **20,** 49 (1969).
54. V. A. Kuznetsov, A. N. Egorochkin, V. A. Savin, E. Ya. Lukevits, and N. P. Erchak, *Dokl. Akad. Nauk SSSR* **221,** 107 (1975).
55. E. Lukevics, A. N. Egorochkin, V. A. Kuznetsov, N. P. Erchak, and O.A. Pudova, in *Vses. Nauchn. Knof. Khim. Tekhnol. Furanovykh Soedin.* (*Tezisy Dokl.*) **3rd,** 69 (1978).
56. A. N. Egorochkin, V. A. Kuznetsov. M. A. Lopatin, N. P. Erchak, and E. Ya. Lukevits, *Dokl. Akad. Nauk SSSR* **258,** 391 (1981).
57. A. N. Egorochkin, M. A. Lopatin, G. A. Razuvaev, N. P. Erchak, L. M. Ignatovich, and E. Ya. Lukevits, *Dokl. Akad. Nauk SSSR* **298,** 895 (1988).
58. V. A. Kuznetsov, A. N. Egorochkin, E. A. Chernyshov, V. I. Savushkina, and V. A. Anisimova, *Tezisy Dokl. Nauchn. Sess. Khim. Tekhnol. Org. Soedin. Sery Sernistykh Neftei, 13th*, 276 (1974).
59. M. A. Lopatin, V. A. Kuznetsov, A. N. Egorochkin, O. A. Pudova, N. P. Erchak, and E. Ya. Lukevits, *Dokl. Akad. Nauk SSSR Ser. Khim.* **246,** 379 (1979).
60. A. N. Egorochkin, M. A. Lopatin, S. E. Skobeleva, N. P. Erchak, L. N. Khokhlov, and E. Ya. Lukevits, *Latv. PSR, Zinat. Akad Vestis, Khim. Ser.*, 220 (1988).
61. G. A. Shvekhgeymer, N. I. Kunavin, and G. V. Gorislavskaya, *Metalloorg. Khim.* **1,** 1334 (1988).
62. N. I. Kunavin, E. A. Baryshnikova, and G. A. Shvekhgeymer, *Zh. Nauchn. Prikl. Fotogr. Kinematogr.*, 183 (1987).
63. G. Briegleb, *Angew. Chem.* **74,** 324 (1964).
64. Z. Du, X. Wen, and J. Lin, Shandong Daxue Zuebao, Ziran Kexueban **22,** 115 (1987).

65. V. O. Reikhsfeld, A. N. Egorochkin, and V. A. Kuznetsov, *Zh. Obshch. Khim.* **50,** 1095 (1980).
66. M. A. Lopatin, A. N. Egorochkin, and V. A. Kuznetsov, *Zh. Obshch. Khim.* **51,** 1086 (1981).
67. A. N. Egorochkin, M. A. Lopatin, S. E. Skobeleva, V. I. Zhun, and V. D. Sheludyakov, *Metalloorg. Khim.* **1,** 350 (1988).
68. J. E. Frey, R. D. Cole, E. C. Kitchen, L. M. Suprenant, and M. S. Sylwestrzak, *J. Am. Chem. Soc.* **107,** 748 (1985).
69. V. F. Traven, M. I. German, and B. I. Stepanov, *Zh. Obshch. Khim.* **45,** 707 (1975).
70. V. F. Traven, M. I. German, and M. Yu. Eismont, *Zh. Obshch. Khim.* **48,** 2232 (1978).
71. A. N. Egorochkin, S. Ya. Khoroshev, N. S. Vyazankin, and E. N. Gladishev, *Izv. Akad. Nauk SSSR, Ser. Khim.*, 1863 (1969).
72. M. J. S. Dewar, "Hyperconjugation." Roland, New York, 1962.
73. H. Bock, G. Wagner, and J. Kroner, *Chem. Ber.* **105,** 3850 (1972).
74. A. Herman, B. Dreczewski, and W. Wojnowski, *J. Organomet. Chem.* **339,** 41 (1988).
75. G. Wagner and H. Bock, *Chem. Ber.* **107,** 68 (1974).
76. H. Sakurai, M. Kira, and T. Uchida, *J. Am. Chem. Soc.* **95,** 6826 (1973).
77. K. Ito and T. Ibaraki, *Bull. Chem. Soc. Jpn.* **61,** 2853 (1988).
78. J. A. A. Ketelaar, C. Van de Stolpe, A. Goudsmit, and W. Dzoubas, *Recl. Trav. Chim. Pays-Bas.* **71,** 1104 (1952).
79. L. M. Julien, W. E. Bennett, and W. B. Person, *J. Am. Chem. Soc.* **91,** 6915 (1969).
80. R. P. Taylor, *J. Am. Chem. Soc.* **90,** 4778 (1968).
81. N. I. Golovanova, N. I. Shergina, N. F. Chernov, and M. G. Voronkov, *Izv. Akad. Nauk SSSR, Ser. Khim.*, 414 (1977).
82. H. Bock and G. Wagner, *Angew. Chem.* **84,** 119 (1972).
83. N. I. Golovanova, N. V. Strashnikova, N. I. Shergina, E. I. Dubinskaya, N. F. Chernov, M. S. Sorokin, V. M. Dyakov, and M. G. Voronkov, *Dokl. Akad. Nauk SSSR* **234,** 625 (1977).
84. N. I. Golovanova, N. I. Shergina, N. F. Chenov, and M. G. Voronkov, *Izv. Akad. Nauk SSSR, Ser. Khim.* 414 (1977).
85. N. I. Golovanova, N. I. Shergina, N. F. Chernov, and M. G. Voronkov, *Zh. Obshch. Khim.* **47,** 1346 (1977).
86. V. B. Knyazhevskaya, V. F. Traven, and B. I. Stepanov, *Zh. Obshch. Khim.* **50,** 606 (1980).
87. H. Bock, P. Mollere, G. Becker, and G. Fritz, *J. Organomet. Chem.* **61,** 113 (1973).
88. V. F. Traven, V. B. Knyazhevskaya, M. Yu. Eismont, and B. I. Stepanov, *Zh. Obshch. Khim.* **51,** 92 (1981).
89. A. Ricci, D. Pietropaolo, and G. Distefano, *J. Chem. Soc., Perkin Trans. 2*, 689 (1977).
90. M. G. Voronkov, V. V. Belyaeva, E. I. Bobrovskaya, and V. O. Reikhsfel'd, *Zh. Obshch. Khim.* **56,** 2095 (1986).
91. H. Bock and W. Esslin, *Angew. Chem., Int. Ed. Engl.* **10,** 404 (1971).
92. Yu. S. Finogenov, V. A. Kuznetsov, A. N. Egorochkin, W. Schmidt, and B. T. Wilkins, *Zh. Obshch. Khim.* **46,** 2517 (1976).
93. V. F. Traven and R. West, *J. Am. Chem. Soc.* **95,** 6824 (1973).
94. G. Briegleb and J. Czekalla, *J. Phys. Chem.* (*Frankfurt am Main*) **24,** 37 (1960).
95. V. F. Traven, V. F. Donyagina, I. G. Makarov, S. P. Kolesnikov, V. M. Kazakova, and B. I. Stepanov, *Izv. Akad. Nauk SSSR, Ser. Khim.*, 1042 (1977).
96. H. Kuroda, M. Kobayushi, and M. Kinoshita, *J. Chem. Phys.* **36,** 457 (1962).
97. G. Ramsey, *J. Organomet. Chem.* **67,** 67 (1974).
98. M. Ishikawa and M. Kumada, *J. Organomet. Chem.*, **42,** 333 (1972).

99. H. Sakurai, Y. Kobayashi, and Y. Nakadira, *J. Am. Chem. Soc.* **96,** 2636 (1974).
100. R. West and A. Indriksons, *J. Am. Chem. Soc.* **94,** 6110 (1972).
101. R. R. McLean and D. W. A. Sharp, *J. Chem. Soc., Dalton Trans.*, 676 (1972).
102. P. R. Hammond, *J. Phys. Chem.* **74,** 647 (1970).
103. V. F. Traven, V. N. Karelskii, V. F. Donyagina, E. D. Babich, B. I. Stepanov, V. M. Vdovin, and N. S. Nametkin, *Dokl. Akad. Nauk SSSR* **224,** 837 (1975).
104. V. F. Traven, V. I. Karelskii, V. F. Donyagina, E. D. Babich, V. M. Volpin, and B. I. Stepanov, *Izv. Akad. Nauk SSSR, Ser. Khim.*, 1681 (1975).
105. H. Sakurai and Y. Komiyama, *J. Am. Chem. Soc.* **96,** 6192 (1974).
106. Yu. G. Bundel, S. I. Bobrovskii, I. D. Novikova, and O. A. Reutov, *Izv. Akad. Nauk SSSR, Ser. Khim.*, 925 (1979).
107. Yu. G. Bundel, S. I. Bobrovskii, V. V. Smirnov, I. A. Novikova, G. B. Sergeev, and O. A. Reutov, *Izv. Akad. Nauk SSSR, Ser. Khim.*, 2129 (1979).
108. C. S. Cundy, M. F. Lappert, J. B. Pedley, W. Schmidt, and B. T. Wilkins, *J. Organomet. Chem.* **51,** 99 (1973).
109. N. S. Nametkin and V. M. Vdovin, *Izv. Akad. Nauk SSSR, Ser. Khim.* 1153 (1964).
110. V. F. Donyagina, V. F. Traven, E. D. Babich, and B. I. Stepanov, *Izv. Akad. Nauk SSSR, Ser. Khim.*, 1038 (1977).
111. R. W. Taft, *in* "Steric Effects in Organic Chemistry," (M. S. Newman, ed.), Chap 13. Wiley, New York, 1955.
112. B. Steiner, C. F. Giese, and M. G. Inghram, *J. Chem. Phys.* **34,** 189 (1961).
113. D. Betteridge and M. Thompson, *J. Mol. Struct.* **21,** 341 (1974).
114. V. F. Traven and R. West, *Zh. Obshch. Khim.* **44,** 1837 (1974).
115. H. C. Gardner and J. K. Kochi, *J. Am. Chem. Soc.* **97,** 5026 (1975); H. C. Gardner and J. K. Kochi, *J. Am. Chem. Soc.* **98,** 2460 (1976).
116. H. C. Gardner and J. K. Kochi, *J. Am. Chem. Soc.* **96,** 1982 (1974); H. C. Gardner and J. K. Kochi, *J. Am. Chem. Soc.* **97,** 1855 (1975).
117. V. F. Traven, V. F. Donyagina, N. S. Fedotov, G. V. Evert, and B. I. Stepanov, *Zh. Obshch. Khim.* **46,** 2761 (1976).
118. Yu. G. Bundel, G. B. Sergeev, V. K. Piotrovskii, V. V. Smirnov, S. I. Bobrovskii, V. I. Rosenberg, and O. A. Reutov, *Izv. Akad. Nauk SSSR, Ser. Khim.*, 1688 (1977).
119. S. H. Hastings, J. L. Franklin, J. C. Schiller, and F. A. Matsent, *J. Am. Chem. Soc.* **75,** 2900 (1953).
120. L. A. Burkardt, P. R. Hammond, R. H. Knipe, and R. R. Lake, *J. Chem. Soc. A*, 3789 (1971).
121. J. M. Dumas, G. Geron, H. Peurichard, and M. Gomel, *Bull. Chem. Soc. Fr.*, 720 (1976).
122. A. Ya. Deich and L. M. Chemokhud, *Latv. PSR, Zinat. Akad. Vestis, Khim. Ser.*, 560 (1974).
123. L. A. Bogdanova, A. Ya. Deich, L. M. Chemokhud, and E. G. Polle, *Zh. Obshch. Khim.* **46,** 655 (1976).
124. J. C. Giordan and J. H. Moore, *J. Am. Chem. Soc.* **105,** 6541 (1983).
125. V. F. Traven, V. V. Redchenko, and B. I. Stepanov, *Zh. Obshch. Khim.* **52,** 358 (1982).
126. E. Carberry and R. West, *J. Am. Chem. Soc.* **91,** 5440 (1969).
127. R. West, *Pure Appl. Chem.* **54,** 1041 (1982).
128. V. F. Donyagina, V. F. Traven, V. G. Zaikin, and B. I. Stepanov, *Izv. Akad. Nauk SSSR, Ser. Khim.*, 2117 (1976).

ADVANCES IN ORGANOMETALLIC CHEMISTRY, VOL. 34

Structure, Color, and Chemistry of Pentaaryl Bismuth Compounds

KONRAD SEPPELT

Institut für Anorganische und Analytische Chemie
Freie Universität Berlin
D-1000 Berlin 33, Germany

I
INTRODUCTION

The first attempt to synthesize pentaaryl bismuth compounds, though unsuccessful, probably dates back to Challenger in 1914. He wrote that "After each addition of the Grignard reagent ($C_6H_5MgBr + (C_6H_5)_3$-$BiBr_2$) a transient but intense purple coloration was produced, and the ether boiled" (*1*). This finding was reproducible, and Gilman and Yablunky used this coloration by $(C_6H_5)_3BiX_2$ as a test for organometallic reagents (*2*). They suggested the presence of $Bi(C_6H_5)_5$, and they also found that $(C_6H_5)_3AsX_2$ and $(C_6H_5)_5SbX_2$ did not produce the color effect.

From the preparative viewpoint it was Wittig and co-workers who first established the existence of $P(C_6H_5)_5$ (*3*), $As(C_6H_5)_5$ (*4*), $Sb(C_6H_5)_5$ (*4*), and $Bi(C_6H_5)_5$ (*5*). The P, As, and Sb compounds are colorless and well-crystallized materials of increasing stability in the given sequence. $Bi(C_6H_5)_5$ is quite unstable both thermally and toward X-rays, so that the structure remained unknown, although the corresponding crystal structures of the lighter homologues have long been known. Wittig did not remark on the color phenomenon. Hellwinkel and co-workers (*6*) later synthesized $Bi(C_6H_5)_3(4\text{-}ClC_6H_4)_2$ and $Bi(C_6H_5)_3(C_{12}H_8)$ ($C_{12}H_8$ = 2,2′-diphenylene). These materials were also intensely colored. Among main group elements, colors are least expected in the highest oxidation states, simply because in these states there are the fewest electrons that can be excited. At this point it was not even proven that the color was not due to an impurity, perhaps of radical nature.

II

PREPARATION OF PENTAARYL BISMUTH COMPOUNDS

Wittigs method of preparing $BiAr_5$ materials in three steps is still useful:

$$BiCl_3 \xrightarrow{ArMgX} BiAr_3 \xrightarrow{Cl_2} Ar_3BiCl_2 \xrightarrow{Ar'Li} BiAr_3Ar'_2 \quad (1)$$

where Ar and Ar′ are phenyl or substituted phenyl. There are normally no difficulties in obtaining the $BiAr_3$ materials. In the second reaction step, chlorination to Ar_3BiCl_2 is sometimes not possible if many electron-withdrawing groups are present on the phenyl rings. Alternatively, fluorination with XeF_2 may give the corresponding Ar_3BiF_2 compounds (*7,8*). For the third step, only lithium reagents have been used. Successful substitution is always obvious from the color of the solution. The pentaaryls are quite often difficult to crystallize and may turn out as viscous liquids. This is possibly due to their quite irregular molecular geometry.

With this general method homoleptic compounds as well as compounds with the composition $BiAr_3Ar_2'$ are readily prepared. Substances with the composition $BiAr_4Ar'$ could probably be prepared by a variation of the method, but such compounds have not yet been described.

$$BiAr_5 \xrightarrow{HCl} BiAr_4{}^+Cl^- \xrightarrow{Ar'Li} BiAr_4Ar' \quad (2)$$

Ligand scrambling in $BiAr_3Ar_2'$ compounds may occur during the preparation, because $Bi(C_6H_5)_5$ is known to take up one more phenyl ligand reversibly to form hexaarylbismuthates (*9,10*).

$$BiAr_5 + Ar^- \rightarrow BiAr_6{}^- \quad (3)$$

III

STRUCTURES OF PENTAARYL BISMUTH COMPOUNDS

Because all known bismuth pentaaryls are pentacoordinated species, the problem of a trigonal bipyramidal versus a square pyrimadal structure needs to be redescribed briefly. The ideal trigonal bipyramid can be described by one parameter, the ratio of axial and equatorial bond lengths. Two parameters are, however, necessary to describe an ideal square pyra-

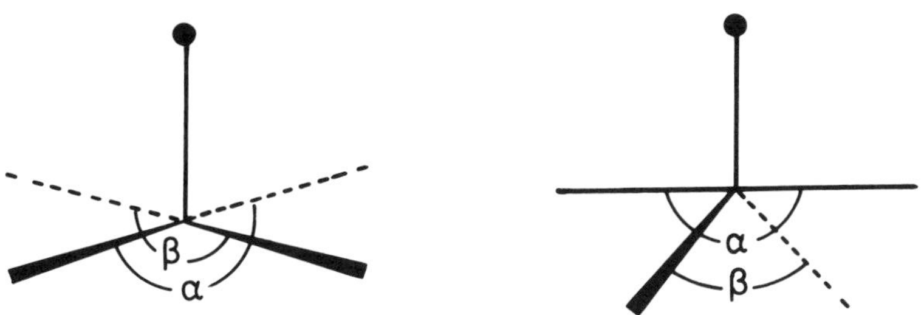

FIG. 1. Definition of trigonal bipyramidal and square pyramidal structures by $\Delta = \alpha - \beta$, where α and β are the two largest angles within the system. This definition holds so long as C_{2v} symmetry is retained; $\Delta = 60°$, trigonal bipyramidal; $\Delta = 0°$, square pyramidal. (Reproduced with permission of Verlag Chemie, Weinheim, New York.)

mid: the ratio of apical and basal bond lengths and one angle (for example, the angle between two opposite basal ligands; see Fig. 1). This angle can be 180° if the central atom is situated exactly in the basal plane, or larger than 180° if the central atom is below the basal plane as in IF_5 (umbrella structure). From the viewpoint of electrostatic interactions and the VSEPR model, this angle should be approximately 130°, so that the central atom lies within the coordination polyhedron.

Trigonal bipyramidal and square pyramidal structures are interconvertible, for example, via the Berry pseudorotation. The reaction coordinate of the pseudorotation may be defined as the difference $\Delta = \alpha - \beta$, where α and β are the two largest angles within the five-coordinate system (*7*); $\Delta = 180° - 120° = 60°$ identifies an ideal trigonal bipyramid; $\Delta = 0°$ identifies an ideal square pyramid. With Δ the deviation from the ideal symmetry can be measured. This approximation will hold so long as C_{2v} symmetry is retained, which qualitatively is the case in all described structures with the single exception of $Bi(C_6H_5)_3(C_{12}H_8)$, in which deviation from C_{2v} symmetry results from the constraints of the chelating ring system.

Table I (*11,12*) shows the values of the known bismuth pentaaryls. The crystal structures of $Bi(C_6H_5)_5$ and $Bi(4\text{-}CH_3\text{–}C_6H_4)_5$ are shown in Figs. 2 and 3. It is obvious that the predominant structure is the square pyramid, most values being closer to 0° than to 60°. This is a surprise. In general, the trigonal bipyramid is considered the principal geometry of a five-coordinated species, certainly based on statistics on the hundreds if not thousands of known cases. Trigonal bipyramidal structures are found for $P(C_6H_5)_5$ ($\Delta = 55.2$°C) (*13*) and $As(C_6H_5)_5$ ($\Delta = 58$°C) (*14*). The first exception here is $Sb(C_6H_5)_5$, $\Delta = 15$°C (*15*), with a structure close to a square pyramid. This finding has caused some excitement, particularly because the structure is the only square pyramidal one known for antimony. Even

TABLE I

List of Known Bismuth Pentaaryls

Compound	Color[a]	λ_{max} (nm)[b]	Δ	Ref.
$Bi(C_6H_5)_5$	Violet	532	13.2°	*5,10*
$Bi(4\text{-}CH_3\text{-}C_6H_4)_5$	Violet	521	26°	*8*
$Bi(C_6H_5)_3(2\text{-}F\text{-}C_6H_4)_2$	Violet	480	19.2°	*12*
$Bi(C_6H_5)_3(4\text{-}Cl\text{-}C_6H_4)_2$	Violet	—	Not known	*6*
$Bi(C_6H_5)_3(2,6\text{-}F_2\text{-}C_6H_3)_2$	Red	—	11.5°, 17.0°	*7*
$Bi(C_6H_5)_3(C_6F_5)_2$	Orange	—	Not known	*11*
$Bi(C_6H_5)_3(C_{12}H_8)$	Red	—	27.2°, 21.7°	*6,8*
$Bi(4\text{-}CH_3\text{-}C_6H_4)_3(2\text{-}F\text{-}C_6H_4)_2$	Orange	465	45.4°	*12*
$Bi(4\text{-}CH_3\text{-}C_6H_4)_3(2,6\text{-}F_2\text{-}C_6F_4)_2$	Red	—	17.5°	*8*
$Bi(4\text{-}CH_3\text{-}C_6H_4)_3(C_6F_5)_2$	Yellow	—	16.0°	*7*
$Bi(4\text{-}CH_3\text{-}C_6H_4)_3(4\text{-}CF_3\text{-}C_6H_4)_2$	Violet	—	Not known	*7*
$Bi(4\text{-}F\text{-}C_6H_4)_3(C_6F_5)_2$	Yellow	—	5.1°	*7*

[a] Color of the solid.
[b] In solution.

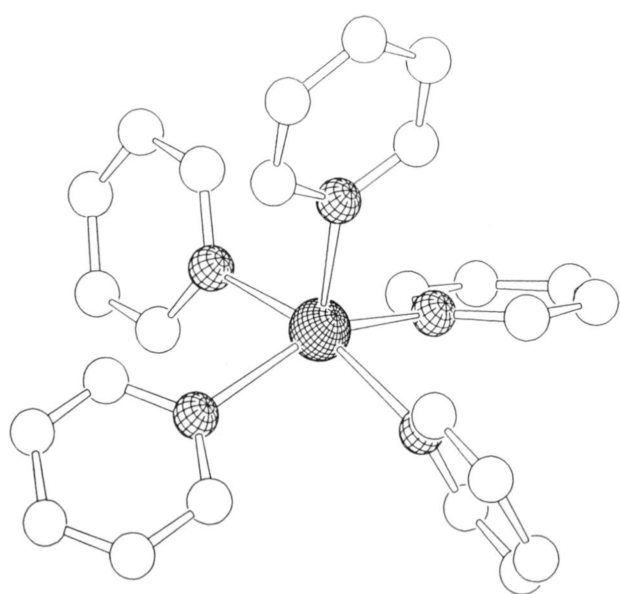

Fig. 2. Crystal structure of $Bi(C_6H_5)_5$, an almost ideal square pyramid.

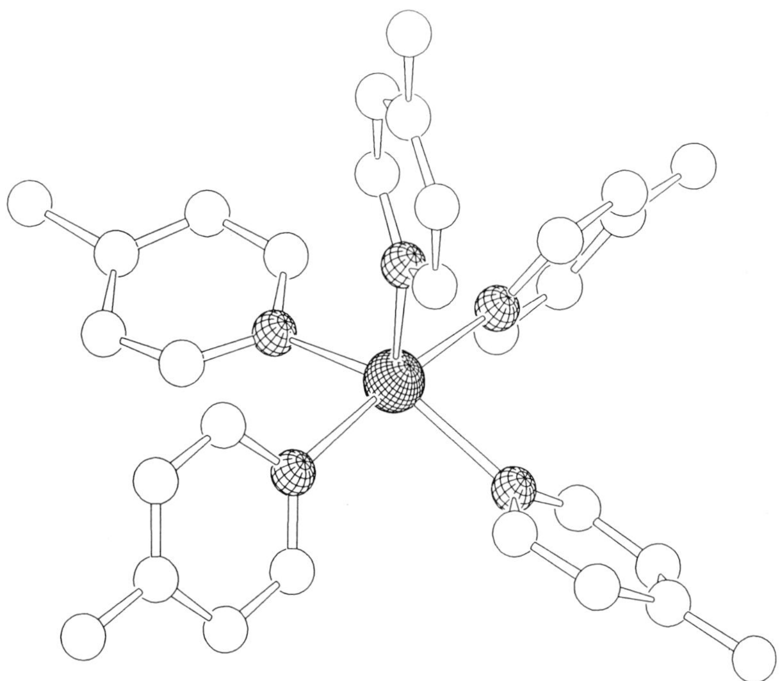

FIG. 3. Crystal structure of $Bi(4\text{-}CH\text{–}C_6H_4)_5$, a structure about halfway between square pyramid and trigonal bipyramid.

the solvate $Sb(C_6H_5)_5 \cdot \frac{1}{2}C_6H_{12}$, $\Delta = 58°$ (*16*), and $Sb(4\text{-}CH_3\text{–}C_6H_4)_5$, $\Delta = 48°$ (*17*), are trigonal bipyramidal.

A closer look at the many $BiAr_5$ structures reveals that phenyl groups carrying electron-withdrawing groups will always occupy basal positions, and if there are only two such rings, they will have a position opposite to each other. The angle between two such phenyl rings is always the largest one in the system. If such a system is considered a very distorted trigonal bipyramid, the two most electron-withdrawing phenyl rings are occupying the axial positions.

Besides the Δ values, bond length systematics also favor the description as square pyramidal: the apical Bi–C bond is >10 pm shorter than the four basal ones. Phenyl groups with fluorine substituents have longer Bi–C bonds, 242–244 pm in the case of C_6F_5 groups. In general, electron withdrawal should tighten a bond; here, however the bond is already electron deficient (Bi^{5+}!), so that the bonds are weakened further.

Electron-donating groups should then be expected to shorten the

Bi–C bonds. The effect of *p*-methyl substitution is visible and in the expected direction, but sometimes hardly larger than the accuracy of the measurement.

The most surprising effect is the tendency of *p*-methyl substitution to drive the structures over to the trigonal bipyramidal side. There are now three pairs of bismuth compounds to compare, and together with the longer known pair of $Sb(C_6H_5)_5$ and $Sb(4\text{-}CH_5\text{–}C_6H_4)_5$ this seems to be a clear but as yet unexplainable effect.

The structures in solution are certainly dynamic. NMR measurements have not revealed any insight about these processes, either because they have not been tried or because at very low temperatures concentrations are very small. As will be shown in the Section IV, the color can give information about structures in solution in special cases.

Sometimes $BiAr_5$ compounds crystallize with solvent molecules. $Bi(C_6H_5)_5$ is the most prominent case. From concentrated solutions, neat $Bi(C_6H_5)_5$ crystallizes about 0°C in large, lumpy, violet crystals. From dilute solutions at lower temperatures, blue needles of $Bi(C_6H_5)_5 \cdot$ tetrahydrofuran are formed. This structure could not be refined because of a disorder of the tetrahydrofuran molecule with two phenyl rings, resulting in an approximate octahedral geometry around the Bi atom. Angles between opposite basal C atoms are close to 180° (*18*).

IV

COLOR OF PENTAARYL BISMUTH COMPOUNDS

Visible color for a main group compound in its highest oxidation state is a surprising phenomenon. This color is due to a broad absorption, for $Bi(C_6H_5)_5$ at $\lambda_{max} = 532$ nm, that is also weak, $\log \varepsilon = 2.4$. The color varies from yellow to blue. Longer wavelength absorption resulting in a green color is not observed. In the case of orange and yellow colors, the UV spectrum shows only a shoulder on the very strong absorption on the phenyl rings. The color depends on the substitution and the structure. From the previous discussion one must be aware that the substitution also guides the structure, so there seems to be a complicated interdependence. However, two simple rules can be derived:

1. Electron-withdrawing groups do not change the structure very much away from square pyramidal, but nevertheless result in a hypsochromic shift. This finding can be explained in terms of a charge-transfer absorption from the ligand to the bismuth atom. This charge transfer is energetically not favored with electron-withdrawing groups.

2. Electron-donating groups, actually so far only the methyl group, direct the geometry toward a trigonal bipyramid for unknown reasons, which also results in a hypsochromic shift.

That the absorption is, in fact, fixed to the square pyramidal geometry is proved by two observations:

1. On well-formed crystals the absorption is dichroic. The color is only observed if the electric vector of the light lies within the basal plane of the square pyramid, otherwise the crystal is—if pure—colorless! (See original photos of oriented crystals in Ref. 11.)
2. The two compounds, $Bi(C_6H_5)_3(2\text{-}F\text{-}C_6H_4)_2$ and $Bi(4\text{-}CH_3\text{-}C_6H_4)_3\text{-}(2\text{-}F\text{-}C_6H_4)_2$, exhibit the same mixed yellow-violet color in solution (see Figs. 4–6). On crystallization, the first solid material is violet and represents a square pyramid; the second is orange and is a trigonal bipyramid. (See original photos of crystals in Ref. 12.) In solution, however, both have the same UV spectrum (see Fig. 4). This shows the Berry pseudorotation at work.

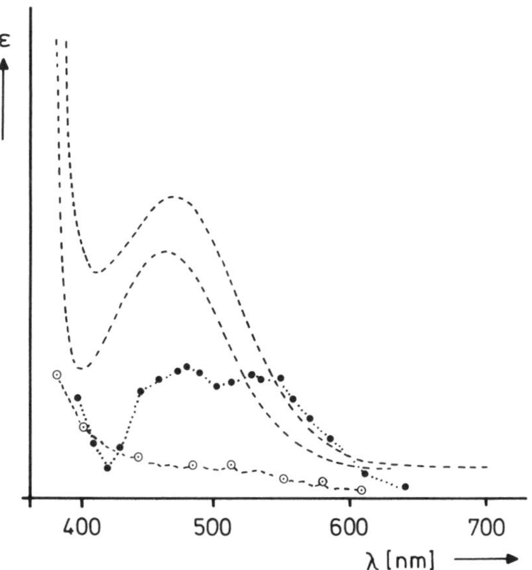

FIG. 4. UV spectra of $Bi(C_6H_5)_3(2\text{-}F\text{-}C_6H_4)_2$ (upper dashed trace) and $Bi(4\text{-}CH_3\text{-}C_6H_4)_3(2\text{-}F\text{-}C_6H_4)_2$ (lower dashed trace) in solution and their respective solids (●···●) (◉···◉). Due to the Berry pseudorotation the solution UV spectra are alike. In the solid state the first material is violet (square pyramidal) and the second is orange (trigonal bipyramidal). (Reproduced with permission of Verlag Chemie, Weinheim, New York.)

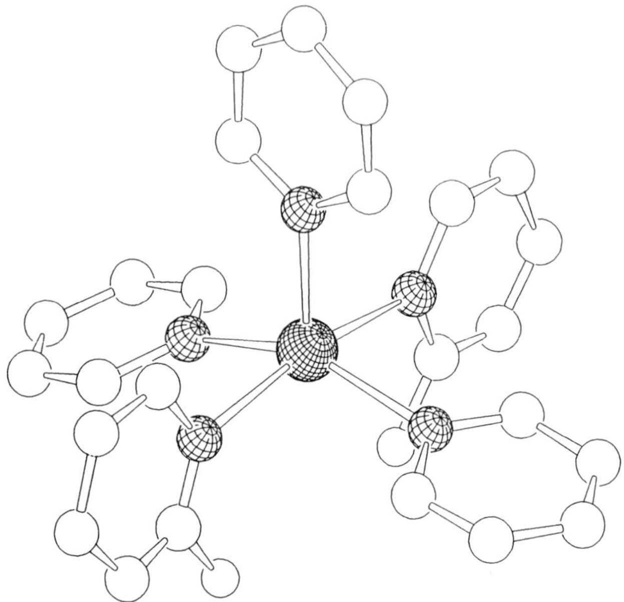

FIG. 5. Crystal structure of violet $Bi(C_6H_5)_3(2\text{-}F\text{-}C_6H_4)_2$, an almost ideal square pyramid. (Reproduced with permission of Verlag Chemie, Weinheim, New York.)

If the empty sixth coordination site of the square pyramid is filled with yet another C_6H_5 ligand as in $Bi(C_6H_5)_6^-$, then the absorption also disappears (*9,10*).

We have simulated these phenomena with simple *ab initio* calculation on BiH_5 (both trigonal bipyramidal and square pyramidal) and on octahedral BiH_6^- (*12*). In order to do this it was necessary to include relativistic effects into the calculations, because Bi is a very heavy atom. The results can be taken from Fig. 7. It is obvious that the highest occupied and lowest unoccupied molecular orbital (HOMO–LUMO) separation for C_{4v} symmetry is only about half as large as that for with D_{3h} or O_h symmetries. Furthermore, in C_{4v} symmetry the HOMO is mostly a ligand-centered orbital, whereas the LUMO has mostly Bi 6*s*, 6*p* character. The description of a charge transfer from ligand to Bi atom seems to be justified and with this the dependence on substitution. It can be shown that it is the relativistic correction in the calculation that lowers particularly the Bi *s* and *p* orbitals (*11*). Here we see, possibly for the first time, the influence of the relativistic correction on an unoccupied orbital: nonrelativistic $Bi(C_6H_5)_5$—if that were possible—would not be violet.

FIG. 6. Crystal structure of orange $Bi(4\text{-}CH_3\text{–}C_6H_4)_3(2\text{-}F\text{–}C_6H_4)_2$, an almost ideal trigonal bipyramid. (Reproduced with permission of Verlag Chemie, Weinheim, New York.)

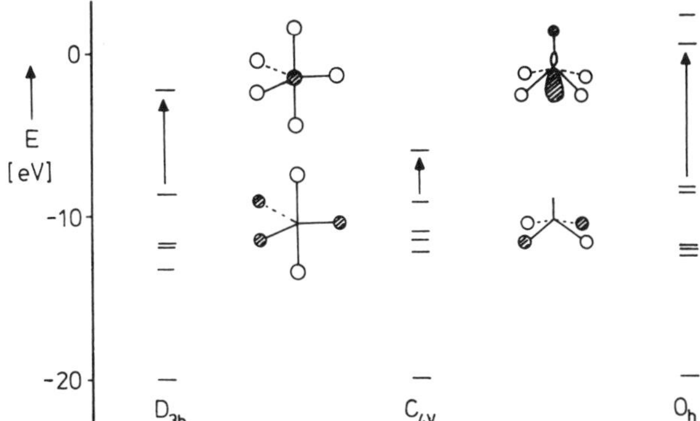

FIG. 7. MO diagram for trigonal bipyramidal BiH_5, square pyramidal BiH_5, and octahedral BiH_6^-, from left to right. The small HOMO–LUMO gap in square pyramidal BiH_5 is a consequence of the lowering of the LUMO by relativistic effects on Bi 6*s*, 6*p* orbitals. Graphs indicate qualitatively the nature of the HOMO and LUMO. (Reproduced with permission of Verlag Chemie, Weinheim, New York.)

The situation in D_{3h} symmetry is only quantitatively different insofar as the LUMO has much less Bi 6*s*, 6*p* character, and in O_h symmetry the LUMO of the C_{4v} symmetry becomes the HOMO. The HOMO–LUMO gap is only half as large in C_{4v} as in D_{3h} symmetry. This agrees well with the difference in energy of the absorbed light. The calculation does not reveal the preference for $BiAr_5$ compounds for the square pyramidal structure. This would indeed need a careful theoretical comparison between (P, As, Sb, Bi)Ar_5 materials. But one can again argue in a similar manner: relativistic effects and/or the lanthanoid contraction lower particularly the Bi 6*s* orbital, so that for bonding mostly *p* orbitals are used, which directs the geometry more toward the rectangular geometries of a square pyramid.

IV

CONCLUSION

The total behavior of pentaaryl bismuth compounds can be summed up as follows: the Bi–C bond is quite nonpolar and in the excited state is polarized toward the bismuth atom. In such cases deviations from the VSEPR geometries can be expected, because the electrostatic repulsion loses the dominance of the ground state geometry. It is also conceivable that reactions of $BiAr_5$ are different than those of pentaaryls of the other five main group elements: the Bi–C bond is cleaved according to $(C_6H_5)_4Bi^-\ C_6H_5^+$. That is why the decomposition of $Bi(C_6H_5)_5$ gives, besides $Bi(C_6H_5)_3$, benzene, polymers and only a little diphenyl (*5*), whereas $Sb(C_6H_5)_5$ decomposes cleanly to $Sb(C_6H_5)_3$ plus diphenyl (*4,19*). Barton has investigated comprehensively the reactions of five valent bismuth organic compounds in organic synthesis. Details of this work have been summarized (*20*). The driving force of all these reactions is the reduction of Bi(V) to Bi(III), thus $(C_6H_5)_3BiCl_2$, $(C_6H_5)_3BiCO_3$, $Bi(C_6H_5)_5$, $(C_6H_5)_4Bi–O–CO–CF_3$, and $(C_6H_5)_3BiCl–O–BiCl(C_6H_5)_3$ are oxidizers. Secondary alcohols are oxidized to ketones and 1,2-glycols are cleaved. These materials serve also as O-phenylating and C-phenylating agents. The latter reaction is particularly of importance (*21,22*). The overall reaction, if $Bi(C_6H_5)_5$ is used, is as follows:

$$\geqq C\text{—}H + BiAr_5 \rightarrow \geqq C\text{—}Ar + BiAr_3 + ArH \tag{4}$$

Particularly ketones, enols, and phenols are phenylated. Without going into the details of the mechanism (*20*), it is obvious that these bismuth compounds act as a rare case of electrophilic arylating agents, which again is a consequence of the special Bi–C bond system.

Finally, it may be said that the excited state of $BiAr_5$ resembles the coordination of three-valent bismuth (for example, in $BiCl_5^{2-}$), and the intramolecular redox process (charge transfer) resembles somewhat the electronic and vibrational exchanges in simple Bi oxide superconductors as in $KB_xPb_{1-x}O_3$, which can be thought of an "intermolecular" Bi^{III}–Bi^{V} redox process coupled with a square pyramidal/octahedral vibrational transition. It is certainly not accidental that a vast number of the modern high-T_c superconductors have the square pyramid as one of their structural principles.

An extension of this chemistry to other central atoms is presently not easy to see. As already described, the faint yellow color of square pyramidal $Sb(C_6H_5)_5$ may develop because of similar chemistry, although so far no dichroism has been detected. The next neighbor, Po, would be expected to have a very exciting chemistry if only it were not radioactive and did not have a short half-life.

References

1. F. Challenger, *J. Chem. Soc.*, 2210 (1914).
2. H. Gilman and H. L. Yablunky, *J. Am. Chem. Soc.* **63,** 839 (1941).
3. G. Wittig and M. Rieber, *Liebigs Ann. Chem.* **562,** 187 (1949).
4. G. Wittig and K. Clauss, *Liebigs Ann. Chem.* **577,** 26 (1952).
5. G. Wittig and K. Clauss, *Liebigs Ann. Chem.* **578,** 136 (1952).
6. D. Hellwinkel and M. Bach, *Liebigs Ann. Chem.* **720,** 198 (1969).
7. A. Schmuck and K. Seppelt, *Chem. Ber.* **122,** 803 (1989).
8. A. Schmuck, D. Leopold, S. Wallenhauer, and K. Seppelt, *Chem. Ber.* **123,** 761 (1990).
9. D. Hellwinkel and G. Kilthan, *Liebigs Ann. Chem.* **705,** 66 (1967).
10. The crystal structure of $Bi(C_6H_5)_6^-$ has been obtained, S. Wallenhauer and K. Seppelt, to be published.
11. A. Schmuck, J. Buschmann, J. Fuchs, and K. Seppelt, *Angew. Chem.* **99,** 1206 (1987); *Angew. Chem., Int. Ed. Engl.* **26,** 1180 (1987).
12. A. Schmuck, P. Pyykkö, and K. Seppelt, *Angew. Chem.* **102,** 211 (1990); *Angew. Chem., Int. Ed. Engl.* **29,** 213 (1990).
13. P. J. Wheatley and G. Wittig, *Proc. Chem. Soc. London*, 251 (1962).
14. P. J. Wheatley, *J. Chem. Soc.*, 2206 (1964).
15. P. J. Wheatley, *J. Chem. Soc.*, 3718 (1964); A. L. Beauchamp, M. J. Bennett, and F. A. Cotton, *J. Am. Chem. Soc.* **90,** 6675 (1968).
16. C. Brabant, B. A. Blanck, and L. Beauchamp, *J. Organomet. Chem.* **82,** 231 (1974).
17. C. Brabant, J. Hubert, and A. L. Beauchamp, *Can. J. Chem.* **51,** 2952 (1973).
18. A. Schmuck, S. Wallenhauer, and K. Seppelt, unpublished.
19. K.-W. Shun, W. E. McEwen, and A. P. Wolf, *J. Am. Chem. Soc.* **91,** 1283 (1969).
20. D. H. R. Barton and J. P. Finet, *Pure Appl. Chem.* **59,** 937 (1987).
21. D. H. R. Barton, N. Y. Bhatnager, J.-C. Blazejewski, B. Charpiot, J.-P. Finet, D. J. Lester, W. B. Motherwell, M. T. Barros Papoula, and S. P. Stanforth, *J. Chem. Soc., Perkin Trans.* **1,** 2657 (1985).
22. D. H. R. Barton, J.-C. Blazejewski, B. Charpiot, J.-P. Finet, W. B. Motherwell, M. T. Barros Papoula, and S. P. Stanforth, *J. Chem. Soc., Perkin Trans. 1*, 2667 (1985).

ADVANCES IN ORGANOMETALLIC CHEMISTRY, VOL. 34

Use of Water-Soluble Ligands in Homogeneous Catalysis

PHILIPPE KALCK and FANNY MONTEIL

Laboratoire de Chimie des Procédés
Ecole Nationale Supérieure de Chimie
31077 Toulouse Cédex, France

I

INTRODUCTION

Homogeneous catalysis provides mild and selective routes to a variety of valuable chemicals from basic organic precursors. An important advantage over heterogeneous systems is that selectivity is achieved by a modulation of the coordination sphere of the metal. However, a recurrent problem is the separation of the organic products from the active catalyst. Supported catalysts have long been shown to combine the advantages of homogeneous and heterogeneous systems. Their performances are, however, diminished by losses of the metal in solution, thereby preventing any application in commercially viable processes. A more attrative approach has been made using water-soluble ligands that are poorly soluble in organic media, allowing the catalysis to be carried out in a two-phase system. Catalyst recovery is thus easily achieved by decantation and separation of the two phases. In addition, catalysis in completely aqueous media is a good way to decrease the toxicity inherent to organic solvents and to modify the kinetics and the stereoselectivities of several reactions. Two reviews have appeared on the synthesis of water-soluble phosphines and their complexes, with potential applications in catalysis (*1,2*). The last one was published in 1987. To the best of our knowledge, no review has appeared more recently, though a short report in 1989 was partially devoted to this topic (*3*). Thus the present review deals with the preparation of water-soluble ligands and their use in the synthesis of water-soluble complexes as well as in catalysis.

II

PREPARATION OF THE LIGANDS AND THEIR TRANSITION METAL COMPLEXES

A. *Preparation of the ligands*

Various methods have been developed to prepare water-soluble ligands. Most of them involve phosphorus-containing ligands.

1. *Phosphine Ligands*

Because numerous complexes with the metal in a low oxidation state are stabilized by phosphorus-containing ligands, it comes as no surprise that much has been done to tailor these ligands with appropriate substituents, including carboxylic, amino, hydroxy, and sulfonate functions, which induce solubilization in water.

a. Sulfonated Phosphines. The first ligand of this type was prepared in 1958 by Chatt and his group (*4*) by addition of diluted oleum (20% SO_3–H_2SO_4) to triphenylphosphine. Under the reported conditions, the monosulfonated ligand, $Ph_2P(m\text{-}C_6H_4SO_3Na)$, L^1, was obtained after neutralization with sodium hydroxide.

Several years later, Kuntz showed that careful control of the reaction conditions afforded the trisubstituted ligand, $P(m\text{-}C_6H_4SO_3Na)_3$, L^3, although it was obtained with the mono- and disulfonated ligands as well as with significant amounts of the corresponding oxides (*5*). Because of the high performance of the highly water-soluble tris(sodium *m*-sulfonatophenyl)phosphine in catalysis (vide infra), it was found of interest to devise high-yield preparative routes. It was shown that the amounts of mono- and disubstitued phosphines can be minimized by increasing the SO_3 : triphenylphosphine molar ratio, provided an oleum with 33% SO_3 in H_2SO_4 is used. A decrease in the reaction temperature reduces the amount of oxides (*6*).

After sulfonation, hydrolysis of the reaction mixture is a key point because large amounts of $OP(m\text{-}C_6H_4SO_3H)_3$ and even $SP(m\text{-}C_6H_4SO_3H)_3$ can be produced. Hydrolysis in very acidic solutions (where the sulfuric acid normality is maintained at least at 12 *N*) reduced significantly the by-products (*7*). Separation of sulfuric acid from the $P(m\text{-}C_6H_4SO_3H)_3$ ligand by liquid–liquid extraction was shown to be a further improvement, giving rise after neutralization to a $P(m\text{-}C_6H_4SO_3Na)_3$ ligand of good quality for catalysis purposes (*8*). This preparation induces substitution in the meta position. However, the SO_3 group can be introduced in the para position by reacting $ClPPh_2$, for example, with $p\text{-}ClC_6H_4SO_3Na$ in the presence of sodium or potassium to prepare $Ph_2P(p\text{-}C_6H_4SO_3Na)$ (*9*).

Whereas current methods of preparing polydentate phosphines usually involve sequential introduction of individual phosphine units into a precursor [either by displacement of halides or tosylates by diphenylphosphide ion, or by direct addition of diphenylphosphine to activated olefins (*10*)], Whitesides and co-workers (*11,12*) developed a new method in which the diphosphine moiety **1** is prepared *in situ* as an intermediate according to Eq. (1).

$$Cl^-H_2N^+(CH_2CH_2Cl)_2 + 2\ HPPh_2 \xrightarrow[\text{2) HCl 2N}]{\text{1) KO-t-Bu, THF, reflux, 16 h}} Cl^-H_2N^+(CH_2CH_2PPh_2)_2 \quad \textbf{1} \quad \text{yield} \sim 90\% \qquad (1)$$

This compound was then acylated with acid chlorides such as trimellitic anhydride acid chloride **2**, tricarballytic anhydride acid chloride **3**, or ethyl oxalyl chloride **4** according to Scheme 1. Further reaction with sodium taurinate, i.e., $H_2NCH_2CH_2SO_3Na$, gave the derivatives **6**, **7**, and **8**, which are diphosphine ligands bearing sulfonate group on a side chain. It is also possible to condense **1** with a sulfobenzoic anhydride such as **5** to obtain **9**.

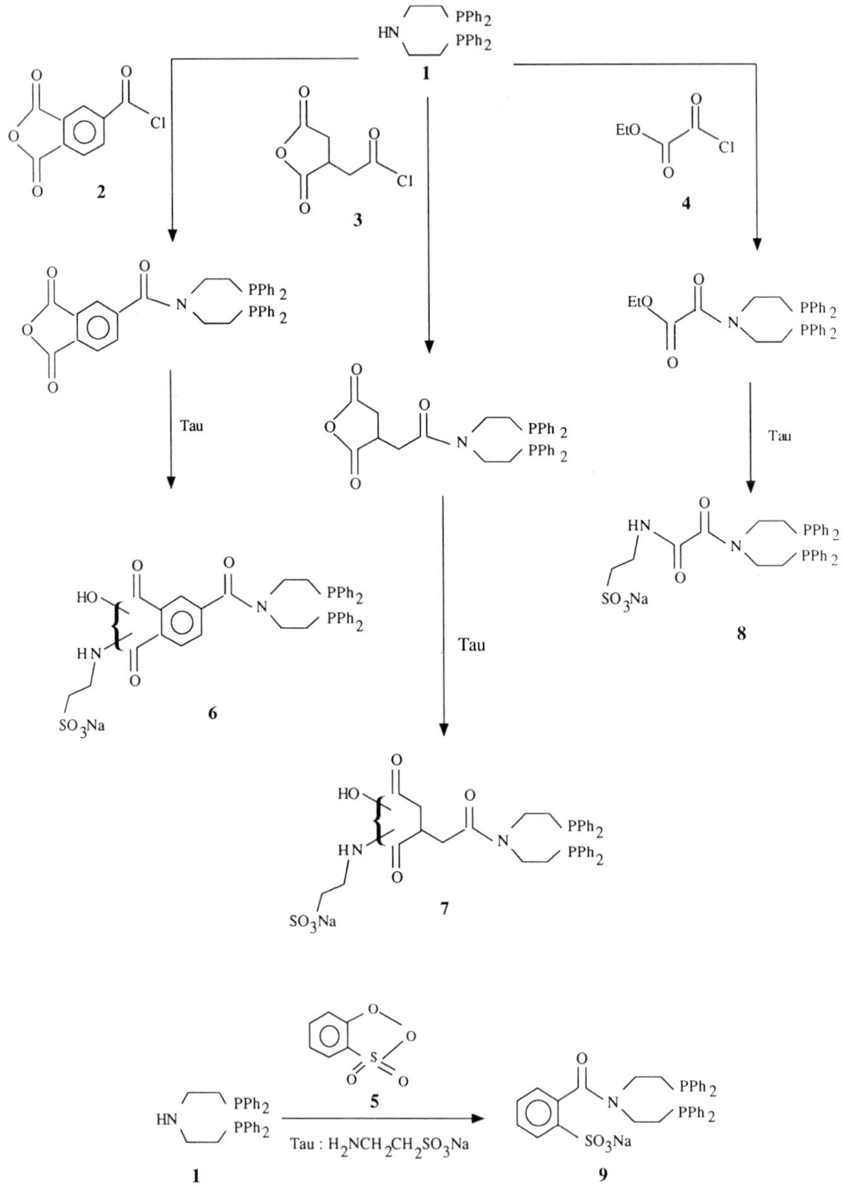

HN PPh2 PPh2
1
2
3
4
Tau
Tau
Tau
6
7
8
5
$Tau : H_2NCH_2CH_2SO_3Na$
1
9

SCHEME 1. Synthesis of sulfonated diphosphines via the Whitesides procedure (*11,12*).

The sulfoalkylation of tris(2-pyridyl)phosphine has been proposed as a new method to bind sulfonate groups to a phosphine. It takes place at mild temperatures with miscellaneous alkyl-1,2-sulfones, which react only with the nitrogen atoms according to Eq. (2) (*13*).

$$\text{P(2-C}_5\text{H}_4\text{N)}_3 + 3\ \text{CH}_3(\text{CH}_2)_n\text{CH}(-\text{O}-\text{SO}_2-)\text{CH}_2 \xrightarrow{0\text{-}20^\circ\text{C}} \text{P}[2\text{-C}_5\text{H}_4\text{N}^+\text{-CH}((\text{CH}_2)_n\text{CH}_3)\text{CH}_2\text{SO}_3^-]_3 \quad (2)$$

A number of interesting studies have provided insight into the properties of the sulfonated monophosphine ligands and have investigated their reactivity. The L[1] ligand is slightly soluble in water (80 g liter^{-1}) whereas L[3] reaches 1100 g liter^{-1} solubility (*14*), thus giving very concentrated aqueous solutions.

Larpent and Patin have been interested in determining the basicity of $P(m\text{-}C_6H_4SO_3Na)_3$ (*15*). By measuring the ^{31}P chemical shift of the oxide and using the Derencsényi's relationship (*16*) between phosphine basicity and the ^{31}P NMR chemical shift of the corresponding oxide, they found a pK_a value of 3.2 ± 0.2, very close to that of PPh_3 (2.85). By comparison of the ν_{co} stretching frequencies in complexes of type $M(CO)_5L$, they also concluded that the two ligands have roughly the same basicities. However, the oxidized ligand OL[3] is easily formed in the presence of metal salts. Indeed, even in the absence of oxygen, water can act as an oxidizing agent for the ligand, giving the metal salt in its reduced form. For instance, it was shown (*17,18*) that $RhCl_3 \cdot 3H_2O$ affords rhodium(I) complexes coordinated by tris(sodium *m*-sulfonatophenyl)phosphine, $[RhCl(L^3)_3]$, and $[RhCl(L^3)_2]_2$ together with OL[3] formed by an oxygen transfer from a hydroxyrhodium(III) intermediate species formulated as $[RhCl_2(OH)(L^3)]$. In addition, when oxygen is present, the ligand is catalytically transformed into the corresponding oxide (*18*).

^{31}P NMR spectroscopy is an attractive tool to follow the reactions and to characterize the complexes (*vide infra*). The chemical shifts of the mono- and trisulfonated triphenylphosphine ligands as well as their oxides are listed in Table I (*19,20*), which shows that the chemical shifts of the ligands

TABLE I

^{31}PNMR DATA FOR $PPh_2(m\text{-}C_6H_4SO_3Na)$ AND $P(m\text{-}C_6H_4SO_3Na)_3$ LIGANDS AND THEIR OXIDES[a]

Ligand	δ (ppm)	Temperature	Ref.
$PPh_2(m\text{-}C_6H_4SO_3Na)$	−5.8	Room temp.	*19*
$P(m\text{-}C_6H_4SO_3Na)_3$	−5.3	5°C	*20*
$OPPh_2(m\text{-}C_6H_4SO_3Na)$	33.4	Room temp.	*19*
$OP(m\text{-}C_6H_4SO_3Na)$	34.8	Room temp.	*20*

[a] NMR spectra were recorded in D_2O (H_3PO_4 as external standard).

and their oxides are, respectively, found in individual ranges. Sinou and co-workers have proposed a new method to separate and analyze mixtures of sulfonated phosphines (*21,22*). Indeed, they have shown that this problem can be solved by using reversed-phase ion-pair chromatography with SAS silica as the stationary phase, a quaternary ammonium salt (usually hexadecyl trimethylammonium bromide) as the counterion, and water/*n*-propanol (5/2) as the eluent. Under these conditions, the sulfonated phosphine ligands are eluted after the corresponding phosphine oxides, and the retention times increase with the number of sulfonic groups in the molecule. Chromatography of mixtures at intervals allowed the authors to determine the kinetics of the sulfonation reaction of several diphosphines (*23*).

Concerning the reactivity of the ligands under investigation, a few studies have been carried out. Patin and co-workers have shown that the mono- and trisulfonated ligands can be involved in nucleophilic additions to activated alkenes and alkynes to afford alkyl- or vinylphosphonium salts (*24–27*). Similarly, α-bromoketones have been quantitatively debrominated by monosulfonated triphenylphosphine (*19*).

b. Aminated Phosphine Ligands. Phosphines for which solubility in water is achieved by quaternary amines as functional groups are also of interest. They are generally obtained through the synthesis of nitrogen-containing phosphine ligands followed by quaternization. According to the general preparative procedure for tertiary aminophosphines (*28*), chlorophosphines and phosphides are essential precursors.

Aminoalkyl- and aminoarylphosphines have been prepared by treating chlorodialkyl- or chlorodiarylphosphines with Grignard reagents derived from ω-bromoamines according to Eq. (3) (*29*), or with a dimethylaminoalkyl (or aryl) lithium compound as summarized in Eq. (4) (*30–32*).

$$R_2PCl + Me_2N\frown MgBr \longrightarrow Me_2N\frown PR_2 + \text{"MgClBr"} \quad (3)$$

$$R'_2(H)C\frown NMe_2 \xrightarrow[2)\ R_2PCl]{1)\ BuLi/Et_2O} R'_2(PR_2)C\frown NMe_2 \quad (4)$$

Alternative preparations involve direct attack by a chlorophenylphosphine on a secondary amine in the presence of triethylamine, as exemplified in Eq. (5) (*33,34*).

$$PhPCl_2 + MeNHCH(Me)Ph \xrightarrow{Et_2O/NEt_3} PhP(N(Me)CH(Me)Ph)_2 \quad (5)$$

Another way to prepare these mixed nitrogen- and phosphorus-containing ligands is to react the appropriate phenylphosphorus Grignard reagent with the corresponding haloalkylamine (*35–37*), for example, as in Eq. (6).

$$PhP(MgBr)_2 + Et_2NCH_2CH_2Br \rightarrow PhP(CH_2CH_2NEt_2)_2 \quad (6)$$

Similarly, condensation of a lithium diarylphosphine with an ω-haloamine gives the expected aminodiarylphosphine [see Eq. (7)] (*38–41*). Secondary phosphines can also be accessed by reacting $NaPH_2$ with a 2-chloroethylamine according to Eq. (8) (*42*). Instead of lithium salts, it is possible to use potassium phosphides as starting materials; these can be prepared *in situ* by the addition of potassium *t*-butoxide (*12,43–45*). The latter salts can be directly condensed with ethyleneimine to afford, after hydrolysis, the desired 2-aminoethylphosphine (*46*).

$$Ph_2PLi + X\frown NR_2 \longrightarrow Ph_2P\frown NR_2 + LiX \quad (7)$$

$$\begin{array}{l} NaPH_2 + ClCH_2CH_2NRR' \xrightarrow[-NaCl]{} H_2PCH_2CH_2NRR' \xrightarrow[-\frac{1}{2}H_2]{Na/NH_3} \\ NaP(H)CH_2CH_2NRR' \xrightarrow[-NaCl]{R''X} R''P(H)CH_2CH_2NRR' \end{array} \quad (8)$$

Another strategy consists in the direct reaction of PH_3 or primary and secondary phosphines with an unsaturated amine (*47–49*). These reactions follow either a free radical pathway or a Michael mechanism. Thus cyanoethylphosphines can be prepared by adding acrylonitrile to phosphines as in Eq. (9).

$$R_2PH + H_2C{=}C(H)CN \rightarrow R_2PCH_2CH_2CN \quad (9)$$

If the ligands present have several donating sites, then bidentate or polydentate compounds are preferred over monodentate ones. Hence, quaternization of the nitrogen atom is required before the ligands are coordinated to the metal center (*50*). It has been shown that quaternization of a tertiary amine group is not achieved when the aminophosphine reacts with an alkylhalide (*29,35*), because only the phosphonium salt is obtained. Nevertheless, Kolodny *et al.* (*39*) have been able to protonate selectively the nitrogen site of $R_2NC_2H_4PPh_2$ with anhydrous hydrogen halide.

The selectivity of this quaternization reaction is uncertain. Indeed, Smith *et al.* reported that simple alkylation of free $Me_2NC_2H_4PPh_2$ occurs only at the phosphorus atom (*50*), whereas it was subsequently recognized that reaction of the species $R_2N–(X)–PPh_2$ with HCl or ICH_3 yields a mixture of phosphonium and ammonium salts (*41*). This is the reason why several researchers currently protect the phosphorus function before quaternizing the ligand. Thus the phosphine oxide may be prepared either by bromination followed by hydrolysis (*41*), or more often by reaction with hydrogen peroxide (*40*) followed by reduction with trichlorosilane after complete alkylation of the nitrogen atom (*40,41,51*). An alternative procedure is to coordinate the phosphosphorus atom to a metal center, then to alkylate the nitrogen atom with the powerful alkylating agent trimethyloxonium tetrafluoroborate (*50,52,53*).

c. Phosphines Involving Hydroxyl or Ether as a Functional Group. Condensation of a primary or secondary alkyl- or arylphosphine with ketones or aldehydes in the presence of hydrochloric acid affords hydroxyalkylphosphines by repeated addition of a carbocation to the phosphine, followed by elimination of HCl from the phosphonium salt being obtained (*54–60*). The reaction can also be performed starting from PH_3. Due to the variable electrophilicity of the hydroxyalkyl groups and their different steric properties, various degrees of alkylation can be achieved. The nature, particularly the basicity, of the solvents plays an important role. For instance, the tetraalkylated phosphonium salt can be obtained in water.

Addition of a phosphide salt to an oxirane, or more generally to a cyclic ether, provides, after hydrolysis, the corresponding hydroxyalkylphosphine, as shown in Eqs. (10) and (11) (*61–63*).

$$LiPR_2 + \underbrace{CH_2—CH_2}_{O} \rightarrow LiOCH_2CH_2PR_2 \xrightarrow{H_2O} HOCH_2CH_2PR_2 + LiOH \tag{10}$$

$$2\,KPPh_2 + ClCH_2—\underbrace{CH—CH_2}_{O} \xrightarrow{H_2O} Ph_2PCH_2CH(OH)CH_2PPh_2 + KOH \tag{11}$$

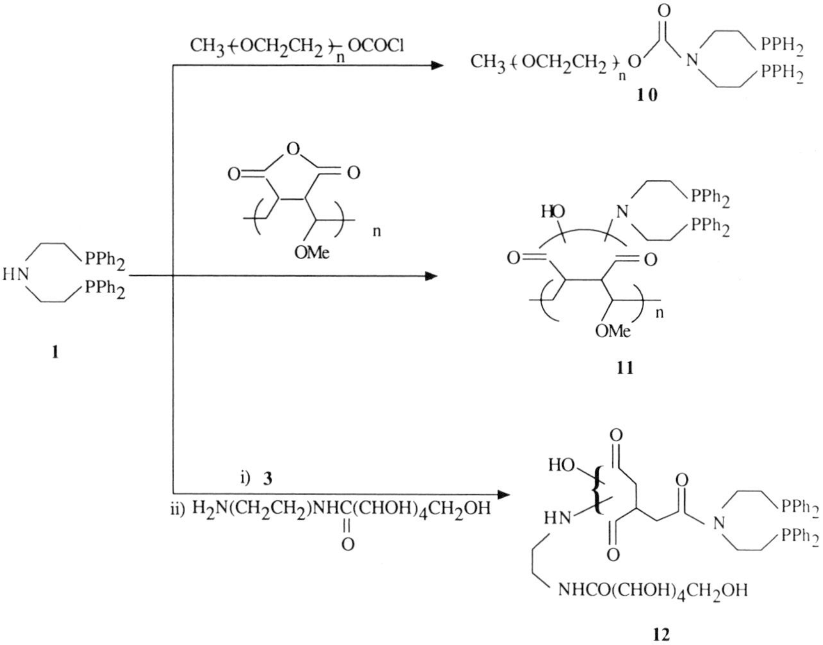

SCHEME 2. Synthesis of water-soluble ligands containing poly(ether) chains.

Like the previously described cyanoethylation, the free radical addition of unsaturated alcohols across the P–H bond allows synthesis of hydroxy-substituted alkylphosphines (*58*). Halophosphines react with Grignard reagents, formed from magnesium and various halopoly(ethers) (*64*). Reaction of bis(2-diphenylphosphinoethyl)amine **1** with acid chlorides or anhydrides that contain poly(ethyleneglycol), poly(ether), or poly(hydroxyalkyl) moieties gives the functionalized ligands shown in Scheme 2 (*12,44*).

d. Carboxylated Phosphine Ligands. Condensation of a phosphido salt with an ω-haloester or ω-halocarboxylate leads to the corresponding ω-phosphinoesters or acids (*65,66*). Similarly, secondary phosphines react directly with ω-bromoesters to yield the ω-phosphinoesters (*67,68*). An indirect procedure has been used to prepare the tris(acetoxymethyl)phosphine ligand by acetylation of tetrakis(hydroxymethyl)phosphonium chloride with acetic anhydride, followed by cleavage with sodium hydroxide, according to Eq. (12) (*69*).

$$[(HOCH_2)_4P]Cl \xrightarrow[H_2SO_4/CH_3COOH]{Ac_2O} [(CH_3COOCH_2)_4P]Cl \xrightarrow{NaOH} (CH_3COOCH_2)_3P \quad (12)$$

Moreover, alkaline hydrolysis of cyanoethylphosphines (vide supra) yields the corresponding carboxyethylphosphines (*48, 49, 70*). Finally, as shown in Scheme 3, the bis(2-diphenylphosphinoethyl)amine **1** appears to be a very useful starting material for the preparation of carboxylic diphosphines **13–15** using compound **2** and camphoric as well as succinic anhydrides (*11, 12, 44*).

e. Chiral Ligands. A number of chiral phosphine ligands have been identified as reagents that achieve asymmetrical induction in catalysis. Most of them are sulfonated. Two main preparative routes have been reported in the literature. They require either the introduction of a polar functional group into a chiral diphosphine or the binding of a chiral moiety to a diphosphine ligand. Most studies have been carried out on 2-(diphenylphosphinomethyl)-4-(diphenylphosphino)pyrrolidine **16** and 2,3-*O*-isopropylidene-2,3-dihydroxy-1,4-bis(diphenylphosphino)butane **17** (DIOP).

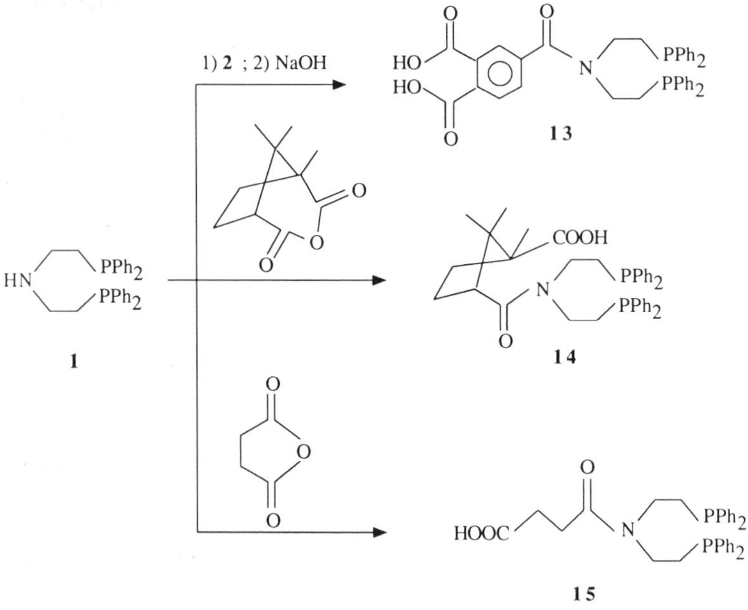

SCHEME 3. Preparation of water-soluble ligands containing carboxylate groups.

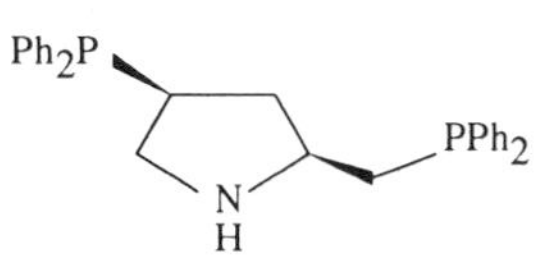

16

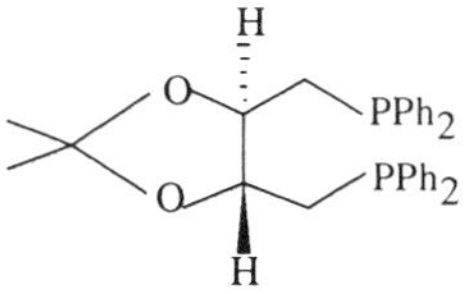

DIOP

17

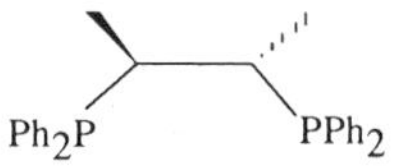

Chiraphos

18

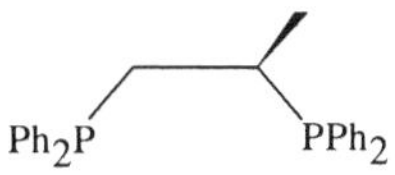

Prophos

19

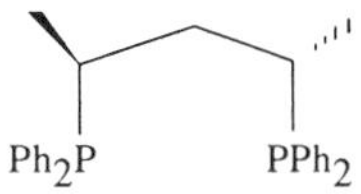

Skewphos

20

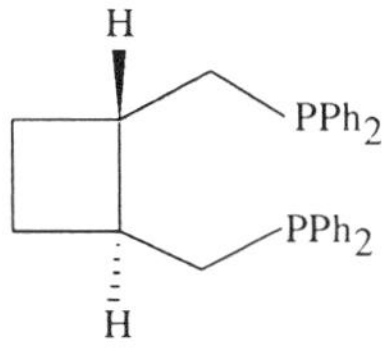

Cyclobutanediop

21

Ligand **16** can be functionalized by acylation of the N–H bond with trimellitic anhydride acid chloride **2**, followed by a treatment with sodium hydroxide or sodium taurinate to give **22** and **23** according to Scheme 4 (*71*). Several reported routes to water-soluble derivatives of the DIOP ligand are summarized in Schemes 5 and 6. Condensation of (*R, R*)-1,4-di-*O*-tosylthreitol with *p*-cyanobenzaldehyde gives an intermediate tosylate, which is then substituted by diphenylphosphide and subsequently functionalized according to the Whitesides procedure (*12*) to give the two ligands **24** and **25** (*72*). The same precursor has been condensed with a poly(ether) to afford the poly(oxadiphosphine) **26** of Scheme 5 (*73*). A similar poly(oxadiphosphine) has been also prepared from (*R*)-isopropylideneglycerol (*73, 74*).

Although the ligands (*S, S*)-chiraphos **18**, (*R*)-prophos **19**, skewphos **20**, and (*S, S*)-cyclobutanediop **21** have been successfully sulfonated in an oleum containing 20% SO_3 by a procedure quite similar to that described for triphenylphosphine (*75–78*), the DIOP ligand with its acetal ring is too sensitive to be subject to reaction with oleum (*53*).

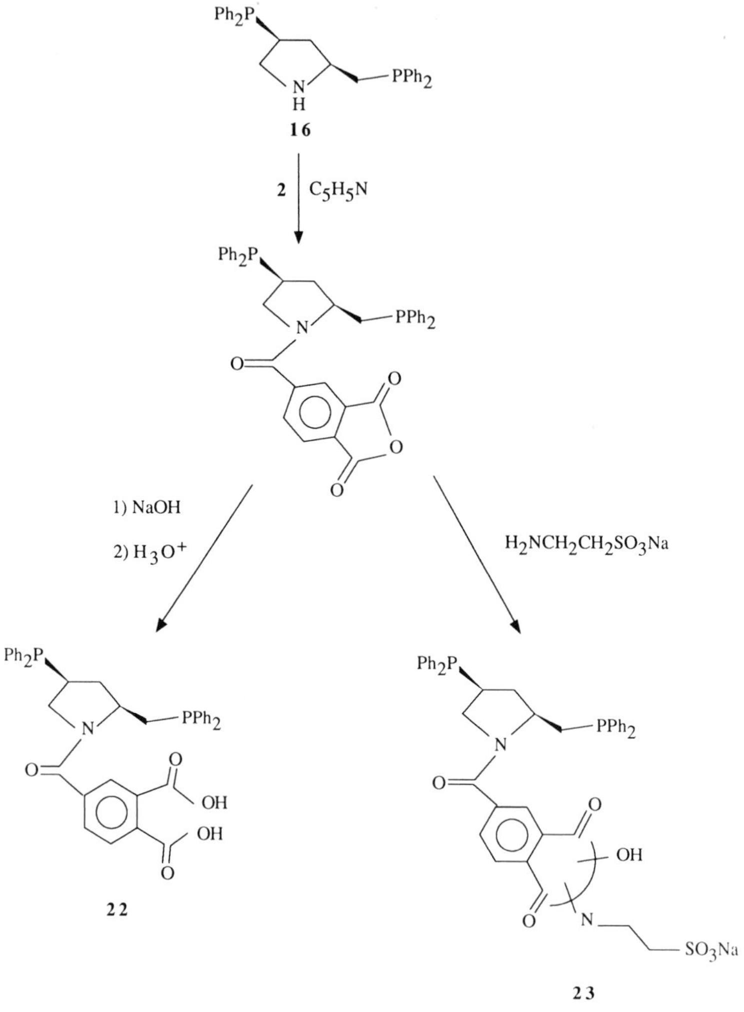

SCHEME 4. Functionalization of ligand **16** to introduce water-soluble moieties.

Tóth and Hanson consider that the amination of the phosphine ligand is a more favorable method to reach a water-solubilized DIOP (*53*). Indeed, as outlined in Scheme 6, potassium bis[*p*-(dimethylamino)phenyl]-phosphide reacts with (*R,R*)-2,3-*O*-isopropylidene-threitol ditosylate to yield the *p*-dimethylamino analog of DIOP **27**. This ligand can be selectively quaternized by trimethyloxonium tetrafluoroborate after protecting the phosphorus atoms through complexation to a rhodium center.

$CH_3(OCH_2CH_2)_nOCH_2CH$ — PPh$_2$ / PPh$_2$ — **26 a** : n = 16; **b** : n = 17

PPh_2Li

$CH_3(OCH_2CH_2)_nOCH_2CH$ — OTs / OTs

$CH_3(OCH_2CH_2)_nOCH_2CH(OCH_3)_2$ / H^+

HO, HO — OTs, OTs

NC — CHO

NC — OTs, OTs

i) PPh_2Li
ii) $LiAlH_4$
iii) **2** / C_5H_5N

H_2N SO_3Na

i) NaOH
ii) H_3O^+

HN — PPH$_2$, PPH$_2$ — COOH, COOH **24**

HN — PPH$_2$, PPH$_2$ — OH, NH SO_3Na **25**

SCHEME 5. Introduction of water-soluble moieties in (R,R),1,4-di-O-tosylthreitol.

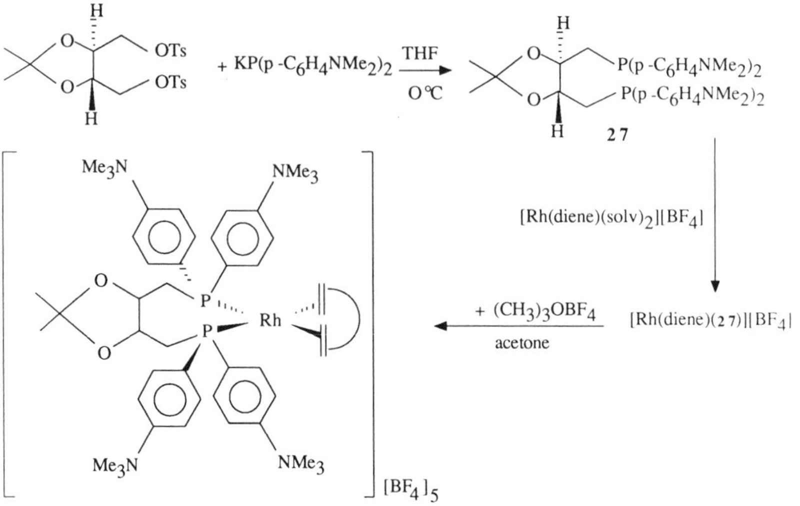

SCHEME 6. Amination and quaternization of DIOP **17**.

Moreover, it is also possible to bind a diphosphine moiety to a chiral arrangement as exemplified by the use of a protein rest (*79*) according to Eq. (13).

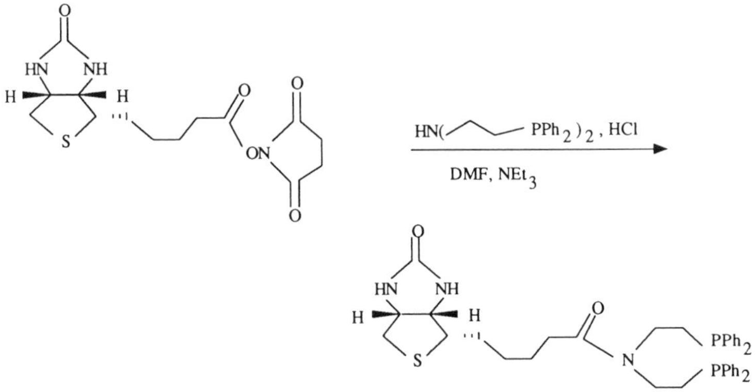

$$\xrightarrow[\text{DMF, NEt}_3]{\text{HN(CH}_2\text{CH}_2\text{PPh}_2)_2\text{, HCl}} \tag{13}$$

Similarly, chiral amines may be used to prepare bis(aminophosphines) $R^*N(PPh_2)CH_2CH_2(PPh_2)NR^*$, as shown in Eq. (14) (*80*).

$$\begin{gathered} 2R^*NH_2 + EtOOCCOOEt \rightarrow R^*NHCOCONHR^* \xrightarrow[\text{THF}]{\text{LiAlH}_4} \\ R^*NHCH_2CH_2NHR^* \xrightarrow[\text{NEt}_3]{\text{Ph}_2\text{PX (X = Cl, Br)}} R^*N(PPh_2)CH_2CH_2N(PPh_2)R^* \end{gathered} \tag{14}$$

where R* = $S(-)$-α-methylbenzyl-$(-)$-menthyl. Finally, in order to enhance the induction effect in catalysis, ferrocenylphosphines have been prepared according to the previously described procedures (*31,32,81,82*).

2. *Nitrogen-Containing Ligands*

The literature relating the preparation of ligands that do not contain a phosphorus atom is relatively scarce and is only concerned with amines. The principal studies deal with the bipyridine ligand. The first sulfonic acid derivatives of 2,2′-bipyridine were obtained from direct reaction of 2,2′-bipyridine with sulfuric acid at a temperature as high as 300°C (*82*), or from the thermolysis of the 1:1 bipy–SO_3 adduct (*83*). A mixture of mono- and disulfonated acids was obtained. Anderson *et al.* have reinvestigated this reaction and have confirmed that temperatures higher than 200°C are required to yield the derivative monosubstituted in the 5-position or the disubstitued (5,5′-position) one (*84,85*). For instance, 2,2′-bipyridine was treated for 24 hours with oleum in the presence of a small amount of mercuric sulfate to give either **28** at 220°C, or a mixture of both ligands, **28** and **29** at 290°C. More recently, Herrmann *et al.* have prepared compound **28** on a larger scale and with improved yield (*86*).

Substitution in the 4-position of the pyridine ring has been achieved by indirect routes. Indeed, after protection of nitrogen atoms by oxidation, the 4- and 4′-positions are sequentially substituted by NO_2, Cl, and SO_3H groups to give the ligand **30** (*85*). By a coupling of γ-picoline, the 4,4′-dimethylbipyridine was obtained; its oxidation by $KMnO_4$ provides the 2,2′-bipyridine-4,4′-dicarboxylic acid **31** (*85,87*).

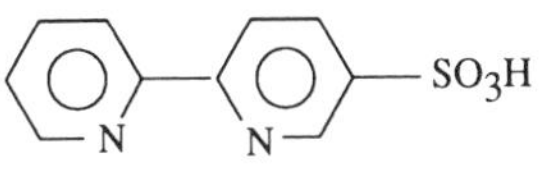

SO_3H N N

28

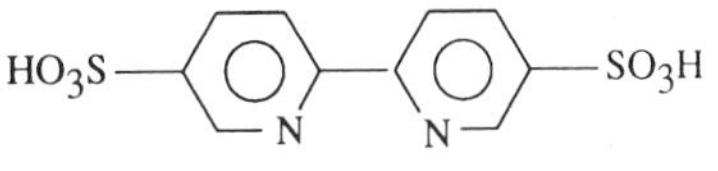

HO_3S SO_3H N N

29

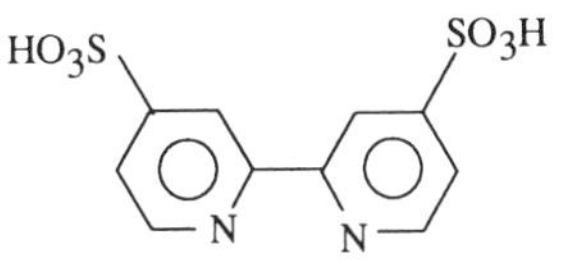

HO_3S SO_3H N N

30

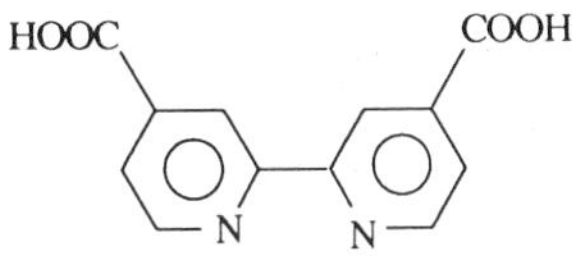

HOOC COOH N N

31

In addition to these water-soluble substituted bipyridine ligands, ethylenediamine *N,N,N′,N′*-tetraacetic acid (EDTA) is one of the most commonly used ligands. EDTA or its β-propionic acid analog is readily prepared by condensing ethylenediamine with either chloroacetic or chloropropionic acid in alkaline solution according to Eq. (15) (*88*). Similar treatment of aniline with sodium chloroacetate is a good method to prepare aniline *N,N*-diacetic acid (*89*).

$$H_2NCH_2CH_2NH_2 + 4ClCH_2COOH \rightarrow \underset{\text{EDTA}}{(HOOCCH_2)_2NCH_2CH_2N(CH_2COOH)_2} \quad (15)$$

Some other nitrogen-containing ligands, including 12,6-diacetylpyridine bis(semicarbazone) ligand (*90*) or the tetrasulfophthalocyanine ligand (*91*), have been designed to solubilize their corresponding metal complexes in water but have not yet been used in catalysis.

B. *Transition Metal Complexes*

Numerous papers describe complexes obtained *in situ* by addition of the ligand to a metal precursor; these complexes have been considered by their discoverers to be analogous to the nonfunctionalized ligand-containing complexes. The following discussion will be devoted only to well-characterized isolated compounds.

1. *Complexes with Phosphine Ligands*

a. Sulfonated Phosphines. The syntheses of complexes containing sulfonated phosphines involve either substitution of labile ligands such as CO, cyclooctadiene (COD), PPh_3, or coordinated solvents (tetrahydrofuran, for instance) by the water-soluble ligand or by reaction with the metal salt, eventually in the presence of additional reagents acting as ligands and/or reducing agents. Moreover, the first method can be applied either by contacting the organic phase containing the parent complex with the sulfonated ligand dissolved in water, or by mixing the complex and the ligand of interest in a solvent in which both are soluble (*92*). Due to the high solubility of the complexes in water, their purification, in particular their separation from free ligand, has long been a tedious problem. The most elegant purification procedure has been that of Anderson *et al.* (*85*), who used gel-permeation chromatography, a procedure extensively further developed by Herrmann *et al.* (*20,93*).

Almost all the complexes fully described and characterized in the literature are based on sodium *m*-sulfonatophenyldiphenylphosphine L^1 and tris(sodium *m*-sulfonatophenyl)phosphine L^3 ligands. Most of the complexes contain ruthenium, rhodium, or more generally noble metals. The infrared data of interest [Tables II (*94–96*) and III (*97*)] are often compared with those of the parent triphenylphosphine complexes. Similarly, the ^{31}P NMR data are listed on Table IV (*98*). The ^{31}P NMR data and

TABLE II

IR Spectral Data for Complexes with L^1 and PPh_3[a]

Complex	ν_{CO}	ν_{M-Cl}	ν_{M-H}	Ref.
$RhCl(L^1)_3$		285^b (298)		*92*
$RhCl(COD)(L^1)$		280^b		*92*
$RhCl(CO)(L^1)_2$	1980 (1960)			*1*
$HRh(CO)(L^1)_3$	1910^b (1918)		2000^b (2041)	*92* (*96*)
$RhH_2Cl(L^1)_3$			2050^b (2065) 2090^b (2099)	*1*
trans-$RuCl_2(CO)_2(L^1)_2$	2005			*94*
cis-$RuCl_2(CO)_2(L^1)_2$	2000^c 2060^c			*95*
$HRuCl(L^1)_3$		275^b (282)	2025^b (2020)	*92*
$HRu(OAc)(L^1)_3$			1996^b	*94*
$HRuCl(CO)(L^1)_3$	1900^b (1903) 1970^b (1922)		2060^b (2020)	*94* (*96*)
cis-$HRuCl(CO)_2(L^1)_2$	1997^c 2060^c		1940^c	*94*
$IrCl(CO)(L^1)_2$	1960 (1961)			*1*
$Ir(O_2)Cl(CO)(L^1)_2$	2010 (2010)			*1*
$HPt(L^1)_3Cl$			2095^b (2102)	*92*
$PdCl_2(L^1)_2$		350^b (357)		*92*

[a] Values given in cm^{-1}. L^1, $PPh_2(m\text{-}C_6H_4SO_3Na)$. Values in parentheses are those found for the analogous PPh_3-containing compounds.

[b] In Nujol.

[c] In KBr.

the spectroscopic data for ligands such as CO, H, and Cl show that the sulfonated phosphine and triphenylphosphine complexes display very close properties, reflecting only minor changes in the electronic and steric behavior of the ligands in the various metal compounds. Thus, the electron-withdrawing sulfonate group in a meta position on the phenyl ring only slightly modifies the basicity of the parent ligand. Because the counterion of the sulfonate group is Na^+, the two ligands involved in the complexes are generally $L^1 = Ph_2P(m\text{-}C_6H_4SO_3Na)$ and $L^3 = P(m\text{-}C_6H_4SO_3Na)_3$. Exceptions include the zwitterionic complex $Pt^+H(L^1)_2(Ph_2PC_6H_4SO_3^-)$ and the neutral species $[C_8H_{12}Rh(Ph_2PC_6H_4SO_3)]_2$ prepared by Borowski *et al.* (*92*). The latter complex includes an oxygen–rhodium σ bond as shown in Scheme 7. Likewise, the complex $Ag(L^3)_3$ contains eight rather than nine Na^+ cations, the ninth positive charge being provided by Ag^+ (*20*).

It is noteworthy that all these ionic species exist in hydrated forms, although this feature is not specified in the formulas listed in Tables II–IV. The reason is that the elemental analyses are generally not sufficiently

TABLE III

IR SPECTRAL DATA FOR COMPLEXES AND CLUSTERS WITH L^3 OR PPh_3[a]

Complexes	ν_{CO} (cm^{-1}; KBr) for L = L^3	ν_{CO} (cm^{-1}; KBr) for L = PPh_3	Ref.
$Fe(CO)_4(L)$	2050 (w) 1977 (w) 1944 (vs)	2050 (w) 1978 (w) 1944 (s)	*93*
$CoH(CO)(L)_3$	1904 (vs)	1910	*93*
$RhCl(CO)(L)_2$	1980 (vs)	1960	*93*
$[Rh(\mu\text{-}Cl)(CO)(L)]_2$	1986 (vs)	1983[b]	*93*
$Ni(CO)_2(L)_2$	2008 (s) 1944 (s)	2000[c] 1941[c]	*93*
		2000[d] 1955[d]	*93*
$IrH(CO)(L)_3$	1927 (m)	1918	*93*
$IrCl(CO)(L)_3$	2005 (m) 1963 (m)		*93*
$Ru_3(CO)_9(L)_3$	2053 (m) 1992 (m) 1973 (s)	2044 (m) 1978 (m) 1967 (s)	*97*
$Ru_3(CO)_{11}(L)$	2080 (sm) 2030 (ms) 2005 (sh) 1990 (vs)	2072 (sm) 2047 (sm) 2019 (s) 1996 (vs)	*97*
$Ru_3(CO)_{10}(L)_2$	2100 (sm) 2050 (ms) 2025 (s) 2010 (sh) 2000 (sh) 1982 (m,br)	2098 (m) 2046 (s) 2038 (sh) 2024 (sh) 2015 (vs) 1985 (sm)	*97*
$Os_3(CO)_{10}(L)_2$	2075 (sm) 2020 (s) 1992 (vs) 1965 (s) 1940 (sh)	2088 (sm) 2035 (s) 2002 (vs) 1963 (s) 1946 (m) 1935 (sh)	*97*
$Ir_4(CO)_9(L)_3$	2049 (m) 2012 (s) 1985 (vs) 1790 (vs) 1784 (sh)	2042 (ms) 2015 (sh) 1982 (vs) 1772 (vs)	*97*

[a] L^3, $P(m\text{-}C_6H_4SO_3^-Na^+)_3$.
[b] In $CHCl_3$.
[c] In Toluene.
[d] In Nujol.

precise to ascertain the number of water molecules. Borowski *et al.* (*92*) and Herrmann *et al.* (*20,93*) stated that one SO_3Na group contains one molecule of water, but Salvesen and Bjerrum reported that the L^1 ligand contains two molecules of H_2O bound strongly enough to resist loss on heating at 400°C (*99*). In the ruthenium complexes synthesized by Tóth *et al.*, an L^1 ligand contains two equivalents of water (*94*). Although the number of water molecules per sulfonate group cannot be ascertained, Borowski *et al.* (*92*) have observed that the infrared spectra may indicate the presence or absence of water in the complex. Indeed, the hydrated form shows a strong band at 1200 cm^{-1} whereas the nonhydrated form presents two absorptions at 1223 and 1192 cm^{-1}, due to the interaction between the sodium cation and the sulfonate group, leading to the perturbation of the C_{3v} symmetry of SO_3^-.

In other respects, the tris(sodium *m*-sulfonatophenyl)phosphine complexes generally present lower coordination numbers than do their triphenylphosphine analogs. Thus, the complexes $[RuCl_2(PPh_3)_4]$, $[Ni(PPh_3)_4]$, and $[Pd(PPh_3)_4]$ afford the undercoordinated species

TABLE IV

^{31}P NMR DATA FOR SOME METAL COMPLEXES WITH L^1 AND L^3 MEASURED IN WATER[a]

Complexes	Temperature (°C)	δ (ppm)[b]	J_{M-P}(Hz)[b]	Ref.
$RhCl(L^1)_3$	28	$\delta P_A{}^c = 50.9$ (48.9)	$J(Rh, P_A)^c = 210$ (192)	*92*
		$\delta P_B{}^c = 32.9$ (32.2)	$J(Rh, P_B)^c = 139$ (146)	
$RhH_2Cl(L^1)_3$	28	$\delta P_A{}^c = 24.0$ (20.7)	$J(Rh, P_A)^c = 96$ (90)	*92*
		$\delta P_B{}^c = 37.2$ (40.3)	$J(Rh, P_B)^c = 104$ (114)	
$RuHCl(L^1)_3$	28	49.8 (59.0)		*92*
$[PtH(L^1)_3]Cl$	28	$\delta P_A{}^c = 21.8$	$J(Pt, P_A)^c = 2205$	*92*
		$\delta P_B{}^c = 21.0$	$J(Pt, P_B)^c = 2812$	
$Fe(CO)_4(L^3)$	5	84.9		*93*
$CoH(CO)(L^3)$	5	45.9		*93*
$Co_2(CO)_4(\mu\text{-alcyne})(L^3)_2$	5	51.8		*93*
$Rh(NO)(L^3)_3$	5	48.4 (48.8)	177 (175)	*93*
$RhCl(L^3)_3$	5	52.8	$J(Rh, P_A) = 196$	*98*
		34.6	$J(Rh, P_B) = 144$	
			$^2J(P_A, P_B) = 41$	
$Rh(OH)(L^3)_3$	21	53.6	$J(Rh, P_A) = 195$	98
		35.2	$J(Rh, P_B) = 144$	
			$^2J(P_A, P_B) = 41$	
$Rh_2(\mu\text{-Cl})_2(COD)(L^3)_2$	5	129	146	*98*
$RhH(CO)(L^3)_3$	5	44.4	156	*98*
trans-$Rh(OH)(CO)(L^3)_2$	21	31.8	129	*98*
trans-$RhCl(CO)(L^3)_2$	5	31.3	128	*98*
$[Rh(\mu\text{-OH})(L^3)_2]_2$	21	58.9	203	*98*
$[Rh(\mu\text{-Cl})(CO)(L^3)]_2$	5	48.2	180	*93*
$[Rh(\mu\text{-S-}t\text{-Bu})(CO)(L^3)]_2$		37.7	152	*149*
$IrCl(CO)(L^3)_3$	5	−2.8		*93*
$IrH(CO)(L^3)_3$	5	19.2		*93*
$IrCl(\eta^4\text{-1,5-COD})(L^3)_2$	5	19.0		*93*
$Ni(CO)_2(L^3)_2$	5	34.9 (32.7)		*93*
$Ni(L^3)_3$	−30[f]	22.7 (23.0)		*93*
$Pd(L^3)_3$	5	22.6 (22.6)		*93*
$Pt(L^3)_4$	5	$\delta P_A{}^d = 24.1$	$J(Pt, P_A)^d = 2210$	*93*
	5	$\delta P_B{}^d = 22.2$	$J(Pt, P_B)^d = 2853$	
	5		$^2J(P_A, P_B)^d = 20$	
cis-$PtCl_2(L^3)_2$	5	13.9 (13.9)	3727 (3678)	*93*
trans-$PtCl_2(L^3)_2$	5	21.7 (18.4)	2597 (2595)	*93*
$RuCl_2(L^3)_2$	5	57.0		*93*
$Ru(NO)_2(L^3)_2$	5	56.0 (56.0)		*93*
$Ru_3(CO)_9(L^3)_3$	Room temp.	38.7		*97*
$Ru_3(CO)_{11}(L^3)$	Room temp.	37.9		*97*
$Ru_3(CO)_{10}(L^3)_2$	Room temp.	35.4		*97*
$Os_3(CO)_{10}(L^3)_2$	Room temp.	−2.3		*97*
$Ir_4(CO)_9(L^3)_3$	Room temp.	$\delta P_A{}^e = -15.0$ (−17.0)		*97*
		$\delta P_B{}^e = 21.3$ (20.0)		

[a] L^1, $PPh_2(m\text{-}C_6H_4SO_3Na)$; L^3, $P(m\text{-}C_6H_4SO_3Na)_3$.

[b] Values in parentheses are those for the corresponding PPh_3-containing compounds.

[c] P_A is the unique phosphorus atom; the P_B are the two equivalent mutually trans phosphorus atoms.

[d] P_A is the unique phosphorus atom; the P_B are the three equivalent phosphorus atoms.

[e] P_A is the phosphorus atom in an axial position; the P_B are the two phosphorus atoms in a radial configuration.

[f] Mixture of ethanol/water.

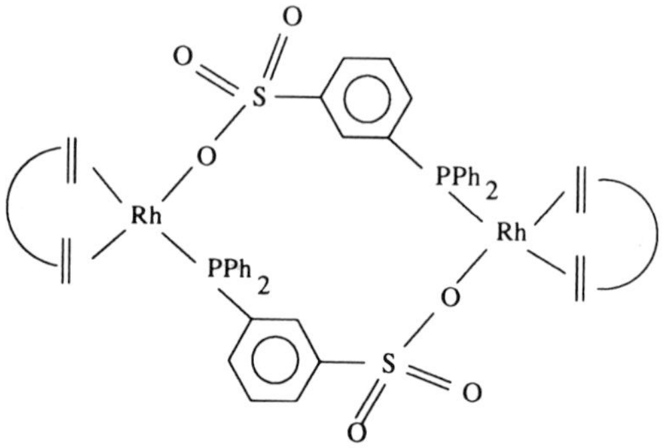

SCHEME 7.

$[RuCl_2(L^3)_2]$, $[Ni(L^3)_3]$, and $[Pd(L^3)_3]$, respectively (*20,93*). It has been previously emphasized that the steric requirements of PPh_3 and L^1 or L^3 are similar (vide supra). These lower numbers of ligands are presumably due to the high charge accumulation within the coordination sphere (*20*).

Except for a few cases, the complexes formed either from L^1 or L^3 ligands are described as particularly air sensitive. An extensive study has been carried out by Patin and co-workers (*15,17,18*) on the L^3 rhodium systems. The presence of oxygen leads to the catalytic oxidation of the ligand to $OP(m\text{-}C_6H_4SO_3Na)_3$, with the latter being unable to coordinate to the metal center, as previously mentioned. In the absence of oxygen, the L^3 ligand can be stoichiometrically oxidized by water through a series of complicated redox mechanisms involving rhodium species.

In addition to the rhodium(III) hydroxo species detected by Patin *et al.*, it has been shown by Herrmann *et al.* that the complex $[HRh(L^3)_4]$ reacts with water to give the dinuclear hydroxo species **33** of Eq. (16) via the intermediate **32**. The species **32** can also be produced by reduction of $[RhCl(L^3)_3]$ with $NaBH_4$, as shown in Eq. (17) (*98*). The isolated complex **32** reacts further with carbon monoxide to produce the two compounds **34** and **35** that are in equilibrium in solution (*20*).

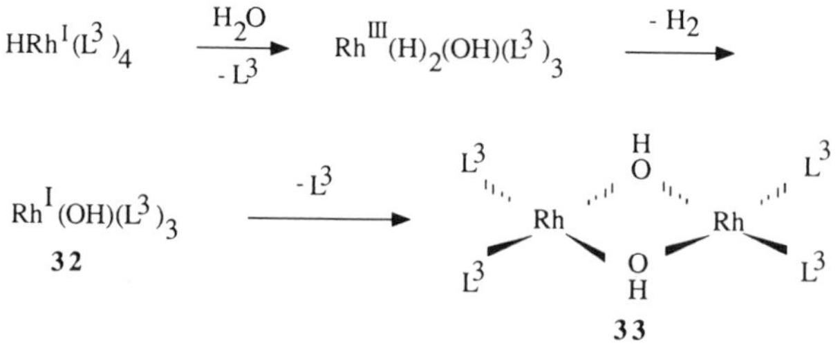

$$HRh^{I}(L^3)_4 \xrightarrow[-L^3]{H_2O} Rh^{III}(H)_2(OH)(L^3)_3 \xrightarrow{-H_2} \underset{\mathbf{32}}{Rh^{I}(OH)(L^3)_3} \xrightarrow{-L^3} \underset{\mathbf{33}}{[(L^3)_2Rh(\mu\text{-}OH)_2Rh(L^3)_2]} \quad (16)$$

$$\mathrm{RhCl(L^3)_3} \xrightarrow{\mathrm{Na[BH_4]/H_2O}} \mathrm{Rh^I(OH)(L^3)_3} \xrightarrow[\mathrm{-TPPTS}]{\mathrm{+CO}}$$

$$\underset{\mathbf{34}}{\mathrm{Rh(OH)(L^3)_2(CO)}} \underset{\Delta T,\ L^3,\ CO,\ H_2}{\overset{\Delta T,\ -L^3,\ -H_2}{\rightleftharpoons}} \underset{\mathbf{35}}{\mathrm{HRh(L^3)_3(CO)}} \qquad (17)$$

Larpent and Patin (*100*; see also *101*) have reported that colloidal dispersions of rhodium stabilized by the oxide $OP(m\text{-}C_6H_4SO_3Na)_3$ can be easily produced from $[RhCl_3 \cdot 3H_2O]/L^3$ mixtures under a hydrogen atmosphere. During the course of their studies on the hydroformylation of alkenes, these authors investigated these aqueous dispersions by introducing deuterium gas or D_2O in a mixture of water, $[RhCl_3 \cdot 3H_2O]$, and L^3. The results were interpreted as resulting from the reduction of colloidal Rh(III) species by H_2 gas followed by oxidative addition of water or molecular hydrogen to the Rh(I) species. The cis hydrogenation (deuteration) of alkenes occurred on the hydrido hydroxo species thus formed, the latter stabilized by ionic repulsions between polyanions OL^3, which prevent the aggregation of rhodium (*102*). Lecomte and Sinou (*78*) have also observed these colloidal particles on addition of chiral diphosphines to $[RhCl_3 \cdot 3H_2O]$, whereas rhodium(I) precursors afford homogeneous catalysts that allow asymmetric hydrogenation and recycling without loss of enantioselectivity.

Similarly, hydrotropic or surface active molecules $R\text{–}C(p\text{-}C_6H_4SO_3Na)_3$ stabilize the hydroxylated rhodium particles by adsorption of the polyanionic head on their polar surface (*103*). This type of trisulfonated surfactant was also used to stabilize Rh(0) particles prepared by reduction of $[RhCl_3 \cdot 3H_2O]$ with $NaBH_4$ (*104*). Thus, these protective colloid agents allow catalytic hydrogenations of alkenes without cosolvent in a purely biphasic system and recycling of colloidal suspensions.

Solutions of rhodium complexes containing L^1 have been shown to be very sensitive to the presence of solvents such as alcohols or ketones, which rapidly induce precipitation of metallic rhodium (*105*).

b. Complexes with Phosphines Containing Hydroxy, Ether, or Carboxylic Functions. The literature related to complexes with phosphines possessing hydroxy groups as substituents is scarce. Bis(hydroxymethyl)-alkylphosphines are described as difficult to obtain in a pure form and their reactions with $[Na_2PdCl_4]$ or with $[Pt_2Cl_4(PR_3)_2]$ generally lead to intractable syrups (*106*). The best way to obtain crystallized complexes is to start from the corresponding acetate and to hydrolyze after coordination of the ligand. It has also been observed that two CH_2OH substituents per phosphine are necessary to gain a significant solubility in water (*106*). More recently, Pringle and co-workers prepared the $P(CH_2OH)_3$ ligand and its nickel, palladium, platinum, ruthenium, rhodium, and iridium complexes by substitution of appropriate ligands in a biphasic medium (*107,108*). Only the first three complexes were described as being tetrahedral zero-valent $[M(P(CH_2OH)_3)_4]$. The X-ray structure of the palladium species has been determined (*107*). The platinum complex is protonated in water to give the $[HPt(P(CH_2OH)_3)_4]^+$ species **36**.

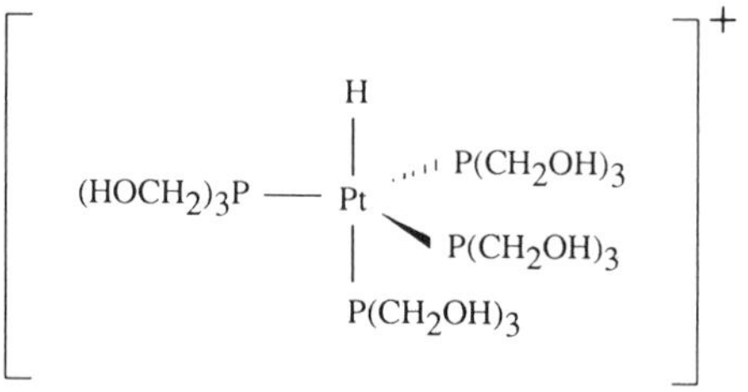

$[HPt(P(CH_2OH)_3)_4]^+$: H, $(HOCH_2)_3P$, Pt, $P(CH_2OH)_3$, $P(CH_2OH)_3$, $P(CH_2OH)_3$

36

Moreover, three ether-containing phosphines have been coordinated to a nickel center, i.e., $Ph_2PCH_2CH_2(OCH_2CH_2)_nPPh_2$, where $n = 1$ (L^4), $n = 2$ (L^5), and $n = 3$ (L^6). Complexes $[L^4NiX_2]$ and $[L^5NiX_2]$ (X = Cl, I) have been analyzed by X-ray diffraction; these studies revealed that only the phosphorus atoms participate in the coordination of the ligands (*109*). These three ligands have also been used in the coordination chemistry of rhodium(I). Their reaction with $[Rh_2(\mu\text{-}Cl)_2(CO)_4]$ produces the mononuclear complex [RhCl(CO)(diphos)], where the two phosphorus atoms occupy trans positions, the oxygen atoms of the bridging chain remaining uncoordinated (*110*). The chlorine atom can be removed by addition of $AgPF_6$ so that in each of the three new complexes, **37** ($n = 1$), **38** ($n = 2$), and **39** ($n = 3$), an oxygen is bonded to the cationic rhodium center. With L^4 and L^5, the oxygen atom comes from the ether chain (except for L^5 with ethanol) whereas in the case of L^6 a water molecule is strongly bound to Rh^+. Hydrogen interactions with two oxygen atoms of the ether chain stabilize the complex (*109–112*). All these species are air stable.

The coordination properties of tertiary phosphines of general formula $(t\text{-}Bu)_2P(CH_2)_nCOOEt$, where $n = 1–3$, thus containing an ester function, have been examined with Pd(II), Pt(II), Rh(II), and Rh(III) metal centers (*68*). Several complexes have been prepared, generally in organic solvents.

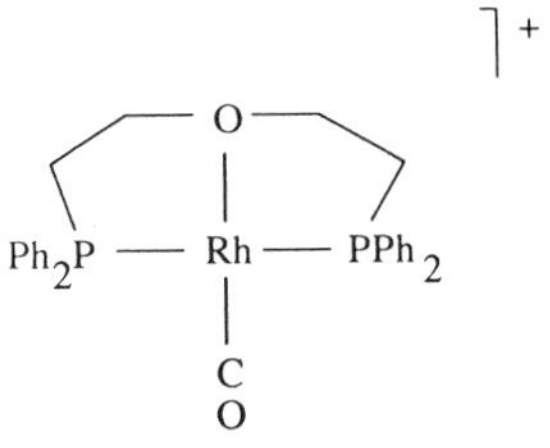

37

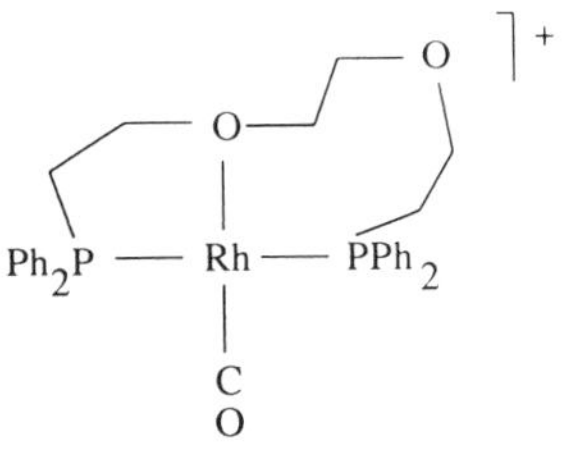

38

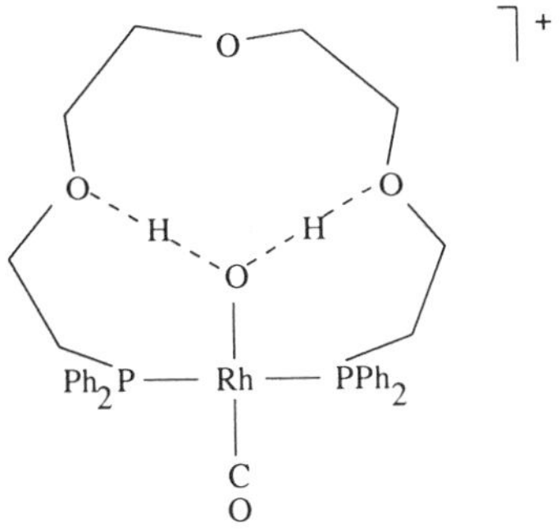

39

Either O or C metallations were observed, according to the chain length. Coordination of the 2,3-bis(diphenylphosphino)maleic anhydride ligand has also been reported. Solvolysis of the anhydride function by methanol afforded the two compounds **40** and **41** according to Scheme 8 (*113*).

c. Complexes with Aminophosphines. The most well-studied water-soluble complexes with aminophosphines as ligands are those related to $[Ph_2PCH_2CH_2NMe_3]^+$, which is referred to as amphos. Because phosphines containing amino groups are only soluble in acidic media, Smith and Baird have introduced the tetraalkylammonium group, which induces solubility in water regardless of the solution pH (*50*).

Two main routes to amphos complexes have been studied. In the first, the precursor $[RhCl(NBD)]_2$ (NBD = norbornadiene) was treated with amphos nitrate to give $[(NBD)RhCl(amphos)](NO_3)$ by chlorobridge splitting (*114,115*). A similar procedure was applied to the reaction of $[H_2N(CH_2CH_2PPh_2)_2]Cl$ with $[Ir_2Cl_2(C_8H_{14})_4]$. However, loss of HCl gave the neutral complex $[IrCl(HN(CH_2CH_2PPh_2)_2)]$. The same behavior was observed for rhodium (*116*). In the second procedure, carbonyl derivatives have been conveniently prepared initially by reacting the parent carbonyls $[M(CO)_x]$ (M = Fe, W; $x = 5, 6$) with amines in the presence of the mild

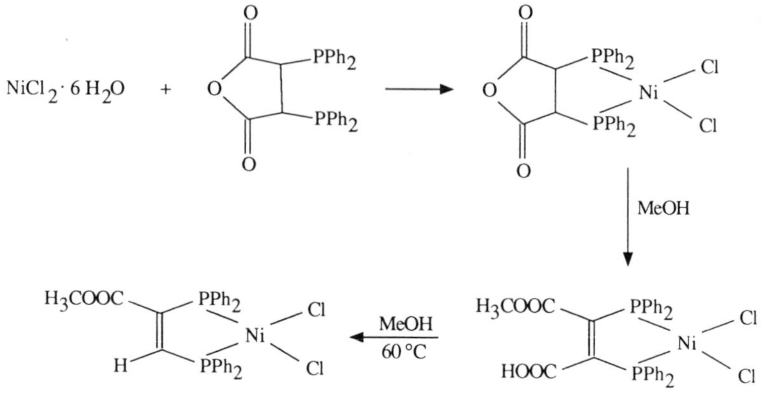

SCHEME 8. Preparation of carboxylated complexes.

decarbonylating agent Me_3NO, then adding methyl iodide in order to alkylate the amine function (*40,50*). The spectroscopic properties of these complexes have been compared with those of the triphenylphosphine analogs. The slight differences measured for the IR data (Table V) as well as for the NMR data (Table VI) suggest that amphos behaves very similarly to triphenylphosphine, the positive charge being attenuated along the aliphatic chain of the ligand. It should be noted that all the amphos complexes reported so far are highly air stable.

2. *Complexes Made from Water-Soluble Ligands that Do Not Contain Phosphorus*

In some cases phosphorus may be replaced by nitrogen. Thus, complexes containing either substituted bipyridine or ethylenediaminetetraacetic acid as ligands have been prepared. Two main preparative routes have been used for bipyridine-containing complexes. Ruthenium complexes of 2,2′-bipyridine-4,4′-disulfonic acid and 2,2′-bipyridine-5-sulfonic acid have been synthesized by reacting the free ligand with blue ruthenium(II) solutions (*85*), which result from the vigorous shaking of a methanolic solution of $[RuCl_3 \cdot 3H_2O]$ and platinum black under 1.5 to 2 bar of hydrogen for several hours (*117*). Complexes of chromium, molybdenum, tungsten, manganese, rhenium, or osmium that contain carbonyl or oxo ligands react with monosubstituted bipyridine ligands to afford the corresponding water-soluble compounds (*86*). Because these complexes have been prepared from the salts **43**, the sulfonamides **44**, or even the acid **42** (see Scheme 9), the solubility of the complexes thus obtained can be

TABLE V

ν_{CO} ABSORPTIONS FOR COMPLEXES DISSOLVED IN CH_2Cl_2[a]

Complex type	Ligand	ν_{CO} (cm^{-1})			
$Fe(CO)_4L$	Amphos	2054 (m)	1983 (w)	1945 (vs)	1932 (vs)
	PPh_3	2049 (m)	1975 (w)	1935 (vs)	
$W(CO)_5L$	Amphos	2073 (w)	1938 (vs)		
	PPh_3	2070 (w)	1935 (vs)		
$Mo(CO)_5L$	Amphos	2074 (w)	1944 (vs)		
	PPh_3	2070 (w)	1940 (vs)		

[a] From Ref. 40.

TABLE VI

^{31}P AND ^{1}H NMR DATA

Complexes	Solvent	δ (NMe)	δ (P)[a]	J_{M-P}[b]	Ref.
$[Fe(CO)_4amphos]^+$	CD_3CN	3.21	64.1		*40*
$[Mo(CO)_5amphos]^+$	CD_3CN	3.13	26.6		*40*
$[W(CO)_5amphos]^+$	CD_3CN	3.08	6.9	244	*40*
$[Rh(NBD)Cl(amphos)]^+$	MeOH		25.3	176	*115*
$[Rh(NBD)Cl(PPh_3)]$			31.6	171	*115*
$[Rh(NBD)(amphos)_2]^{3+}$	MeOH		15.7	152	*115*
$[Rh(NBD)(PPh_3)_2]^+$			29.8	157	*115*
$[RhH_2(amphos)_2(MeOH)_2]^{3+}$	MeOH		31.5	122	*115*
$[RhH_2(PPh_3)_2(MeOH)_2]^+$			41.8	121	*115*
$[Rh(amphos)_2(MeOH)]^{3+}$	MeOH		46.4	205	*115*
$[Rh(PPh_3)_2(MeOH)_2]^+$			57.2	207	*115*
$[RhH_2(amphos)_3X]^{3+\text{ or }4+}$	MeOH		11.9	95	*115*
			31.2	117	
$[RhH_2Cl(PPh_3)_3]$			20.7	90	*115*
			40.3	114	

[a] Chemical shift in parts per million relative to external H_3PO_4.
[b] Coupling constant in hertz.

significantly different. Sodium salts are soluble in water, whereas tetra-*n*-butylammonium and tetrapenylphosphonium salts and the sulfonamides are soluble in polar organic solvents (*86*). Nevertheless, the polar groups and their position on the 2,2′-bipyridine have no significant influence on the chemistry and the reactivity of the complexes obtained (*118*).

Many transition metal complexes have been synthesized, in which EDTA employs various coordination sites. A nonexhaustive list is provided in Table VII according to the coordination number for the EDTA

SCHEME 9. Precursors for substituted bipyridine-containing complexes.

TABLE VII

EXAMPLES OF EDTA COMPLEXES SHOWING VARIOUS COORDINATION NUMBERS[a]

Ligand EDTA	Complexes	Ref.
Tridentate	$[Cr^{III}(H_3EDTA)Cl_2]\cdot H_2O$	*119*
Tetradentate	$[Cr^{III}(EDTA)(H_2O)_2]^-$	*120*
	$K[RuCl_2(H_2EDTA)]\cdot H_2O$	*121*
Pentadentate	$[Cr^{III}(HEDTA)(H_2O)]$	*120*
	$[Rh(EDTA)(H_2O)]^-$	*119*
	$K[RuCl(HEDTA)]\cdot 2H_2O$	*121*
	$[Ru^{III}(EDTA)L]^-$	*119*
	$[Ru^{II}(EDTA)L]^{2-}$	*119*
Hexadentate	$NH_4[Co^{III}(EDTA)]\cdot 2H_2O$	*119*
	$NH_4[Fe(EDTA)]\cdot H_2O$	*119*

[a] $L = H_2O$, SCN^-, C_5H_5N, $C_3H_4N_2$, $C_4H_4N_2$, $C_5H_7N^+$, or $C_6H_6N_2O$.

ligands. For many transition metals, especially rhodium and ruthenium, the tetra- and pentadentate forms are most frequently encountered. IR spectroscopy is generally used to detect free acid that remains ($\nu_{COOH} = 1720\ cm^{-1}$) in addition to the coordinated acid groups ($1650\ cm^{-1}$). Attempts to prepare hexadentate EDTA complexes of chromium, ruthenium, and rhodium failed (*120,121*), whereas hexadentate iron(III) and cobalt(III) complexes have been prepared (*119,120*). Compounds containing EDTA as a tridentate ligand are scarce (*121*). The chemistry of EDTA–ruthenium complexes has been extensively explored, due to their high reactivity toward water, dioxygen, dinitrogen, carbon monoxide, and related small molecules. Their activity in catalysis will be considered in the following sections.

Aqueous complexes containing semicarbazones, such as 2,6-diacetylpyridine bis(semicarbazone) (DAPSC) (*122,123*), semicarbazides (*90*), and tetrasulfophthalocyanines (*91*) have also been prepared; reports on these complexes deal essentially with their crystalline structures and magnetic properties.

Stability constants have been determined for a variety water-soluble complexes that have not as yet been used in catalysis. Chatt and co-workers were the first to examine stability constants for silver and cadmium complexes containing sulfonated aromatic ligands (*4,124*). Moreover, the L^1 ligand has been reinvestigated by Bjerrum and co-workers (*99,125*) in mercury, bismuth, copper, gold(I), palladium(II), and platinum(II) complexes. Other water-soluble ligands containing carboxylic groups have also been studied from a stability point of view for nickel, copper, cadmium, zinc, and cobalt (*88,89*). More recently, the water-soluble cluster $Au_{55}(L^1 \cdot 2H_2O)_{12}Cl_6$, successfully obtained by exchanging PPh_3 by L^1 in $Au_{55}(PPh_3)_{12}Cl_6$, was the first stable M_{55} cluster that could be imaged with its intact ligand by means of transmission electron microscopy (*126*). Finally, Aquino and Macartney substituted the water molecule of the compound $[Rh_2(OOCH_3)_4(H_2O)_2]$ with sulfonated, carboxylated, or aminated water-soluble phosphines in order to study precisely the kinetics and the mechanism of the exchange (*127*). The present review is restricted to the use of such compounds in catalysis.

III

WATER GAS SHIFT CATALYSIS

Numerous studies have been concerned with the production of dihydrogen from carbon monoxide and water according to Eq. (18), especially since the 1973 petroleum crisis.

$$H_2O + CO \leftrightharpoons H_2 + CO_2 \tag{18}$$

Because this reaction is exothermic under ambient conditions ($\Delta H^\circ_{298} =$ 0.68 kcal), it was expected that homogeneous systems operating at low temperature could replace the classical heterogeneous catalysts that work at temperatures of 400–450°C, wherein the equilibrium is only slightly shifted to the right. For instance, at 673 and 400 K, the equilibrium constants are 11.7 and 1.5×10^3, respectively. Several metal complexes and clusters, for example, $[RhI_2(CO)_2]^-$, $[Ru_3(CO)_{12}]$, $[Fe(CO)_5]$, have been shown to catalyze this reaction at slow rates. These systems have been recently reviewed (*128–130*). The present article is devoted to the species containing water-soluble ligands.

The metallic center must accommodate several oxidation states in such a way that the two main reactions involved can produce CO_2 then H_2, as shown by Eqs. (19) and (20), (*131, 132*). In order to improve the contact between the catalyst and water, complexes containing water-soluble ligands have been designed and their reactivity examined. Decomposition of the complexes, side reactions, and low kinetics, problems encountered with classical ligands, were expected to be eliminated.

$$ML_n + CO + H_2O \leftrightharpoons ML_n^{2-} + CO_2 + 2H^+ \tag{19}$$

$$ML_n^{2-} + 2H^+ \leftrightharpoons L_nMH^- + H^+ \leftrightharpoons ML_n + H_2 \tag{20}$$

The most efficient transition metals generally used are rhodium and ruthenium. The complex $[Fe(CO)_4(amphos)]I$ was shown by Smith and Baird to be 200 times less active than $[Fe(CO)_5]$ under the same basic conditions (140°C, $P_{CO} = 34$ bar, aqueous methanol). Moreover, the two complexes $[Mo(CO)_5(amphos)]I$ and $[W(CO)_5(amphos]I$ do not show any catalytic activity (*40*). In addition, no reactivity was noted for these three complexes under neutral conditions. We shall consider in turn the activity of rhodium and ruthenium complexes.

A. *Rhodium Complexes*

By adding the amphos ligand to $[Rh_2(NBD)_2Cl_2]$ in a molar ratio $P:Rh = 3$, the resulting aqueous mixture was shown to hydroformylate hex-1-ene moderately effectively at 40 bar in the absence of hydrogen (*115*). In fact, these hydroformylation reactions as well as hydrogenations are often a good means to detect and measure the water gas shift reaction (WGSR), because water is the only source of hydrogen. The formation of molecular hydrogen is generally confirmed by carrying out the reaction in the absence of olefin. Thus, Smith *et al.* (*115*) observed that the amounts of hydrogen produced were similar to those of heptanals and carbon dioxide formed in the latter experiment [Eq. (21)].

$$RCH{=}CH_2 + 2CO + H_2O \rightarrow RCH_2CH_2CHO + CO_2 \tag{21}$$

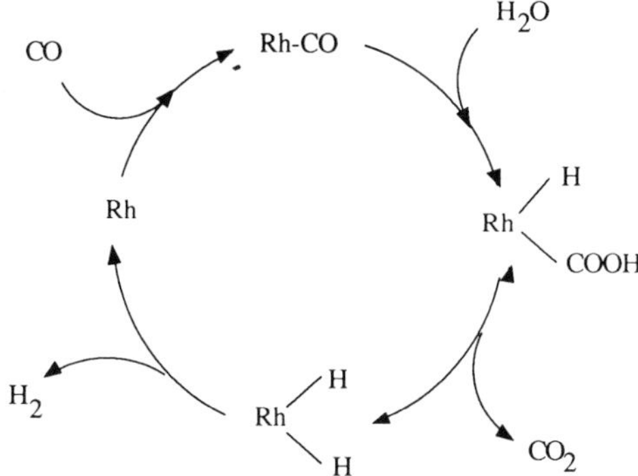

SCHEME 10. Simplified catalytic cycle for the WGGSR. Adapted from Ref. 133.

Presumably, the WGSR controls the reaction rate, produces dihydrogen according to Scheme 10 (*133*), and another rhodium activates H_2, CO, and the alkene to convert it into the corresponding aldehyde.

We have examined the efficiency of dinuclear rhodium(I) complexes $[Rh_2(\mu\text{-}SR)_2(CO)_2(L^3)_2]$ in these two reactions under mild conditions (8 bar, 80°C) (*134,135*). It was found that the presence of alkene in the medium significantly increases the production of hydrogen from water. In this case, it can be assumed that the dihydrido species generated by the WGSR is involved in the catalytic hydroformylation cycle according to Scheme 11.

The WGSR is often performed under neutral or basic conditions. However, a basic medium leads to extensive oxidation of the sulfonated L^3 ligand to the corresponding OL^3 compound. We have observed that acidic media exert a good influence on the activity and selectivity of the reaction. Interesting performances were achieved at pH 4.8, with a turnover frequency of 40 $hour^{-1}$ and selectivities in linear aldehyde as high as 96% (*134–136*). At similar concentrations, the mononuclear species $[Rh(COD)(L^3)_2][BF_4]$ and $[RhH(CO)(L^3)_3]$ exhibit lower activity.

The synergism between the two metallic centers has been proposed by Tikkanen *et al.* (*137*), who studied the ability of the dinuclear precursor $[Rh_2(\mu\text{-}CH_3COO)_3(bpnp)][PF_6]$ **45** to catalyze the WGSR in a 20% water/pyridine solution. Nevertheless, the achieved turnover frequency remains quite low because it is around 5 mol CO_2 (mol catalyst)$^{-1}$ hour^{-1} at 100°C and 1 bar CO. Rarely has the WGSR been studied in its own right. Similarly, Nuzzo *et al.* (*11*) showed that the complex [RhCl(CO)**10**] displays a catalytic activity for the same reaction. Under basic conditions (pH 13.8) and at 85°C, 4 bar, these authors obtained a good turnover frequency of 32 hour$-^{1}$. Moreover, the stoichiometric WGSR has been

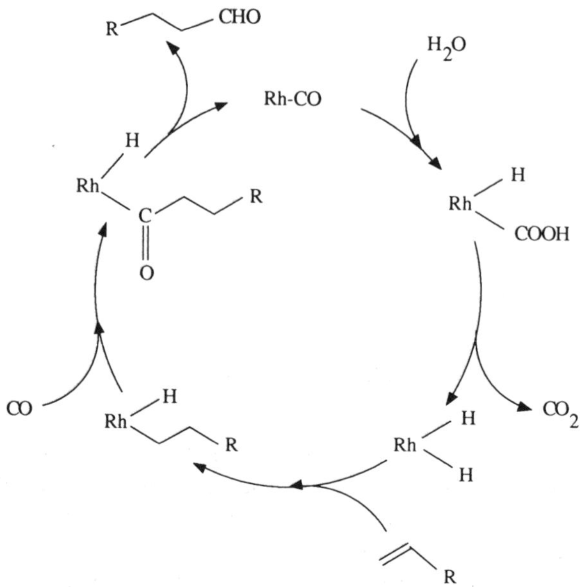

SCHEME 11. Simplified catalytic cycle involving the direct coordination of the alkene on the dihydro species.

recognized to account for the formation of $[HRh(CO)(L^3)_3]$ from $[RhCl(L^3)_3]$ (*20*) or from $[Rh(CO)_2(acac)]$ and $[Rh(C_2H_4)(acac)]$ under a CO atmosphere in the presence of an excess of the water-soluble L^3 ligand (*138*).

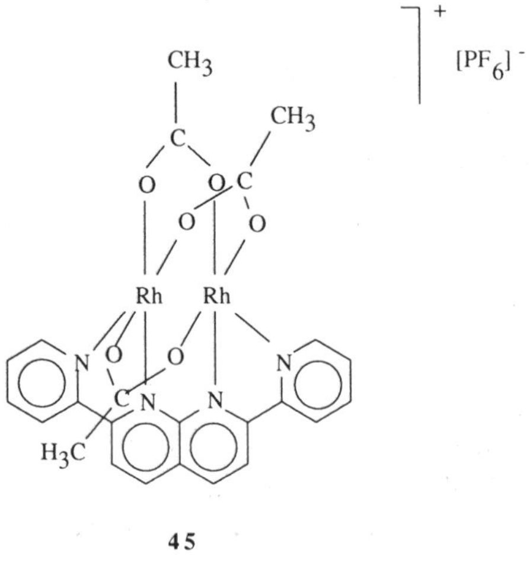

45

B. *Ruthenium Complexes*

During the past 10 years, significant results have been obtained by Taqui Khan *et al.*, using the classical but robust aforementioned ethylenediaminetetraacetic ligand, which induces high solubility in water. These workers have described the most active precursor yet reported for the WGSR in homogeneous catalysis (*139–141*), namely, a ruthenium(II)/EDTA complex. Turnover frequencies of 350 hour^{-1} were obtained. The authors have proposed that the active species $[Ru(EDTA-H)(CO)]^-$ induces the oxidative addition of a water molecule to afford a hydrido hydroxo carbonyl species. The OH migration on the coordinated CO ligand leads to the key complex **46**, from which H_2 and then CO_2 are evolved

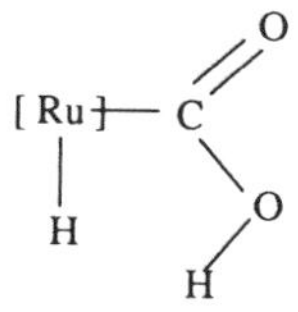

46

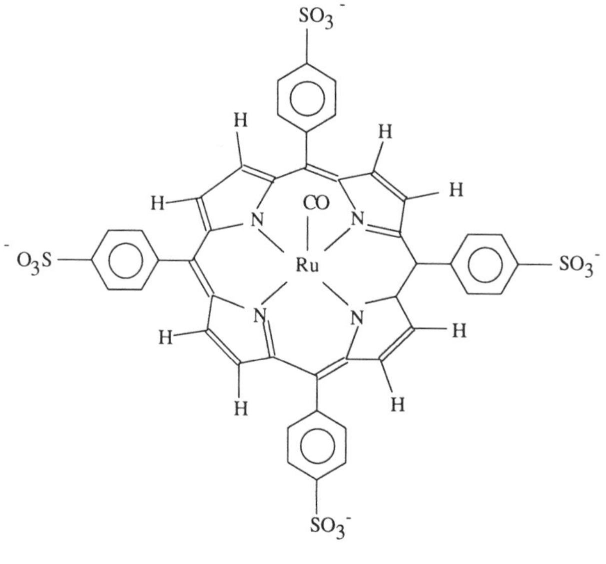

47

through a concerted mechanism. These authors have also observed the formation of formaldehyde and formic acid (*140*). The latter compound results from a rearrangement of complex **46** into a new species in which formic acid is coordinated to ruthenium through the oxygen atom. The formaldehyde formation implies the insertion of a CO ligand into the Ru–H bond instead of the Ru–OH bond in the hydrido hydroxo carbonyl intermediate.

Another approach to water-soluble ruthenium complexes for the WGSR has been to introduce sulfonate groups in the four phenyl rings of a ruthenium tetraphenylporphyrin. Under a CO atmosphere, the species $[Ru(TPPS)(CO)]^{4-}$ **47** was characterized and was shown to produce 6.3-fold more hydrogen and carbon dioxide than the classical $[Ru_3(CO)_{12}]$ precursor previously studied by Ford and Rokicki (*130*). The peculiar role of the porphyrin ligand was underlined because the complex can accommodate the two ligands, CO and OH^-, being in cis positions on one side of the porphyrin plane (*142*).

IV
HYDROFORMYLATION

The hydroformylation reaction is widely used to convert an alkene to an aldehyde by addition of hydrogen and carbon monoxide in the presence of a catalyst [Eq. (22)].

$$RCH{=}CH_2 + H_2 + CO \xrightarrow{(cat)} RCH_2CH_2CHO + RCH(CHO)CH_3 \qquad (22)$$

Cobalt and rhodium are active in this reaction. Cobalt requires severe temperature and pressure conditions (180°C, 200 bar). However, rhodium associated with triphenylphosphine allows mild conditions (120°C, 20 bar) to operate, resulting in a "low-pressure oxo" process, which was developed by Union Carbide, Davy Mc Kee, and Johnson Matthey; 12 plants have been constructed worldwide to convert propene into butanal (*143, 144*). The major disadvantage of this process is that it is restricted to light alkenes because the corresponding aldehydes are separated by distillation. For heavier alkenes, the temperatures required for the separation induce extensive catalyst degradation. Numerous attempts to anchor the complex on a support have been made to overcome this problem. However, due to losses of activity and selectivity, it was found that introduction of the active species in water with the appropriate ligands is a good means to retain a high level of activity with a simple way of recycling. Kuntz, at Rhône-Poulenc Industries, was the first to succeed and to patent a Rh/L^3 process, which has been developed by Ruhr Chemie Co. (*5,145,146*).

Three metals have been used for the hydroformylation in a biphasic medium: cobalt, ruthenium, and the most widely studied metal, rhodium.

A. *Rhodium Complexes*

1. *L^1 and Amphos Ligands*

By using the monosulfonated triphenylphosphine ligand L^1, the complex $[HRh(CO)(L^1)_3]$ has been prepared by Borowski *et al.* from $[HRh(CO)(PPh_3)_3]$ by an exchange method (*92*). This compound was employed in the hydroformylation reaction and was shown to be active above 70°C. However, at the end of the catalytic runs, the organic layer was found to be orange colored, due to the presence of some soluble rhodium species. No further work has been published on this system.

Leaching of rhodium into the organic phase, giving rise to yellow solutions, was also observed in the case of amphos/Rh systems, although it was claimed that less than 0.1% Rh was lost (*114,115*). The complex generated by addition of six equivalents of amphos to $[Rh_2(\mu\text{-Cl})_2(NBD)_2]$ was found to hydroformylate hex-1-ene into heptanals at 40 bar and 90°C, with 90% yield and an $n:i$ ratio of about 4 (this ratio is solvent dependent). Borowski *et al.* added an excess of L^1 to stabilize the catalyst and improve the $n:i$ ratio; in contrast, an amphos : rhodium ratio of 10 was found to induce a lower selectivity in linear aldehyde (3.4 instead of 4). It was also noted that acidic media favor hydrogenation of hex-1-ene into hexane (16% at pH 5).

2. *L^3 Ligand*

This highly water-soluble ligand provides a very interesting way to maintain the catalyst in the aqueous phase; all researchers using this compound have reported colorless organic layers. Most of the early studies concern propene, as evidenced by the first patents (*5,145,146*). In general, a large excess of ligand L^3, as high as 340 g $liter^{-1}$ L^3/0.4 g $liter^{-1}$ Rh, was added to $[Rh_2(\mu\text{-Cl})_2(COD)_2]$ in order to reach a 95% selectivity in *n*-butanal and a productivity of 150 g of C_4 aldehydes per hour and per liter of solution (corresponding to a turnover frequency 95.5 $hour^{-1}$) (*5,14,145,146*). A selectivity in butanals of 96% at 30 bar and 125°C was characteristic of this system (*145*).

Complex $[HRh(CO)(L^3)_3]$ is highly efficient in transforming hex-1-ene into C_7 compounds at 5 bar and 80°C with selectivities into linear aldehyde of up to 93% (*147*). The same complex was prepared starting from $[Rh(CO)_2(acac)]$ or $[Rh(C_2H_4)_2(acac)]$ and its catalytic behavior has been compared with that of $[HRh(CO)(PPh_3)_3]$ (*138*). Both ^{13}C and ^{31}P NMR studies have shown that the species $[HRh(CO)(L^3)_3]$ remains intact under 200 bar of CO/H_2, whereas the dicarbonyl species $[HRh(CO)_2(PPh_3)_2]$ is formed with triphenylphosphine. Thus, absence of the dicarbonyl complex in the case of L^3 ligand accounts for the high selectivity in linear aldehyde (*138*).

Higher performances are obtained with the dinuclear rhodium complexes $[Rh_2(\mu\text{-}SR)_2(CO)_2(L^3)_2]$, provided that a slight excess of ligand is added (*148*). For example, at 5 bar and 80°C, with a P : Rh molar ratio of 3, a 100% conversion of hex-1-ene into heptanal and a selectivity of 94% in linear aldehyde is reached. On increasing to a slight extent the P : Rh ratio (6, for instance, and at 10 bar), the selectivity in *n*-heptanal is raised to 97% (*148*). As mentioned in all patents related to L^3/Rh systems (*5,145,146,149*), simple decantation after catalysis allows separation of the organic and aqueous layers, thereby permitting easy catalyst recycling. For propene (*149*) or hex-1-ene (*148*), the linearity is not affected by recycling, the same level of activity being measured. The organic solutions remain colorless and analysis shows that the amount of leached rhodium is 0.1 ppm (w/w).

3. *Monocarboxylated Triphenylphosphine*

When used in catalysis, the L^3 ligand discussed above is characterized by slow reaction rates, due to mass transfer control, as opposed to kinetic control of the homogeneous hydroformylation systems. In order to improve the transfer, Russell and Murrer from Johnson Matthey have shown that addition of the water-soluble diphenyl (*p*-carboxylatophenyl)-phosphine is effective provided a phase transfer agent is added (*150,151*). High selectivities to normal aldehyde, relatively good rates, and minimized rhodium losses were obtained when poor mixing of the liquid phases was adjusted. In fact, low mixing rates allow the formation of micelles, which are the key to the high selectivity observed. Such organized systems explain why heavy alkenes such a hexadec-1-ene can be transformed relatively quickly (73% in 1 hour instead of 0.5% in 3 hours if $C_{12}H_{25}NMe_3Br$ is absent). The reaction occurs under mild conditions (10 bar and 100°C).

4. *New Surface Active Ligands*

The sulfobetaine derivatives of tris(2-pyridyl)phosphine described in Section II,A appeared to be an attractive material to achieve the biphasic hydroformylation of higher olefins. Indeed, the micellar hydroformylation of tetradec-1-ene gave up to 79% yield when carried out at 125°C and 75 bar CO/H_2 for 3 hours in the presence of the water-soluble catalyst

$[Rh_4(CO)_{12}]$ and tris(2-pyridyl)phosphine sulfoalkylated by octane-1,2-sultone. Thus this surfactant phosphine improves the contact between the catalyst and the organic layer, which contains substrates and products, allowing a quantitative recovery of the active species by simple decantation (*13*).

5. *Hydroformylation under Water Gas Shift Conditions*

In the absence of added dihydrogen, hydroformylation of hex-1-ene has been shown to occur only from carbon monoxide and water. Amphos/rhodium systems gave low yields (~5% at 40 bar and 90°C for 24 hours) in aldehydes with a poor linearity ($n:i = 1.1$). Presumably, the twofold CO pressure used for the WGS experiment, when compared with the CO/H_2 runs, induces the formation of the $HRh(CO)_2(amphos)_2$ species, which is known to decrease the linearity (*138*).

With L^3, the three complexes $[HRh(CO)(L^3)_3]$, $[Rh(COD)(L^3)_2]$-$[ClO_4]$, and $[Rh_2(\mu\text{-S-}t\text{-Bu})_2(CO)_2(L^3)_2]$ were shown to selectively convert hex-1-ene into the corresponding aldehydes. The best yields were obtained with the dinuclear complex in an acetic buffer ($P_{CO} = 8$ bar, 80°C, turnover frequency 40 hour^{-1}, selectivity to normal aldehyde 96%) (*134–136*).

B. *Ruthenium Complexes*

During their preliminary exploration of the catalytic activity of L^1 complexes, Borowski *et al.* noted that $[RuHCl(L^1)_3]$ is a precursor for hydroformylation of hex-1-ene at 60 bar and 90°C (*92*). However, extensive isomerization of the terminal alkene occurred so that only 30% conversion to aldehydes was obtained with a linear selectivity of 75%.

The $[Ru(EDTA)]^-$ complex studied by Taqui Khan *et al.* was investigated for the hydroformylation of allyl alcohol as well as hex-1-ene (*152,153*). The reactions were carried out at 50 bar within the range of 90 to 130°C. Kinetic studies led the authors to propose the mechanism given in Scheme 12. The $[Ru(EDTA–H)(CO)]^-$ complex is assumed to be the precursor, which reacts in a first step with dihydrogen to generate the active species $[Ru(EDTA\text{-}H)(CO)(H)]^{2-}$. Coordination of the alkene occurs very quickly; the rate-determining step is the hydride transfer to the coordinated alkene giving the alkyl species. For hex-1-ene, very high linearities were gained for the C_7 aldehyde (98 to 100%) (*141,153*). However, for allyl alcohol, only 35% hydroxybutyraldehyde was produced along with its cyclization products γ-butyrolactone (25%), dihydrofuran (25%), and polymeric compounds (*152*).

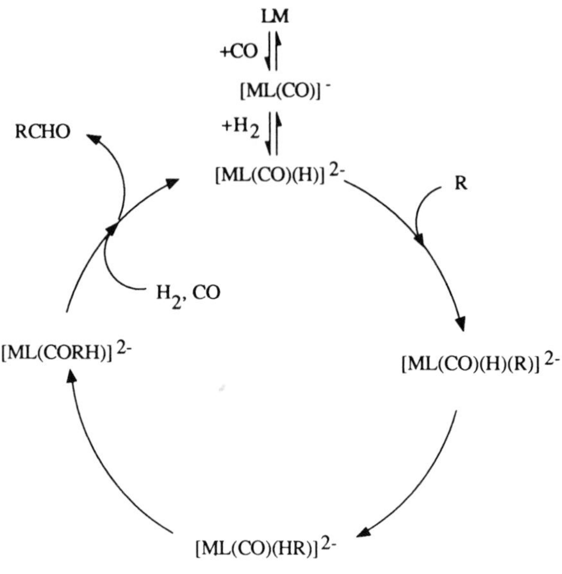

L = EDTA-H ; M = Ru

SCHEME 12. Catalytic cycle for the hydroformylation of allylic alcohol or hex-1-ene.

C. *Cobalt Complexes*

Given that studies of the catalytic performances of water-soluble complexes that operate in biphasic systems are very recent, and because rhodium appeared to be the most appropriate metal to achieve hydroformylation, papers related to cobalt systems are at present rather scarce. The hydroformylation of terminal or internal alkenes by $[Co_2(CO)_8]/L^3$ has been patented by Rhône-Poulenc Industries (*154*). For instance, at 140°C and 130 bar (H_2:CO = 2:1) this catalytic system gives 89% heptanals after 3 hours, with a selectivity in linear aldehyde of 62%. The same reaction starting from hex-2-ene affords aldehydes after 18 hours, with 68% selectivity, the linearity being 59%, as usually observed for cobalt catalysts. In all these cases the organic solutions were slightly yellow colored and the cobalt leaching was about 1% of the initial metal amounts (*154*). Thus cobalt appears to be less attractive than rhodium for the selective hydroformylation of alkenes, except the internal ones, because the operating conditions are generally more severe, the loss in organic solution is higher, and the selectivities are poorer.

These results have been confirmed by Markiewicz and Baird, who explored the activity of $[Co(CO)_3(amphos)_2][PF_6]_2$ in a two-phase system (*155*). Indeed, at 100°C, 80 bar, during 24 hours, a turnover frequency of 6 $hour^{-1}$ was achieved; the selectivities were, respectively, 35% for *n*-heptanal, 33% for 2-methylhexanal, and 32% for *n*-heptanol. Hydrogenation of aldehydes can be avoided by addition of benzene as cosolvent.

V
HYDROGENATION

Catalytic hydrogenation tests appear from the literature to be very popular. This may be due to the fact that active systems work usually at 1 bar. In order to separate the products from the catalyst, the first application of two-phase systems involved the hydrogenation of olefins catalyzed by rhodium complexes containing the monosulfonated triphenylphosphine L^1 (*105*). In addition, water-soluble substrates were hydrogenated in aqueous phase (*95*). Since the early investigations, particularly by Wilkinson *et al.* on $[RhCl(PPh_3)_3]$, it was known that varying the coordination sphere of the metal center can induce high selectivities (*156*). These studies were extended to asymmetric hydrogenations using water-soluble chiral ligands, thereby revealing a peculiar role played by water on the enantiomeric excess with regard to that observed in the homogeneous phase.

A. *Hydrogenation in Two-Phase Systems*

1. *Hydrogenation of Water-Immiscible olefins*

Dror and Manassen have shown that the hydrogenation of cyclohexene in a two-phase system can be catalyzed by a complex prepared *in situ* from $[RhCl_3 \cdot 3H_2O]$ and an excess of ligand L^1, provided a cosolvent is added (*105*). The activity of the catalyst increases in the order dimethylacetamide < dimethoxyethane < ethanol < methanol, the conversions being as high as 90% in the last case.

Recently, ternary diagrams have been established for the water/oct-1-ene/cosolvent systems (*157*). Such calculations have shown that methanol and ethanol are particularly convenient to solubilize oct-1-ene in the aqueous phase, whereas the solubility of water in the organic phase is maintained at a very low level. Accordingly, when higher catalytic activities are observed in the presence of better cosolvents, it can be concluded that catalysis occurs in the aqueous phase and not at the interface. High stirring rates are required to be maintained in order to ensure a good gas–liquid transfer.

Borowski *et al.* have reported the hydrogenation of hex-1-ene, *cis/trans*-hex-2-ene, and cyclohexene in a two-phase medium without cosolvent (*92*). Among the two species, $[RhCl(L^1)_3]$ and $[RuHCl(L^1)_3]$, the former was shown to be the most effective and the most selective catalyst for terminal alkenes. However, significant amounts of internal alkene were obtained from hexenes (*92*), and we interpret these results as being due to the absence of cosolvent. Indeed, the solubility of substrate in water is much reduced so that the isomerization process competes with hydrogenation. Mild conditions have been adopted by the authors (25–80°C, 3 bar), who observed that the two phases separate readily after reaction. Yet, the main disadvantage of such a process is that air must be carefully excluded, due to the air sensitivity of the two Rh(I) and Ru(II) complexes.

Another analog of the Wilkinson complex, prepared by Larpent *et al.* by mixing of $[RhCl_3 \cdot 3H_2O]$ and L^3, has been shown to hydrogenate numerous olefins with 100% conversion and complete selectivity on the C–C double bond, regardless of the identity of the various functional group (*101*). Nevertheless, ^{31}P NMR analyses showed that full oxidation of L^3 occurred very quickly so that colloidal dispersions of rhodium stabilized by the oxide OL^3 become the catalytically active species (*101*). Such a reduction to rhodium(0) with simultaneous formation of phosphine oxide has not been observed in the case of α,β-unsaturated aldehydes (*158*). The rhodium(I) complex, presumably $[RhCl(L^3)_3]$, prepared by *in situ* addition of L^3 to $[Rh_2(\mu\text{-}Cl)_2(COD)_2]$, produced the corresponding saturated aldehydes with no decarbonylation at all. At 20 bar, hydrogenation of the aldehyde function occurred sufficient to afford the saturated alcohols, but generally selectivities higher than 95% were reached for the aldehydes (*158,159*). However, the absence of reduction of the latter complex is in agreement with recent experiments of Lecomte and Sinou. In order to determine the homogeneous or heterogeneous nature of the catalyst, they used chiral phosphines and measured enantioselectivities after recycling the catalyst. Thus they showed that the Rh(I)/chiral sulfonated phosphine system is a homogeneous catalyst whereas Rh(III) induces oxidation of phosphine ligand, leading finally to a heterogeneous system (*78*).

Further studies carried out by Larpent and Patin have shown that the dihydride species $[RhH_2Cl(L^3)_3]$ (cis-mer and cis-fac isomers) have disappointing catalytic activity. Indeed, no hydrogenation took place if sufficient amounts of phosphine oxide had not been produced (*160*). Thus when Rh(III) complexes were introduced into the medium, catalytic activity was attributed to metallic rhodium particles dispersed and stabilized by sulfonate anions, which prevent metal aggregation, as shown by hydrogenation of cycloheptene followed by ^{31}P NMR and light-scattering measurements (*100*).

In contrast with the rhodium catalysts containing the ligand L^1, the corresponding amphos systems are much more stable to oxidation by air and can even be handled in air and recycled without any special precautions (*115*). The species $[Rh(NBD)(amphos)_2]^{3+}$ behaves as its triphenylphosphine analog so that hex-1-ene is hydrogenated faster than styrene. Isomerization of hex-1-ene into internal olefins (10–20%) was observed. The activity of the catalyst is strongly dependent on the organic solvent; diethyl ether is more efficient than *n*-heptane, possibly due to its greater solubility in water, *n*-heptane displaying higher efficiency as solvent than dichoromethane, which can even present an inhibiting effect. The aqueous phase can be reused because only 0.1% of the catalyst leaks into the organic layer.

Ruthenium complexes such as $[RuCl_2(L^1)_2]$, $[RuHCl(L^1)_3]$, and $[RuH(OAc)(L^1)_3]$ have been shown by Joó and co-workers to hydrogenate hex-1-ene and styrene under similar conditions but with much slower rates than $[RhCl(L^1)_3]$ (*94*).

As a result of improvements in the synthesis of hydrido or polyhydrido complexes of ruthenium in recent years, Taqui Khan *et al.* have focused on water-soluble ruthenium complexes containing EDTA to catalyze the hydrogenation of cyclohexene under mild conditions (10–40°C, 0.41 bar) (*161,162*). The four complexes $[K\{Ru(EDTA{-}H)Cl\}\cdot 2H_2O]$, $[Ru(EDTA{-}H)(PPh_3)]$, $[K\{Ru(EDTA{-}H)(CO)\}]$, and $[K\{Ru(EDTA{-}H)(SnCl_3)\}]$ have been shown to follow the same rate equation, Eq. (23):

$$-\frac{d[H_2]}{dt} = \frac{k_1K_1[H_2][C][S] + k_2K_2[H_2][C][S]}{(1 + K_1[H_2])(1 + K_2[S])} \tag{23}$$

where $[H_2]$ is the dihydrogen partial pressure, [C] and [S] are the initial catalyst and substrate concentrations, respectively, K_1 and K_2 are the equilibrium constants for the formation of the hydrido complex [Ru(EDTA–H)(L)(H)] and olefin complex [Ru(EDTA–H)(L)(S)], respectively ($L = Cl^-$, CO, PPh_3, $SnCl_3^-$), and k_1 and k_2 are the constants corresponding to the hydride transfer to the olefin by the two preceding routes. This equation accounts for two competing pathways. The first involves the coordination of cyclohexene after formation of the hydride species (hydride route) and the second is the opposite (olefin) route. Whereas the olefin route is largely faster than the hydride one, because the olefin complex is dramatically less stable than the hydride ($K_1 > K_2$), the reaction proceeds preferentially through the hydride pathway. Thermodynamic data and activation parameters correlate quite well with the influence of the π acidity of the ligands on the stability and the reactivity of the various species as determined by the ^1H NMR shift of the hydride ligand. The two complexes, [Na{Ru(EDTA–H)N_3}·$2H_2O$] and [{Ru(EDTA–H)(NO)}-BF_4], were shown to afford similar behavior (*162*).

2. *Hydrogenation of Unsaturated Aldehydes*

We have emphasized that rhodium/L^3 systems are very efficient for hydrogenation of the carbon–carbon double bond of unsaturated aldehydes, such as crotonaldehyde, but-2-eneal, or cinnamaldehyde (see Section V,1,a). However, reduction of the aldehyde function can be achieved selectively with ruthenium complexes under two-phase conditions. Following selections of several hydrido ruthenium complexes for their catalytic activity in hydride transfer reactions, Joó and Bényei found that aromatic and aliphatic aldehydes could also be reduced to the corresponding unsaturated alcohols with complexes containing the L^1 ligand and using sodium formate as hydrogen donor in the aqueous phase (*163,164*). Except for the crotonaldehyde, which gave only 18% yield of crotyl alcohol, numerous aromatic as well as allylic alcohols were highly selectively obtained with the most efficient catalyst [$RuCl_2(L^1)_2$] at 80°C. Although the reaction mechanism has not been defined, the [$RuH(O_2CH)(L^1)_3$] species appears to be a key intermediate of the process. High rates and conversions could be obtained provided an excess of L^1 ligand was added. Contrary to the formation of an inactive species, in this case the bis(aldehydo)-ruthenium complex [$RuH(O_2CH)(RCHO)_2(L^1)_2$] (*164*), the tensioactive properties of the ligand may be involved in the acceleration of the catalytic process.

Grosselin *et al.* have shown that the deactivation phenomenon that prevents efficient recycling of the aqueous phase could be overcome by using the L^3 ligand (*158,165,166*). An active ruthenium species, generated *in situ* by addition of the salt [$RuCl_3 \cdot 3H_2O$], was found to convert several unsaturated aldehydes into the corresponding unsaturated alcohols, with selectivities as high as 99%. The reduction could be carried out either with hydrogen (generally at 20 bar) or with sodium formate. Recycling the complex twice after reduction of substrate [Eq. (24)] gave 97% selectivity both times. Interestingly, the presence of a phosphate buffer to maintain neutral pH conditions inhibited the side effect of C=C hydrogenation favored by an acidic medium (*158*).

O ⟶ OH (99 %) + OH (1 %) (24)

B. *Aqueous Phase Catalysis*

Whereas the water insolubility of transition metal complexes containing conventional ligands restricts their application to catalysis in organic solutions, the introduction of water-soluble ligands allows hydrogenation of many olefinic or keto acids in aqueous media.

1. *Ruthenium Complexes*

Following the early disappointing report by Chatt *et al.* that hydroxymethylphosphine complexes of rhodium, palladium, and platinum were inactive in catalysis (*106*), Joó *et al.* carried out pioneering investigations on active water-soluble complexes. Their preliminary experiments showed that the Ru/L^1 system catalyzes the hydrogenation of pyruvic acid to lactic acid with low conversion rates, that is, the turnover number does not exceed 30 (95). These poor performances were attributed initially to the conversion of the catalytic species to inactive ruthenium(II) complexes, but later the effect was ascribed to catalyst deactivation due to the presence of small amounts of dioxygen in the medium (*94*).

This study was extended to several keto and other unsaturated acids. The ruthenium complexes $[HRu(OOCCH_3)(L^1)_3]$ **48**, $[RuCl_2(L^1)_2]$ **49**, and $[HRuCl(L^1)_3]$ **50** were prepared and characterized and their catalytic activity was investigated. Only complexes **48** and **50**, which contain three phosphine ligands, catalyzed the hydrogenation of 2-keto acids (pyruvic, phenylpyruvic, 2-ketopentanoic, and 2-ketooctanoic acids). It was found that when only two phosphines are present in the coordination sphere of the metal, the keto acid coordinates strongly enough to produce a catalytically inactive species. On the other hand, complex **50** is a very effective catalyst for the hydrogenation of ketonic function—1300 turnovers were obtained for pyruvic acid, giving lactic acid selectively [Eq. (25)] (*167*).

$$CH_3COCOOH + H_2 \xrightarrow[\text{1 bar, 60°C, pH 1}]{\textbf{50}} CH_3CH(OH)COOH \qquad (25)$$

It is necessary to start from complex **50** to gain a high activity. Otherwise, reduction of the $[RuCl_3 \cdot 3H_2O]/L^1$ system in the presence of pyruvic acid generates a complex that contains the pyruvato ligand and two L^1 ligands and thus leads to an induction period. The kinetic study of hydrogenation of pyruvic acid led the authors to propose that $[HRuCl(L^1)_3]$ or $[HRu(OAc)(L^1)_3]$ was the active species. For crotonic acid, however, the active species is likely to be $[HRuCl(L^1)_2]$ or $[HRu(OAc)(L^1)_2]$ (*94*).

Very recently, Basset and co-workers have shown that superior activity occurs in water rather than in classic organic solvents, provided that an iodide salt is added (*168*). Starting from either $[RuCl_2(L^3)_3]$, $[HRuCl(L^3)_3]$, $[HRu(OAc)(L^3)_3]$, or $[RuH_2(L^3)_4]$, the active species would be $[HRuI(L^3)_2]$ in the presence of NaI. Moreover, NaI would assist the rapid formation of a metal–carbon bond to give with propionaldehyde the intermediate Ru–CH(Et)ONa, which reacts with water to give a Ru–CH(OH)Et species. This peculiar effect is due to the polarity of water.

2. *Rhodium Complexes*

Joó and co-workers have also studied the kinetics of hydrogenation in the presence of complex $[RhCl(L^1)_3]$ (*169*), the preparation of which had been previously reported (*95*). From studies with crotonic, maleic, and fumaric acids, two innovating concepts were proposed, though they have not yet been confirmed by detailed mechanistic investigations. As with the homogeneous systems, the reduction of crotonic acid involves dissociation of a ligand and formation of a dihydride species. However, studies with the two other acids imply instead the dissociation of a chloro ligand, leading to the $[Rh(L^1)_{2\text{ or }3}]^+$ species. Moreover it has been claimed that carboxylato species $[Rh(RCOO)(L^1)_3]$ are probably formed in aqueous medium, regardless of the identity of the substrate. Water in this system could play a determinant role as the proton acceptor, a phenomenon that cannot occur in an organic solvent. Thus, displacement of Cl^- by $RCOO^-$ would give the catalytically active carboxylato species, each R group exhibiting its characteristic dissociation constant.

The highly air-stable complex $[Rh(NBD)(amphos)_2]^{3+}$, obtained from the reaction of $[Rh(NBD)Cl]_2$ with four equivalents of amphos nitrate, catalyzes the hydrogenation of maleic and crotonic acids, at 1 bar, with higher activities than the analogous rhodium complexes containing L^1(*114,115*). Moreover, the reaction rates are higher in methanol than in water, as a consequence of either the greater solubility of dihydrogen in methanol or the lower stability of $[RhH_2(amphos)_2(H_2O)_2]^{3+}$ as compared with $[RhH_2(amphos)_2(MeOH)_2]^{3+}$. An excess of amphos is undesirable because it prevents olefin coordination.

Addition of three equivalents of the L^3 ligand to $[RhCl_3 \cdot 3H_2O]$ affords an air-stable catalyst that converts numerous unsaturated carboxylic acids to their saturated analogs. It was shown by ^{31}P NMR spectroscopy that the system is in fact a mixture of the oxide OL^3 (75%), the dimeric complex $[RhCl(L^3)_2]_2$ (15%), and the monomer $[RhCl(L^3)_2(H_2O)_x]$ (10%) (*170*).

Several rhodium(I) complexes containing water-soluble bidentate ligands derived from bis[2-(diphenylphosphino)ethyl]amine **1** have been investigated as catalyst precursors in the homogeneous hydrogenation of acids (*11,12*). The rhodium complex made from **13** was the easiest to prepare. It was the most effective catalyst for the hydrogenation of α-acetamidoacrylic acid—turnover frequencies higher than 275 $hour^{-1}$ were found at 2.3 bar and at room temperature. Accordingly, it was used for other substrates.

The advantageous addition of surfactant must also be emphasized. Indeed, whereas the rhodium(I) complex containing $HOOCCH_2CH_2CON(CH_2CH_2PPh_2)_2$ gave a turnover frequency lower than 5 $hour^{-1}$ for the hydrogenation of α-acetamidoacrylic acid, addition of sodium dodecyl sulfate induced the formation of micelles, which stabilize the catalyst and certainly increase the local concentration of the complex and presumably that of the substrate within the small volume formed; turnover frequencies as high as 120 $hour^{-1}$ were thus reached (*12*).

Performances of $[HRu(OOCCH_3)(L^1)_3]$, $[RuCl_2(L^1)_2]$, $[HRuCl(L^1)_3]$, $[RhCl(L^1)_3]$, and $[\mathbf{6}\cdot Rh(NBD)][CF_3SO_3]$ in the hydrogenation of several unsaturated substrates are listed in Table VIII.

C. *Asymmetric Hydrogenation*

During the past decade, numerous studies have been devoted to the hydrogenation of prochiral substrates in the presence of rhodium complexes containing chiral ligands. Among interesting precursors for amino acids, α-amidoacrylic acid **51a** and α-amidocinnamic acid **51b** have often been selected. Thus hydrogenation can occur either in aqueous solution or in two-phase systems.

R_1–CH=C(COOR$_2$)(NHCOR$_3$)

51

CH_2=C(COOH)–CH_2–COOH

52

R_1	R_2	R_3	
H	H	Me	**51a**
Ph	H	Me	**51b**
Ph	Me	Me	**51c**
Ph	H	Ph	**51d**
Ph	Me	Ph	**51e**
3-OMe-4-OAc-C_6H_3-	H	Me	**51f**

1. *Asymmetric Hydrogenation in Aqueous Solutions*

Because unsaturated acids are generally reduced in alcoholic solutions, attempts were made to vary the solvent composition by adding water and using water-soluble chiral ligands. Enantioselectivity fluctuations were

TABLE VIII

INITIAL RATES FOR HYDROGENATION OF UNSATURATED SUBSTRATE IN AQUEOUS SOLUTION

Substrate	Catalyst[a]	Temp. (°C)	pH	H_2 pressure (bar)	Initial rates[b]	Ref.
trans-HOOC—CH=CH—COOH	D	60	—	0.8	53	*169*
	A	60	4.8	1	92	*167*
	C	60	1	1	49	*167*
	E	60	7	2.2	144	*11*
	E	25	7	2.2	48	*11*
	E	25	7	8.4	130	*11*
cis-HOOC—CH=CH—COOH	A	60	4.8	1	92	*167*
	C	60	1	1	23	*167*
	D	60	—	0.8	1270	*169*
trans-H_3C—CH=CH—COOH	A	60	4.8	1	198	*167*
	B	60	6-7	1	164	*167*
	C	60	1	1	23	*167*
	D	60	—	0.8	180	*169*
	E	25	7	2.2	13	*11*
Ph—CH=CH—COOH	A	60	4.8	1	92	*167*
	C	60	1	1	30	*167*
	D	60	—	0.8	46	*169*
trans-4(OH)C_6H_4—CH=CH—COOH	E	25	7	2.2	13	*11*
CH_2=C(COOH)—CH_2—COOH	C	60	1	1	84	*167*
CH_2=CH—CH=CH—COOH	A	60	4.8	1	175	*167*
CH_2=C(NHAc)COOH	E	25	7	2.2	>200	*11*
(Z)—Ph—CH=C(NHAc)COOH	E	25	7	2.2	>30	*11*
CH_3—CO—COOH	A	60	4.8	1	46	*167*
	C	60	1	1	157	*167*
	D	60	—	0.8	35	*169*
	E	25	7	2.2	14	*11*
C_6H_5—CO—COOH	A	60	4.8	1	46	*167*
	C	60	1	1	115	*167*
CH_3—$(CH_2)_2$—CO—COOH	A	60	4.8	1	74	*167*
	C	60	1	1	70	*167*
	D	60	—	0.8	18	*169*
CH_3—CO—$(CH_2)_2$—COOH	A	60	4.8	1	25	*167*
	C	60	1	1	13	*167*
	D	60	—	0.8	9	*169*
HOOC—$(CH_2)_2$—CO—COOH	A	60	4.8	1	67	*167*
	C	60	1	1	95	*167*
HO—CH_2—CO—CH_2OH	C	60	—	1	60	*167*
	D	60	1	0.8	12	*169*
CH_2=CH—CH_2—OH	D	60	—	0.8	111	*169*
	E	25	7	2.2	19	*11*

[a] A, $[HRu(CH_3COO)(L^1)_3]$; B, $[RhCl_2(L^1)_2]$; C, $[HRuCl(L^1)_3]$; D, $[RhCl(L^1)_3]$; E, $[\mathbf{6} \cdot Rh(I)(NBD)]^+Tf^-$.

[b] Expressed as moles of H_2 per mole of catalyst per hour.

generally observed and were attributed not only to the modifications on the ligand but also to solvent effects.

The first application of the use of mixed solvents was made on chiral ferrocenylphosphines (*171*), which proved to be very effective for asymmetric hydrogenation in methanol or ethanol (*81,82*). The (*S*)-α-[(*R*)-1′,2-bis(diphenylphosphino)ferrocenyl] ethyldimethylamine **53**, abbreviated (*S*)-(*R*)-BPPFA, coordinated to rhodium(I), induced higher optical yields in water than in alcohols. Such an improvement in enantioselectivity was interpreted as a consequence of the interactions between the amino substituent of **53** and the carboxylic group of the substrate (*171*).

In contrast to these results, the carboxylated and sulfonated diphosphines **22** and **23**, derived from **16**, exhibit a slightly lower enantioselectivity in water than that obtained in ethanol for the hydrogenation of **51a** and **51b** (*71*). The same trend but with a smaller variation between the two kinds of solvents was observed when **51a** was reduced with the pyrrolidinium–rhodium species **54** (*52*). The influence of the composition of the

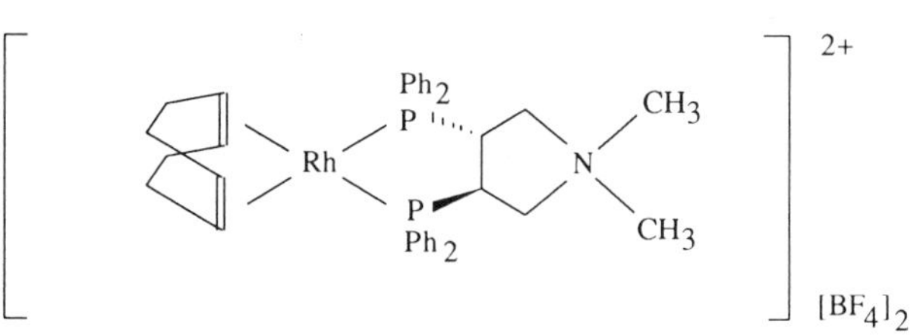

54

water/alcohol mixture was studied by Sinou *et al.* (*172,173*) and extended by Tóth *et al.* with the object of explaining the lower optical yields in water (*174*). The results are summarized in Table IX (*172–176*).

It appears that the enantiomeric excess obtained in water with a 1,2-diphosphine is generally close to that observed in ethanol, whereas greater differences are observed with 1,3- and 1,4-diphosphines. Indeed, when using DIPAMP **55**, norphos **56**, or chiraphos **18**, the enantioselectivity is

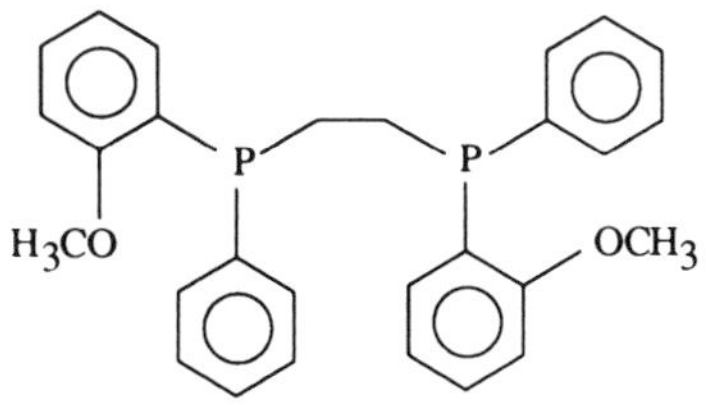

DIPAMP

55

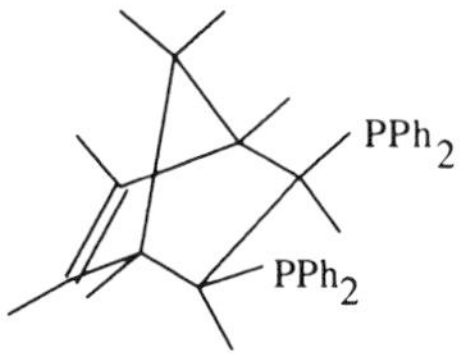

NORPHOS

56

TABLE IX

INFLUENCE OF SOLVENT ON THE ASYMMETRIC HYDROGENATION OF SEVERAL UNSATURATED ACIDS[a]

Substrate	Ligand	Solvent	Enantiomeric excess (%)	Configuration	Ref.
51a	$\mathbf{26_{16}}$[b]	EtOH	43	*R*	*73*
		$EtOH/H_2O$ (75/25)	39	*R*	*73*
		$EtOH/H_2O$ (50/50)	36	*R*	*73*
		$EtOH/H_2O$ (25/75)	33	*R*	*73*
		H_2O	11	*R*	*73*
	$\mathbf{26_{42}}$[b]	EtOH	46	*R*	*73*
		H_2O	10	*R*	*73*
51b	**55**[b]	EtOH	98		*175*
		$EtOH/H_2O$ (50/50)	98		*175*
	56[b]	EtOH	95		*176*
		$EtOH/H_2O$ (50/50)	94		*176*
	18d[c]	MeOH	90	*R*	*174*
		H_2O	90	*R*	*174*
	20a[b]	EtOH	73		*173*
		$EtOH/H_2O$ (85/15)	70		*173*
		$EtOH/H_2O$ (75/25)	68		*173*
		$EtOH/H_2O$ (50/50)	63		*173*
		$EtOH/H_2O$ (20/80)	43		*173*
		H_2O	33		*173*
	20d[c]	MeOH	91	*R*	*174*
		$MeOH/H_2O$ (50/50)	73	*R*	*174*
		$MeOH/H_2O$ (20/80)	70	*R*	*174*
	21a[b]	EtOH	52		*173*
		$EtOH/H_2O$ (75/25)	46		*173*
		$EtOH/H_2O$ (50/50)	42		*173*
		$EtOH/H_2O$ (25/75)	36		*173*
		H_2O	28		*173*
	$\mathbf{26_{16}}$[b]	EtOH	69	*R*	*73,172*
		H_2O	30	*R*	*73,172*
	$\mathbf{26_{42}}$[b]	EtOH	65	*R*	*73*
		H_2O	30	*R*	*73*
51c	**18d**[c]	MeOH	75	*R*	*174*
		H_2O	74	*R*	*174*
	$\mathbf{26_{16}}$[b]	EtOH	60		*73*
		$EtOH/H_2O$ (50/50)	46		*73*
	$\mathbf{26_{42}}$[b]	EtOH	50		*73*
		$EtOH/H_2O$ (50/50)	43		*73*
52	$\mathbf{26_{16}}$[b]	EtOH	39	*S*	*73*
		H_2O	20	*S*	*73*
	$\mathbf{26_{42}}$[b]	EtOH	47	*S*	*73*
		H_2O	9	*S*	*73*

[a] Conditions: T = 25°C; 1 bar of H_2 partial pressure.
[b] Complex involved in the reaction is $[Rh(COD)(ligand)][ClO_4]$.
[c] Complex involved in the reaction is $[Rh(diene)(ligand)][BF_4]_5$.

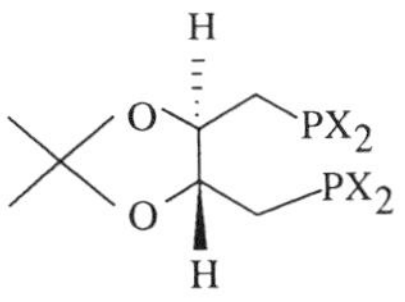

17 (DIOP)

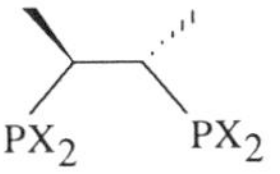

18 (CHIRAPHOS)

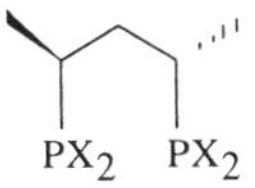

20 (SKEWPHOS)

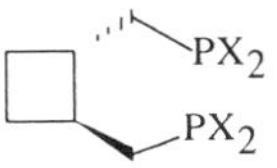

21 (CBD : cyclobutanediop)

X	
C_6H_5	
$[m\text{-}C_6H_4SO_3Na]$	a
$[p\text{-}C_6H_4NMe_2]$	b
$[p\text{-}C_6H_4NMe_3]^+[BF_4]^-$	c
$[p\text{-}C_6H_4NHMe_2]^+[BF_4]^-$	d

not affected by water addition (*73,175,176*), whereas the polyoxadiphosphines derived from DIOP **17** (*73,172*), the tetrasulfonated CBD or skewphos ligands (*173*), and the aminated skewphos (*174*) induce large decrease in enantiomeric excess (e.e.) when water is added (see Table IX). This solvent effect has been discussed by Landis and Halpern (*177*). Enantiodiscrimination is achieved by the reactivity difference between the two diastereomeric forms of $[RhL_2(\mathbf{51b})]^+$ toward dihydrogen addition. Indeed, it was shown that the less stable isomer reacts much faster than the major one. The resulting enantioselectivity is governed by the difference between the free energies of the two transition states. The stabilities of the two diastereomeric species $[RhH_2L_2(\mathbf{51b})]^+$ thus obtained are strongly dependent on solvation effects. Because water significantly reduces this difference, enantioselectivity is decreased (*73*).

More recently, a detailed investigation carried out by Sinou and coorkers has revealed that the lowering in enantioselectivity for the reduction of acids in water must be attributed to a modification of the solvophobic parameter rather than the solvent polarity (*173*). However, in some cases, the solvent effect is low and introduction of a polar function is sufficient to alter the enantioselectivity, as shown in Table X.

TABLE X

INFLUENCE OF LIGANDS ON THE REDUCTION OF ACIDS CARRIED OUT IN ALCOHOLS[a]

Substrate	Ligand	Solvent	Enantiomeric excess (%)	Configuration	Ref.
51a	**20**	MeOH	90	*R*	*174*
	20b	MeOH	88	*R*	*174*
	17	EtOH	81		*73*
	26_{16}	EtOH	43		*73*
	26_{42}	EtOH	46		*73*
	17	MeOH	73	*R*	*174*
	17b	MeOH	8	*S*	*174*
51b	**18**	MeOH	89	*R*	*174*
	18b	MeOH	87	*R*	*174*
	18d	MeOH	90	*R*	*174*
	20	MeOH	96	*R*	*174*
	20b	MeOH	92	*R*	*174*
	20d	MeOH	91	*R*	*174*
	17	EtOH	81		*73*
	26_{16}	EtOH	69		*73*
	26_{42}	EtOH	65		*73*
	17	MeOH	82	*R*	*174*
	17b	MeOH	59	*S*	*174*
	21	EtOH	82		*173*
	21a	EtOH	52		*173*

[a] Reaction conditions: T = 25°C; 1 bar of H_2 partial pressure.

Amino groups have been introduced by Tóth *et al.* in DIOP **17**, skewphos **20**, and chiraphos **18** ligands. Only functionalization in DIOP resulted in a decreased enantiomeric excess (*174*). It was postulated that enantioselection is not generated by the substituents but by the ligand chelate ring conformation that derives from them. DIOP, with other 1,4-diphosphines, leads to flexible seven-membered rings that adopt different chelate conformations according to the substituents. The chair or twisted boat conformations induce opposite optical yields. Tri- or tetrasulfonated chiraphos give the same optical yields (*77*) and this phenomenon can be interpreted as being due to the formation of a rather rigid five-membered ring.

The introduction of a metal center in the specific chiral site of a protein is an attractive way to obtain chiral diphosphine rhodium(I) complexes. The only efficient one for the hydrogenation of **51a** is avidin, which is attached to rhodium by means of biotin, previously converted into a chelating diphosphine by the Whitesides procedure (*12, 79*). With an easier synthetic method, this moiety exhibits enantiomeric excess higher than 40% provided that an excess of one equivalent of avidin is added. However, lower enantioselectivity (e.e. 30%) was obtained with a rhodium complex containing the synthetically prepared ligand **14** (*44*).

Because hydrogen is less soluble in water than in organic solvents such as alcohols, it was of great interest to explore the hydrogenation of α,β-unsaturated carboxylic acids in an aqueous solution with formate as the hydrogen source and with rhodium complexes containing water-soluble ligands (*178*). Hydrogenation of α-acetamidocinnamic acid **51b** with ammonium formate and a **21a**/Rh complex yielded a higher enantiomeric excess (43%) than that obtained using molecular dihydrogen (28%) (*173*), showing the usefulness of modifying these systems.

2. *Two-Phase Asymmetric Hydrogenation*

In order to separate easily the reaction products from the catalyst, asymmetric hydrogenation of the same substrates has been conducted in aqueous–organic two-phase solvent systems. Results are reported in Table XI (*179*). Of these experimental values, substrate **51d** is noteworthy in that it is not water soluble and is not reduced when slurried in aqueous solution. Moreover, the optical yields obtained in two-phase systems are close to those observed in the absence of organic solvents, so that it can be confirmed that catalysis occurs in the water bulk. However, it has been shown by Laghmari and Sinou that water acts not only as a solvent but also as a reactant. Indeed, they proved that hydrogenation of **51a** and **51c** in an AcOEt/D_2O two-phase system involved an appreciable amount of deuterium incorporation (*180*). From Table XI, it appears that ligand **18** leads to the most efficient catalysis, which induces enantiomeric excesses up to 87%, regardless of the nature of the substituents. In addition to the high selectivities observed, the aqueous solutions can be reused several times without any loss of enantioselectivity (*77,174,179*).

D. *Special Application to Hydrogenation of Lipids*

Because cell membrane fluidity is connected with biochemical and biological processes, several attempts have been made to modulate it, e.g., genetic manipulations, drug treatments, nutritional supplements, or changes in environmental temperature (*181,182*). Nevertheless, homogeneous hydrogenation of unsaturated fatty acids associated with the phospholipids of biomembranes appears to be more attractive than other methods, due to its ability to change membrane fluidity at a mild temperature, with the selectivities generally provided by homogeneous catalysis (*182*).

TABLE XI

TWO-PHASE AND SLURRY REDUCTION OF ENAMIDES WITH $[Rh(diene)(ligand)]^+$ AT AMBIENT TEMPERATURES

Substrate	Ligand	Solvent	P_{H_2} (bar)	Enantiomeric excess (%)	Configuration	Ref.
51b	**21a**	$AcOEt/H_2O$ (2/1)	1	51		*75*
		$AcOEt/H_2O$ (1/1)	1	34	*S*	*77*
			1	37[a]	*S*	*77*
		$AcOEt/H_2O$ (1/1)	5	21		*76*
	18a	$AcOEt/H_2O$ (2/1)	10	87	*R*	*77*
	18c	H_2O, slurry	14	94	*R*	*174*
	18d	H_2O, slurry	14	90	*R*	*174*
	17c	H_2O, slurry	14	25	*S*	*174*
	17d	H_2O, slurry	14	34	*S*	*174*
	20a	$AcOEt/H_2O$ (1/1)	15	65	*R*	*77*
	20c	$AcOEt/C_6H_6/H_2O$ (0.5/0.5/1)	14	67	*R*	*179*
		H_2O, slurry	14	67	*R*	*174*
	20d	$AcOEt/C_6H_6/H_2O$ (0.5/0.5/1)	14	73	*R*	*174*
		H_2O, slurry	14	71	*R*	*174*
51c	**17c**	H_2O, slurry	14	8	*S*	*174*
	17d	$AcOEt/C_6H_6/H_2O$ (0.5/0.5/1)	14	25	*S*	*174*
	18a	$AcOEt/H_2O$ (1/1)	5	82		*77*
			5	86[a]		*77*
		CH_2Cl_2/H_2O (1/1)	10	89		*77*
		C_6H_6/H_2O (1/1)	10	70		*77*
		$AcOEt/H_2O$ (1/1)	10	81	*R*	*77*
		$AcOEt/H_2O$ (2/1)	10	80		*75*
	18c	$AcOEt/C_6H_6/H_2O$ (0.5/0.5/1)	14	75		*174*
			14	77[a]		*174*
			14	77[b]		*174*
		H_2O, slurry	14	68	*R*	*174*
	18d	$AcOEt/C_6H_6/H_2O$ (0.5/0.5/1)	14	74	*R*	*174*
	20a	$AcOEt/H_2O$ (1/1)	15	45	*R*	*77*
	20c	H_2O, slurry	14	40	*R*	*174*
		$AcOEt/C_6H_6/H_2O$ (0.5/0.5/1)	14	44		*179*
			14	43[a]		*179*
			14	41[b]		*179*
			14	42[c]		*179*
	20d	$AcOEt/C_6H_6/H_2O$ (0.5/0.5/1)	14	50	*R*	*174*
	21a	CH_2Cl_2/H_2O (1/1)	1	12	*S*	*77*
		C_6H_6/H_2O (1/1)	1	12	*S*	*77*
		$AcOEt/H_2O$ (1/1)	1	20	*S*	*77*
			1	23[a]	*S*	*77*
51d	**18a**	$AcOEt/H_2O$ (2/1)	10	86	*R*	*77*
	20a	$AcOEt/H_2O$ (1/1)	15	44	*R*	*77*
	20c	H_2O, slurry	14	—[d]		*179*
	21a	$AcOEt/H_2O$ (2/1)	1	13	*S*	*77*

TABLE XI (*continued*)

Substrate	Ligand	Solvent	P_{H_2} (bar)	Enantiomeric excess (%)	Configuration	Ref.
51e	**17c**	AcOEt/C_6H_6/H_2O (0.5/0.5/1)	14	9	*S*	*174*
	17d	AcOEt/C_6H_6/H_2O (0.5/0.5/1)	14	11	*S*	*174*
	18c	AcOEt/C_6H_6/H_2O (0.5/0.5/1)	14	65	*R*	*174*
	18d	AcOEt/C_6H_6/H_2O (0.5/0.5/1)	14	58	*R*	*174*
	20c	AcOEt/C_6H_6/H_2O (0.5/0.5/71)	14	54	*R*	*174*
	20d	AcOEt/C_6H_6/H_2O (0.5/0.5/1)	14	67	*R*	*174*
51f	**17c**	H_2O, slurry	14	42	*S*	*174*
	17d	H_2O, slurry	14	67	*S*	*174*
	18c	H_2O, slurry	14	93	*R*	*174*
			14	95[a]	*R*	*174*
			14	90[b]	*R*	*174*
	18d	H_2O, sluury	14	88	*R*	*174*
	20c	H_2O, slurry	14	76	*R*	*174*
	20d	H_2O, slurry	14	79	*R*	*174*
	21a	AcOEt/H_2O (2/1)	1	37	*S*	*77*

[a] Catalyst reused.
[b] Catalyst reused twice.
[c] Catalyst reused three times.
[d] No conversion was observed.

In 1976, Chapman and Quinn showed that unsaturated membrane lipids could be hydrogenated with [RhCl(PPh_3)$_3$], but the complex needed an organic vector, such as tetrahydrofuran or dimethyl sulfoxide, to be introduced into the liposomes or the biomembranes. Once trapped in the hydrophobic membrane structure, the catalyst cannot be removed without disruption of the membrane structure (*181,182*). The use of [RhCl(L^1)$_3$] (*183–185*), [HRh(CO)(L^1)$_3$] (*186*), [RhCl(dpup)$_{2\text{ or }3}$] [dpup = bis(sodium-diphenylphosphinoundecylphosphate)] (*184*), and [RuCl_2(L^1)$_2$] (*185*) allowed for an easier catalyst removal. For instance, after the complete hydrogenation of liposomes, Madden *et al.* removed up to 70% of [RhCl(L^1)$_3$] with an anion-exchange resin (*183*). Ruthenium works faster than rhodium at "physiological" temperatures. In addition, for each catalyst, a decrease in reaction rates is observed as the substrate is hydrogenated until it reaches a plateau. This phenomenon has been attributed to the increase of the lipid phase viscosity, thus hampering the diffusion of the catalyst (*183*).

Owing to the amphiphilic properties of the catalyst, the reaction was shown to occur at the interface of the hydrocarbon substrate and the polar aqueous phase (*183*). Some 85% of [RhCl(L^1)$_3$] proved to be associated with the lipid fraction, the remainder staying in the aqueous phase, presumably in micellar association. Because a change in the catalyst concentration had no significant effect, it was assumed that the critical micellar concentration was exceeded providing a constant monomer concentration so that no more catalyst was found in the bilayer (*184*). As the sulfonate or the charged groups remain in close proximity to the polar substrate terminations, the catalyst cannot penetrate entirely into the bilayer, and the double bonds, which are located at the surface, are preferentially reduced (*186*). Thus, hydrogenation of fatty acids can be achieved with great selectivity.

VI

MISCELLANEOUS STUDIES

A. *Addition to Unsaturated Compounds*

Preparation of the water-soluble $P(CH_2OH)_3$ ligand has been achieved by direct addition of formaldehyde to PH_3 catalyzed by the complexes $[Pt(P(CH_2OH)_3)_4]$ or $[Pd(P(CH_2OH)_3)_4]$ (*107*). However, addition reactions catalyzed by water-soluble complexes have been essentially confined to olefinic compounds. The Michael reaction especially has been shown to be catalyzed cleanly by rhodium/L^3 systems. Indeed, activated CH_2 groups have been selectively added to isoprene, myrcene, or β-farnesene, which have the general framework $H_2C{=}C(R){-}CH{=}CH_2$. The reaction was carried out in Na_2CO_3 solutions (*187,188*). Presumably, the rhodium species activates the C–H bond of the CH_2 moiety to afford a hydride complex that gives an allylic group with the conjugated diene. Thus, 1,2- and 1,4-additions were obtained. High selectivities were observed, but optimal conversions (91.5% for isoprene, 76.3% for pentane-2,4-dione) were reached when addition of these two reactants was carried out at 80°C, during 6.5 hours, with a L^3:Rh molar ratio greater than 3. Interestingly, addition of a cosolvent such as alcohol improved the kinetics via the transfer rate between the two phases, as illustrated by addition of methanol for the condensation of methyl acetyl acetate with myrcene (*188*). For example, the rate proceeds twice as fast when going from a medium containing 20 ml of water to one containing 15 ml of H_2O/5 ml CH_3OH. In regard to the mechanism, it was assumed that the $CH_2Z^1Z^2$ substrates generate the active species $[Rh(L^3)_3CHZ^1Z^2]$ (*189*). A catalytic cycle such as that proposed in Scheme 13 involving an η^3-allylic intermediate accounts for the high selectivity of the reaction.

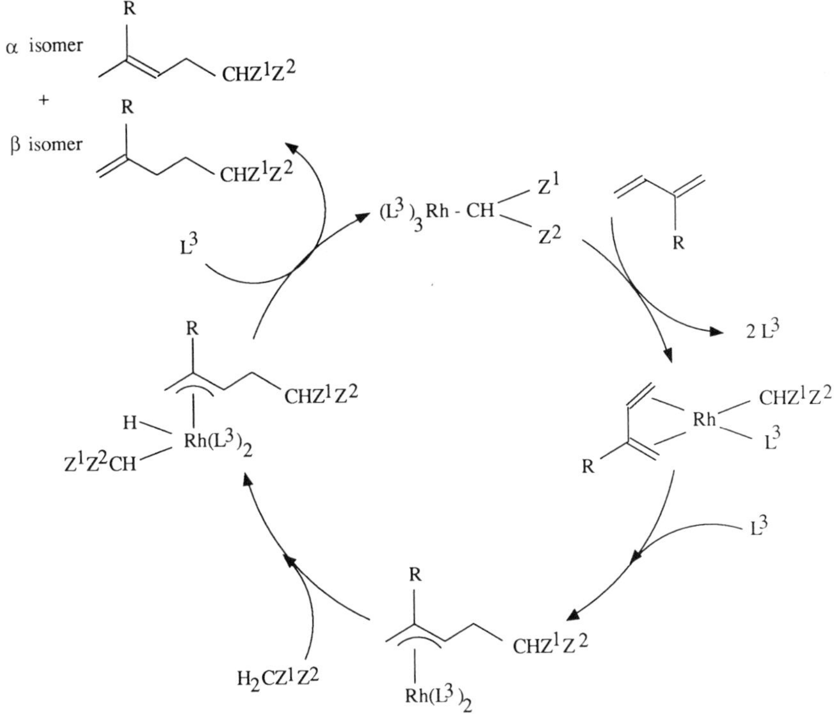

SCHEME 13. Addition of active methylene-containing compounds to conjugated dienes.

Interestingly, compounds that have a terminal 1,3-diene structure substituted in the 3-position could be selectively extracted from a C_{15} fragment according to this reaction. For instance, using acetyl methyl acetate, only the two products, **58** and **59**, were obtained from **57**, the other dienes being

$$C_{10}H_{17}-CH_2-\underset{\underset{CH_2}{||}}{C}-CH=CH_2$$

57

$$C_{10}H_{17}-CH_2-\underset{\underset{CH_2}{||}}{C}-CH_2-CH_2-CH\begin{matrix} COOCH_3 \\ COOCH_3 \end{matrix}$$

58

$$C_{10}H_{17}-CH_2-\underset{\underset{CH_3}{|}}{C}=CH-CH_2-CH\begin{matrix} COOCH_3 \\ COOCH_3 \end{matrix}$$

59

untouched. This chemical purification method is important for the preparation of intermediates in vitamin E synthesis (*190*). Moreover, the same Rh/L^3 systems also catalyze the condensation of myrcene, isoprene, and butadiene with secondary cyclic amines (*191*). The reaction takes place in water and at 100°C with high selectivities, giving the three isomers $R(Me)C{=}C(R)CH_2NR^1R^2$, $RC({=}CH_2)CH(R)CH_2NR^1R^2$ (resulting from the 1,4- and 1,2-additions, respectively) and $R(Me)CHCR{=}CHNR^1R^2$ in low yield.

Hydrocyanation of alkenes occurs readily in water at 80°C in the presence of Ni^0/L^3 systems (*192*). Beside obtaining yields greater than 80% with respect to HCN and high reaction rates, as well as high selectivity in linear nitrile up to 92%, catalyst recycling could be achieved without any loss in activity. Finally, selective hydration of acetylene to acetaldehyde was carried out by Taqui Khan *et al.* by using the catalyst precursor K[Ru(EDTA–H)Cl] at 80°C and 1 bar (*193*).

B. *Carbonylations*

Allyl chloride and benzyl chloride appear to be versatile starting materials for hydroxycarbonylation or alkoxycarbonylation [Eq. (26)], either in aqueous alkaline solution or in organic two-phase systems, in the absence of phase transfer cataryst (*194–196*).

$$RCH_2Cl + CO + R'OH \rightarrow RCH_2COOR' + HCl \tag{26}$$

where $R = CH_2{=}CH$, C_6H_5 and $R' = H, CH_3, C_2H_5$. Whereas this carbonylation requires drastic reaction conditions in homogeneous media, it occurs under atmospheric pressure and at mild temperatures when $[PdCl_2(L^1)_2]$ is used in a two-phase system (*194,196*). In this case, any excess of L^1 ligand is undesirable because it results in long induction periods by hampering the reduction of Pd(II) to the active Pd(0) species.

Having previously shown that reduction of allylic compounds with sodium formate and palladium complexes containing water-soluble phosphines takes place in the aqueous phase (*197*), Okano *et al.* proposed that transfer of substrate occurs from the organic to the aqueous phase subsequent to its coordination to a $Pd(L^1)_n$ species (*197*). For our part, we have carried out hydroxycarbonylation of bromobenzene in two-phase systems, but using $Pd(L^3)_n$ complexes, which are insoluble in organic solvents (*198*). To account for the high yields obtained, we consider that the transfer of the substrate in the aqueous phase takes place prior to coordination to the palladium complex. Okano *et al.* have also studied the influence of organic solvents and showed that polar solvents favor faster carbonylation of benzyl chloride (*196*). Nevertheless, *n*-butanol as solvent was not convenient because it led to benzyl *n*-butyl ether (42%) by the Williamson reaction. Surprisingly, no ester formation was observed, whereas significant amounts of methyl benzoic ester were obtained with water/methanol mixtures. The latter results can be interpreted as resulting from a competition between the alcoholysis and hydrolysis of the aroyl species $[(L^3)_nPdCOPh]$ (*198*). Moreover, complete conversion of benzyl chloride was achieved with [Ru(III)–EDTA] systems in ethanol/water solvents, leading to phenylacetic acid and ethyl phenyl acetate with a turnover frequency of 44 $hour^{-1}$ (*195*). A mechanism quite similar to that generally accepted for palladium complexes has been proposed, but no precise details about the exact role of water or ethanol have been given.

Interestingly, carbonylation of ammonia, diethylamine, and triethylamine is catalyzed by $[Ru(EDTA–H)(CO)]^-$ in aqueous media under a CO partial pressure of 5–30 bar and at temperatures in the range 80–100°C (*199,200*). Triethylamine gave 100% diethylpropionamide whereas diethylamine gave 80% *N,N*-diethylformamide in addition to 20% *N,N*-tetraethylurea, and NH_3 yielded 70% urea, 5% formamide, and by-products. Formation of urea requires a N–H bond in the substrate. From the two proposed mechanisms, we believe that oxidative addition of the N–H bond is likely to occur.

C. *Telomerizations*

Telomerization of butadiene with methanol was patented in 1976 by Rhône-Poulenc Industries. Palladium salts were used as precursors with the L^3 ligand (*201*). Reactions carried out in basic media at 30°C led to 56% conversion of butadiene and 95% selectivity in 1-methoxy-2,7-octadiene. Several examples were given using other dienes or metal precursors. The main results are the high linearities of the telomer.

Quaternized aminophosphines have been prepared by Peiffer *et al.* and studied in the telomerization of butadiene and isoprene (*41*). Satisfactory results were obtained for butadiene; at 80°C and for 20 hours 80% conversion and 78% selectivity in linear product were reached. However, isoprene gave conversions no higher than 52% with only 25% of 1-methoxy-2,7-dimethyl-2,7-octadiene. Large amounts of high-molecular-weight products were formed. Moreover, the catalytic activity was maintained by recycling the aqueous phase and the selectivity was slightly improved.

D. *Oxidations*

Until the last decade, catalysts generally used for oxidations were cobalt(II) complexes, porphyrinic iron(II), or manganese(II) complexes. The hard donor EDTA ligand can also give dioxygen complexes. Whereas the Ti(IV) peroxo compound $[Ti(EDTA)(O_2)]^{2-}$ was formed either from the addition of dioxygen to the species $[Ti(EDTA)(H_2O)]^-$ or from the addition of H_2O_2 to titanium(IV) complexes (*202*), Ru(III) was the first platinum metal to be reported to form dioxygen compounds with EDTA-type ligands in aqueous media. Indeed, equilibria between numerous hydroxo or peroxo compounds formed by addition of dioxygen to ruthenium complexes containing aminocarboxylic acids were identified by potentiometry, spectrophotometry, electrochemical, and gas absorption techniques (*202–206*). The first application of such species was the oxidation of triphenylphosphine to its oxide (*207*).

Ruthenium has also been coordinated to Schiff bases, leading to oxygen complexes. Thus, oxidation of cyclohexene could be achieved in water–dioxane mixtures by complexes containing the bis(salicylaldehyde)-o-phenylenediiminato ligand (saloph) (*208*). However, the oxidation of hydrazine, hydroxylamine, and cysteine occurred in aqueous solution by molecular oxygen in the presence of the complex $[CO^{II}(TSPC)]$, where TSPC represents the 3,10,17,24-tetrasulfophthalocyanine ligand (*209*).

E. *Substitutions*

Palladium complexes maintained in aqueous phase by L^3 ligands are particularly attractive for the allylic nucleophilic substitution reactions because the separation of the catalyst from heavy organic products is very easily achieved. This reaction has been applied to the substitution of an allylic hydrogen in cinnamyl acetate, ethyl cinnamyl carbonate, and ethyl geranyl carbonate by acetyl ethyl acetate (*210*). In the latter case, geranylacetone, a key intermediate for the synthesis of vitamin A, was prepared (*210*). Other nucleophiles, such as acetylacetone, morpholine, and sodium *p*-sulfonatobenzyl were then studied, while substrates were extended to the epoxy compounds (*211*).

C-alkylation of phenols, essentially in ortho and para positions, can be performed using rhodium/L^3 systems. Examples are restricted to reactions of isoprene or myrcene with many phenols (*212*). Moreover, alkylations of numerous aryl and heteroaromatic halides with aryl or vinyl boronic acids, alkynes, an alkene, and a dialkyl phosphite are catalyzed by the complex $[Pd(L^1)_3]$, prepared and characterized by single-crystal

X-ray diffraction. The reactions took place in aqueous solutions at temperatures that did not exceed 80°C, and in some cases complete conversions could be observed (*213*). Thus this water-soluble Pd(0) catalyst is very promising because C–C coupling reactions have previously been developed only in organic media.

F. *Immobilization of Water-Soluble Complexes on Supports*

Because they contain polar functional groups that are easily ionized, water-soluble ligands are particularly suited to the immobilization of transition metal complexes on ion-exchange resins. Due to their thermal stability, cation-exchange resins are generally preferred to anionic exchangers, although the latter are used, owing to the presence of the sulfonate or carboxylate functions on the ligands. With cation-exchange materials, amino groups are linked either by the quaternary ammonium function to the resin in its sodium form or by direct reaction of the amine with the resin in the acidic form. Numerous resins have been successfully applied to miscellaneous catalytic reactions, for example, hydrogenation, oxidation, and hydroformylation (*115,214–218*).

For hydroformylation, aminophosphines have been grafted onto macroreticular sulfonated styrene–divinylbenzene resins. The anchored ligands were then treated with cobalt, rhodium, or platinum precursors (*214*). Rhodium catalysts were used to hydroformylate 1,5-cyclooctadiene at around 100 bar and 80°C giving ~55% cyclooct-4-enecarboxaldehyde and 45% 1,3-cyclooctadiene, resulting from the isomerization of the substrate. With cobalt catalysts obtained from $[Co_2(CO)_8]$, only hydroformylation occurred, but with low rates, in contrast to the $[HCo(CO)_3(PBu_3)]$ system, which affords alcohols. Starting from hex-1-ene, linearities of heptanal were near 70%. Precursors of the type $[Pt(H)(SnCl_3)L_2]$ converted hex-1-ene into C_7 aldehydes with selectivities as high as 98% and linearities of 91%. The rates here also dramatically decreased with respect to analogous homogeneous systems (*214*).

The *p*-(dimethylphosphino)-*N,N*-dimethyl or diethylaniline ligands are protonated on the nitrogen atom by the strongly acidic resins (*216*). These supported ligands are reacted with $[Rh_2Cl_2(CO)_4]$ to generate $[RhCl(CO)L_2]$-type precursors, which were then evaluated in the hydroformylation of oct-1-ene at ~60 bar and 80°C. This substrate was converted into 99.9% *n*- or *iso*-C_9 aldehydes. However, the linearity was a deceptive 70% (*216*). Similar results were obtained at slower rates starting from *p*-(diphenylphosphino) benzoic acid anchored on basic exchange resins (*216*).

Contrary to these results, hydrogenation of classic substrates such as hex-1-ene, cyclohexene, styrene, or crotonic acid and related unsaturated acids shows results very similar to the homogeneous systems (*115,217–219*). It was generally found that the ionic substituents of the ligands far enough from the metal center do not exert a significant influence on the catalytic performances. The main advantage is the catalyst recovery by a simple filtration process. In addition, the metal can be reextracted from the support by elution with classic reagents. Several recycling experiments were identified with almost no appreciable loss of activity. Moreover, during the asymmetric hydrogenation of methyl (Z)-*N*-acetylamino-cinnamic acid methyl ester, catalyzed by rhodium quaternized chiral ditertiary phosphines supported on cationic resins, it was shown that the conversion was unchanged until six recyclings, and the enantiomeric excess was essentially maintained (*219*).

Oxidation reactions have also been demonstrated by autoxidation of thiols catalyzed by cobalt(II)-supported chelates [Eq. (27)]. Cobalt(I) was obtained after reaction so that regeneration of the active species required an oxidation by air or, alternatively, treatment with aqueous iron(III) solutions (*215*).

$$2Co(II) + 2RSH \rightarrow RSSR + 2H^+ + 2Co(I) \qquad (27)$$

Mineral clays, which present a layer structure consisting of alternating layers of cations and negatively charged silicate sheets, are also attractive heterogenizing materials. Indeed, active ionic species can be intercalated by ion metathesis. Whereas simple metal cations were previously introduced in these clays, coordination through positively charged ligands, such as $Ph_2PCH_2CH_2P^+Ph_2(CH_2Ph)$ (PP^+), is an interesting way to immoblize the catalyst. This PP^+ ligand was added to $[Rh_2Cl_2(COD)_2]$ in the presence of triphenylphosphine ($PP^+:PPh_3:Rh$ molar ratio of 1:1:1). This system hydrogenated hex-1-ene with activities very similar to those obtained in the homogeneous phase (*220*).

Recently, a novel family of immobilized catalysts has been developed in order to improve chemical reactions at the interface of the two liquid phases. This new method, called "supported aqueous phase catalysis" (SAPC), consists in dissolving a water-soluble organometallic complex in a thin film of water, which is then supported on a high-surface-area hydrophilic solid such as silica (*221*). The solid thus impregnated is contacted with the immiscible liquid organic reactants, which diffuse from the bulk phase into the porous solid.

This novel supported catalysis presents great advantages when compared with the two-phase systems. Indeed, in contrast with the latter systems, reaction was shown to occur at the aqueous–organic interphase. Thus the solubility of reactants in water is not a limiting factor, as shown by similar activities and selectivities obtained when hex-1-ene, oct-1-ene, and dec-1-ene were hydroformylated with $[HRh(CO)(L^3)_3]/L^3$ in a silica-supported aqueous phase (*222*). Moreover, a large enhancement in activity was observed by immobilization of the catalysts. Similarly, this system dramatically improves the hydroformylation of oleyl alcohol (96.6% conversions to aldehydes after 5.5 hours), which did not occur after 13 hours in a two-phase system (*223*). The same trends were observed with $[Co_2(CO)_6(L^1)_2]$, which is twice as active when immobilized on a hydrophilic support (*224*). This behavior was attributed to the increase in surface area between the two phases or, in some cases, to the stabilization of the complex, which decomposes in aqueous phase. However, immobilization induces changes in active species, so that the $n:i$ ratio can be decreased or improved. This method is very attractive because loss of cobalt can be dramatically minimized and significant leaching of rhodium is avoided (*222–224*).

G. *Photocatalytic Reactions*

Photocatalytic hydrogenations and hydroformylations have been achieved in aqueous media by using water as a hydrogen source. An early method involved $[Ru(bipy)_3]^{2+}$ as a photosensitizer, ascorbate (or more frequently, EDTA) as a sacrificial electron donor to reduce the Ru(II) complex in $[Ru(bipy)_3]^+$, methyl viologen as an electron relay, and a transition metal catalyst (*225,226*). The latter was added in metal colloid form to hydrogenate ethylene and acetylene, whereas hydrogenation and hydroformylation of these substrates were achieved, respectively, via $[ClRh(H)_2(L^1)_3]^{3-}$ and $[RhH(CO)(L^1)_3]^{3-}$, both derived from $[Rh^ICl(L^1)_3]^{3-}$.

The photosplitting of water was also observed in the presence of semiconducting CdS powders previously loaded with Pt or Rh and RuO_2 (*227–231*). Their irradiation in aqueous solution excites an electron from the valence band to the conduction band (band gap of 2.4 eV). Electrons of the valence band are supplied by RuO_2-catalyzed photooxidation of water to $2H^+$ and $\frac{1}{2}O_2$ or by oxidation of a sacrificial electron donor such as EDTA. At the conductance band, although ethylene and acetylene were shown to be directly reduced (*227*), water-soluble Ru complexes are generally competing favorably for electrons with H^+ (*230,231*), due to their negative potentials. Indeed, for the reduction of dinitrogen to ammonia, electron transfer to $[Ru^{II}(EDTA–H)N_2]^-$ gives the intermediate $[Ru^I(EDTA–H)N_2]^{2-}$, which is then subject to protonation and electron transfer to afford NH_3 and $[Ru^{II}(EDTA{-}H)(H_2O)]^-$ (*229*). The hydrogenation of cyclohexene requires electron transfer to $[Ru^{III}(EDTA{-}H)(H_2O)]$, thus leading to $[Ru^{II}(EDTA{-}H)(H_2O)]^-$, and the H_2O ligand is then displaced by the olefin (*231*).

VII

CONCLUSION

Catalysis employing water-soluble complexes appears to be very useful. The principal advantage is that the organic products can be separated easily from the aqueous phase containing the active species. In addition, high selectivities are often obtained. These systems are generally well adapted to light substrates that are sufficiently soluble in water. However, they are less efficient for heavy organic compounds due to their lower solubility, thereby reducing both the reaction rates and the product yields. This problem may be solved by the introduction of micellar systems or supported aqueous phase catalysts, because catalysis takes place mainly at the interface. In our opinion, these two promising methods should be widely developed in the future.

References

1. F. Joó and Z. Tóth, *J. Mol. Catal.* **8,** 369 (1980).
2. D. Sinou, *Bull. Soc. Chim. Fr.*, 480 (1987).
3. T. G. Southern, *Polyhedron* **8,** 407 (1989).
4. S. Ahrland, J. Chatt, N. R. Davies, and A. A. Williams, *J. Chem. Soc.*, 276 (1958).
5. E. G. Kuntz, Fr. Patent 2,314,910 to Rhône-Poulenc Industries (06-20-1975).
6. D. Morel and J. Jenck, Fr. Patent 2,550,202 to Rhône-Poulenc Recherches (08-03-1983).
7. C. Varre, M. Desbois, and J. Nouvel, Fr. Patent 2,561,650 to Rhône-Poulenc Recherches (03-26-1984).
8. J. L. Sabot, Eur. Patent 0,104,967 to Rhône-Poulenc Chimie de base (08-31-1982).
9. H. Schindlbauer, *Monatsch. Chem.* **96,** 2051 (1965).
10. B. W. Bangerter, R. P. Beatty, J. K. Kouba, and S. S. Wreford, *J. Org. Chem.* **42,** 3247 (1977).
11. R. G. Nuzzo, D. Feitler, and G. M. Whitesides, *J. Am. Chem. Soc.* **101,** 3683 (1979).
12. R. G. Nuzzo, S. L. Haynie, M. E. Wilson, and G. M. Whitesides, *J. Org. Chem.* **46,** 2861 (1981).
13. B. Fell and G. Pagadogianakis, *J. Mol. Catal.* **66,** 143 (1991).
14. E. G. Kuntz, *Chemtech*, 570 (1987).
15. C. Larpent and H. Patin, *Appl. Organomet. Chem.* **1,** 529 (1987).
16. T. T. Derencsényi, *Inorg. Chem.* **20,** 665 (1981).
17. C. Larpent, R. Dabard, and H. Patin, *Inorg. Chem.* **26,** 2922 (1987).
18. C. Larpent, R. Dabard, and H. Patin, *New J. Chem.* **12,** 907 (1988).
19. C. Larpent, G. Meignan, J. Priol, and H. Patin, *C. R. Acad. Sci. Ser. 2* **310,** 493 (1990).

20. W. A. Herrmann, J. A. Kulpe, J. Kellner, H. Riepl, H. Bahrmann, and W. Konkol, *Angew. Chem., Int. Ed. Engl.* **29,** 391 (1990).
21. L. Lecomte, J. Triolet, D. Sinou, J. Bakos, and B. Heil, *J. Chromatogr.* **408,** 416 (1987).
22. L. Lecomte and D. Sinou, *J. Chromatogr.* **514,** 91 (1990).
23. L. Lecomte and D. Sinou, *Phosphorus, Sulfur, Silicon Relat. Elem.* **53,** 239 (1990).
24. C. Larpent and H. Patin, *C. R. Acad. Sci. Ser. 2* **305,** 1427 (1987).
25. C. Larpent and H. Patin, *Tetrahedron* **44,** 6107 (1988).
26. C. Larpent and H. Patin, *Tetrahedron Lett.* **29,** 4577 (1988).
27. C. Larpent, G. Meignan, and H. Patin, *Tetrahedron* **46,** 6381 (1990).
28. W. Kosolapoff and L. Maier, "Organic Phosphorus Compounds," Vol. 1. Wiley (Interscience), New York, 1972.
29. F. G. Mann and H. R. Watson, *J. Chem. Soc.*, 3945 (1957).
30. H. P. Fritz, I. R. Gordon, K. E. Schwarzhans, and L. M. Venanzi, *J. Chem. Soc.*, 5210 (1965).
31. T. Hayashi, K. Yamamoto, and M. Kumada, *Tetrahedron Lett.* **49-50,** 4405 (1974).
32. T. Hayashi and M. Kumada, *Acc. Chem. Res.* **15,** 395 (1982).
33. M. Fiorini, G. M. Giongo, F. Marcati, and W. Marconi, *J. Mol. Catal.* **1,** 451 (1975/1976).
34. G. Pracejus and H. Pracejus, *Tetrahedron Lett.* **39,** 3497 (1977).
35. F. G. Mann and I. T. Millar, *J. Chem. Soc.*, 3039 (1952).
36. G. R. Dobson, R. C. Taylor, and T. D. Walsh, *Inorg. Chem.* **6,** 1929 (1967).
37. R. C. Taylor, G. R. Dobson, and R. A. Kolodny, *Inorg. Chem.* **7,** 1886 (1968).
38. K. Issleib and R. Rieschel, *Chem. Ber.* **98,** 2086 (1965).
39. R. A. Kolodny, T. L. Morris, and R. C. Taylor, *J. Chem. Soc., Dalton Trans.*, 328 (1973).
40. R. T. Smith and M. C. Baird, *Inorg. Chim. Acta* **62,** 135 (1982).
41. G. Peiffer, S. Chhan, and A. Bendayan, *J. Mol. Catal.* **59,** 1 (1990).
42. K. Issleib, R. Kümmel, H. Oehme, and I. Meissner, *Chem. Ber.* **101,** 3612 (1968).
43. L. Sacconi and R. Morassi, *J. Chem. Soc. A*, 2997 (1968).
44. M. E. Wilson, R. G. Nuzzo, and G. M. Whitesides, *J. Am. Chem. Soc.* **100,** 2269 (1978).
45. I. Tóth, B. E. Hanson, and M. E. Davis, *Organometallics* **9,** 675 (1990).
46. K. Issleib and D. Haferburg, *Z. Naturforsch. B: Anorg. Chem. Org. Chem. Biochem. Biophys. Biol* **20b,** 916 (1965).
47. K. Issleib, A. Kipke, and V. Hahnfeld, *Z. Anorg. Allg. Chem.* **444,** 5 (1978).
48. M. M. Rauhut, I. Hechenbleikner, H. A. Currier, F. C. Schaefer, and V. P. Wystrach, *J. Chem. Soc.*, 1103 (1959).
49. M. M. Rauhut, H. A. Currier, A. M. Semsel, and V. P. Wystrach, *J. Org. Chem.* **26,** 5138 (1961).
50. R. T. Smith and M. C. Baird, *Trans. Met. Chem.* **6,** 197 (1981).
51. K. Naumann, G. Zon, and K. Mislow, *J. Am. Chem. Soc.* **91,** 7012 (1969).
52. U. Nagel and E. Kinzel, *Chem. Ber.* **119,** 1731 (1986).
53. I. Tóth and B. E. Hanson, *Tetrahedron: Asymmetry* **1,** 895 (1990).
54. K. A. Petrov and V. A. Parshina, *Zh. Obshch. Khim.* **31,** 2729 (1961).
55. K. A. Petrov, V. A. Parshina, and V. A. Gaidamak, *Zh. Obshch. Khim* **31,** 3411 (1961).
56. K. A. Petrov and V. A. Parshina, *Zh. Obshch. Khim.* **31,** 3417 (1961).
57. S. Trippett, *J. Chem. Soc.*, 2813 (1961).
58. E. Steininger, *Chem. Ber.* **95,** 2541 (1962).
59. K. A. Petrov, V. A. Parshina, and M. B. Luzanova, *Zh. Obshch. Khim.* **32,** 553 (1962).

60. K. A. Petrov, V. A. Parshina, and A. F. Manuilov, *Zh. Obshch. Khim.* **35,** 2062 (1965).
61. K. Issleib and H.-M. Möbius, *Chem. Ber.* **94,** 102 (1961).
62. K. Issleib and K. Rockstroh, *Chem. Ber.* **96,** 407 (1963).
63. K. Issleib and H. R. Roloff, *Chem. Ber.* **98,** 2091 (1965).
64. T. Okano, M. Yamamoto, T. Noguchi, H. Konishi, and J. Kiji, *Chem. Lett.*, 977 (1982).
65. K. Issleib and G. Thomas, *Chem. Ber.* **93,** 803 (1960).
66. K. Issleib, R. Kümmel, and H. Zimmermann, *Angew. Chem., Int. Ed. Engl.* **4,** 155 (1965).
67. K. Issleib and G. Thomas, *Chem. Ber.* **94,** 2244 (1961).
68. H. D. Empsall, E. M. Hyde, D. Pawson, and B. L. Shaw, *J. Chem. Soc., Dalton Trans.*, 1292 (1977).
69. Z. N. Mironova, E. N. Tsvetkov, A. V. Nikolaev, and M. I. Kabachnik, *Zh. Obshch. Khim.* **37,** 2747 (1967).
70. F. G. Mann and I. T. Millar, *J. Chem. Soc.*, 4453 (1952).
71. R. Benhamza, Y. Amrani, and D. Sinou, *J. Organomet. Chem.* **288,** C37 (1985).
72. Y. Amrani, Ph.D. Thesis, University of Lyon, Lyon (1986).
73. D. Sinou and Y. Amrani, *J. Mol. Catal.* **36,** 319 (1986).
74. Y. Amrani, D. Lafont, and D. Sinou, *J. Mol. Catal.* **32,** 333 (1985).
75. T. P. Dang, J. Jenck, and D. Morel, Fr. Patent 2,549,840 to Rhône-Poulenc Santé (07-28-1983).
76. F. Alario, Y. Amrani, Y. Colleuille, T. P. Dang, J. Jenck, D. Morel, and D. Sinou, *J. Chem. Soc., Chem. Commun.*, 202 (1986).
77. Y. Amrani, L. Lecomte, D. Sinou, J. Bakos, I. Tóth, and B. Heil, *Organometallics* **8,** 542 (1989).
78. L. Lecomte and D. Sinou, *J. Mol. Catal.* **52,** L21 (1989).
79. M. E. Wilson and G. M. Whitesides, *J. Am. Chem. Soc.* **100,** 306 (1978).
80. M. Fiorini, F. Marcati, and G. M. Giongo, *J. Mol. Catal.* **4,** 125 (1978).
81. T. Hayashi, T. Mise, and M. Kumada, *Tetrahedron Lett.* **48,** 4351 (1976).
82. W. R. Cullen and E.-S. Yeh, *J. Organomet. Chem.* **139,** C13 (1977).
83. Y. V. Kurbatov, O. S. Otroshchenko, and A. S. Sadykov, *Nauchm. Tr. Tashk. Gos. Univ.* **263,** 36 (1964), quoted by S. Anderson *et al.*, Ref. *85*.
84. O. S. Otroshchenko, Y. V. Kurbatov, and A. S. Sadykov, *Nauchm. Tr. Tashk. Gos. Univ.* **263,** 27 (1964), quoted by S. Anderson *et al.*, Ref. *85*.
85. S. Anderson, E. C. Constable, K. R. Seddon, J. E. Turp, J. E. Baggott, and M. J. Pilling, *J. Chem. Soc., Dalton Trans.*, 2247 (1985).
86. W. A. Herrmann, W. R. Thiel, and J. G. Kuchler, *Chem. Ber.* **123,** 1953 (1990).
87. G. Sprintschnik, H. W. Sprintschnik, P. P. Kirsch, and D. G. Whitten, *J. Am. Chem. Soc.* **99,** 4947 (1977).
88. H. Irving, R. Shelton, and R. Evans, *J. Chem. Soc.*, 3540 (1958).
89. L. D. Pettit and H. Irving, *J. Chem. Soc.*, 5336 (1964).
90. S. O. Sommerer and G. J. Palenik, *Organometallics* **10,** 1223 (1991), and references therein.
91. J. H. Weber and D. H. Busch, *Inorg. Chem.* **4,** 469 (1965).
92. A. F. Borowski, D. J. Cole-Hamilton, and G. Wilkinson, *Nouv. J. Chim.* **2,** 137 (1978).
93. W. A. Herrmann, J. Kellner, and H. Riepl, *J. Organomet. Chem.* **389,** 103 (1990).
94. Z. Tóth, F. Joó, and M. T. Beck, *Inorg. Chim. Acta* **42,** 153 (1980).
95. F. Joó and M. T. Beck, *React. Kinet. Catal. Lett.* **2,** 257 (1975).
96. N. Ahmad, J. J. Levison, S. D. Robinson, and M. F. Uttley, *Inorg. Synth.* **15,** 45 (1986).

97. B. Fontal, J. Orlewski, C. C. Santini, and J. M. Basset, *Inorg. Chem.* **25,** 4320 (1986).
98. W. A. Herrmann and J. A. Kulpe, *J. Organomet. Chem.* **389,** 85 (1990).
99. B. Salvesen and J. Bjerrum, *Acta Chem. Scand.* **16,** 735 (1962).
100. C. Larpent and H. Patin, *J. Mol. Catal.* **44,** 191 (1988).
101. C. Larpent, R. Dabard, and H. Patin, *Tetrahedron Lett.* **28,** 2507 (1987).
102. C. Larpent and H. Patin, *J. Mol. Catal.* **61,** 65 (1990).
103. C. Larpent, F. Brisse-Le Menn, and H. Patin, *New J. Chem.* **15,** 361 (1991).
104. C. Larpent, F. Brisse-Le Menn, and H. Patin, *J. Mol. Catal.* **65,** L35 (1991).
105. Y. Dror and J. Manassen, *J. Mol. Catal.* **2,** 219 (1977).
106. J. Chatt, G. J. Leigh, and R. M. Slade, *J. Chem. Soc., Dalton Trans.*, 2021 (1973).
107. K. N. Harrison, P. A. T. Hoye, A. Guy Orpen, P. G. Pringle, and M. B. Smith, *J. Chem. Soc., Chem. Commun.*, 1096 (1989).
108. P. G. Pringle and M. B. Smith, *Platinum Met. Rev.* **34,** 74 (1990).
109. P. Dapporto and L. Sacconi, *J. Chem. Soc. (A)*, 1914 (1971), and references therein.
110. N. W. Alcock, J. M. Brown, and J. C. Jeffery, *J. Chem. Soc., Chem. Commun.*, 829 (1974).
111. N. W. Alcock, J. M. Brown, and J. C. Jeffery, *J. Chem. Soc., Dalton Trans.*, 583 (1976).
112. N. W. Alcock, J. M. Brown, and J. C. Jeffery, *J. Chem. Soc., Dalton Trans.*, 888 (1977).
113. H. J. Becher, W. Bensmann, and D. Fenske, *Chem. Ber.* **110,** 315 (1977).
114. R. T. Smith, R. K. Ungar, and M. C. Baird, *Trans. Metal. Chem.* **7,** 288 (1982).
115. R. T. Smith, R. K. Ungar, L. J. Sanderson, and M. C. Baird, *Organometallics* **2,** 1138 (1983).
116. M. M. Taqui Khan, B. Taqui Khan, Safia, and K. Nazeeruddin, *J. Mol. Catal.* **26,** 207 (1984).
117. S. Anderson and K. R. Seddon, *J. Chem. Res.* S74 (1979).
118. W. A. Herrmann, W. R. Thiel, J. G. Kuchler, J. Behm, and E. Herdtweck, *Chem. Ber.* **123,** 1963 (1990).
119. T. Matsubara and C. Creutz, *Inorg. Chem.* **18,** 1956 (1979).
120. R. N. F. Thorneley, A. G. Sykes, and P. Gans, *J. Chem. Soc. (A)*, 1494 (1971).
121. N. A. Ezerskaya and T. P. Solovykh, *Russ. J. Inorg. Chem.* **11,** 991 (1966).
122. G. J. Palenik, D. W. Wester, U. Rychlewska, and R. C. Palenik, *Inorg. Chem.* **15,** 1814 (1976).
123. J. E. Thomas and G. J. Palenik, *Inorg. Chim. Acta* **44,** L303 (1980).
124. S. Ahrland, J. Chatt, N. R. Davies, and A. A. Williams, *J. Chem. Soc.*, 264 (1958); S. Ahrland, J. Chatt, N. R. Davies, and A. A. Williams, *J. Chem. Soc.*, 1403 (1958).
125. G. Wright and J. Bjerrum, *Acta Chem. Scand.* **16,** 1262 (1962); R. George and J. Bjerrum, *Acta Chem. Scand.* **22,** 497 (1968); C. J. Hawkins, O. Monsted, and J. Bjerrum, *Acta Chem. Scand.* **24,** 1059 (1970); J. C. Chang and J. Bjerrum, *Acta Chem. Scand.* **26,** 815 (1972).
126. G. Schmid, N. Klein, L. Korste, U. Kreibig, and D. Schönauer, *Polyhedron* **7,** 605 (1988).
127. M. A. S. Aquino and D. H. Macartney, *Inorg. Chem.* **27,** 2868 (1988).
128. R. Eisenberg and D. Henderson, *Adv. Catal.* **28,** 79 (1979).
129. P. Escaffre, A. Thorez, and P. Kalck, *J. Mol. Catal.* **33,** 87 (1985).
130. P. C. Ford and A. Rokicki, *Adv. Organomet. Chem.* **28,** 139 (1988).
131. E. C. Baker, D. E. Hendriksen, and R. Eisenberg, *J. Am. Chem. Soc.* **102,** 1020 (1980).
132. J. Halpern, *Comments Inorg. Chem.* **1,** 3 (1981).

133. J. Kaspar, R. Spogliarich, G. Mestroni, and M. Graziani, *J. Organomet. Chem.* **208,** C15 (1981).
134. P. Escaffre, A. Thorez, and P. Kalck, *New J. Chem.* **11,** 601 (1987).
135. P. Escaffre, A. Thorez, and P. Kalck, *J. Chem. Soc., Chem. Commun.*, 146 (1987).
136. P. Kalck, *Polyhedron* **7,** 2441 (1988).
137. W. R. Tikkanen, E. Binamina-Soriaga, W. C. Kaska, and P. C. Ford, *Inorg. Chem.* **22,** 1147 (1983).
138. I. T. Horváth, R. V. Kastrup, A. A. Oswald, and E. J. Mozeleski, *Catal. Lett.* **2,** 85 (1989).
139. M. M. Taqui Khan, S. B. Halligudi, and S. Shukla, *Angew. Chem., Int. Ed. Engl.* **27,** 1735 (1988).
140. M. M. Taqui Khan, S. B. Halligudi, N. Nageswara Rao, and S. Shukla, *J. Mol. Catal.* **51,** 161 (1989).
141. M. M. Taqui Khan, *Platinum Met. Rev.* **35,** 70 (1991).
142. M. Pawlik, M. F. Hoq, and R. E. Shepherd, *J. Chem. Soc., Chem. Commun.*, 1467 (1983).
143. J. Falbe, "New Synthesis with Carbon Monoxide." Springer-Verlag, Berlin, 1980.
144. I. Tkatchenko, *in* "Comprehensive Organometallic Chemistry" (G. Wilkinson, F. G. A. Stone, and E. W. Abel, eds.), Vol. 8, p 101. Pergamon, Oxford, 1982.
145. J. Jenck, Fr. Patent 2,478,078 to Rhône-Poulenc Industries (03-12-1980).
146. E. Kuntz, Fr. Patent 2,349,562 to Rhône-Poulenc Industries (04-29-1976).
147. A. Thorez, Thèse d'Etat, Institut National Polytechnique de Toulouse, Toulouse (1985).
148. P. Kalck, P. Escaffre, F. Serein-Spirau, A. Thorez, B. Besson, Y. Colleuille, and R. Perron, *New J. Chem.* **12,** 687 (1988).
149. B. Besson, P. Kalck, and A. Thorez, U. S. Patent 4,778,905 to Rhône-Poulenc Chimie de Base (10-16-1985).
150. M. J. H. Russell and B. A. Murrer, U. S. Patent 4,399,312 to Johnson Matthey Public Limited Company (08-27-1981).
151. M. J. H. Russell, *Platinum Met. Rev.* **32,** 179 (1988).
152. M. M. Taqui Khan, S. B. Halligudi, and S. H. R. Abdi, *J. Mol. Catal.* **45,** 215 (1988).
153. M. M. Taqui Khan, S. B. Halligudi, and S. H. R. Abdi, *J. Mol. Catal.* **48,** 313 (1988).
154. J. Jenck and E. Kuntz, Fr. Patent 2,473,504 to Rhône-Poulenc Industries (10-31-1979).
155. M. K. Markiewicz and M. C. Baird, *Inorg. Chim. Acta* **113,** 95 (1986).
156. F. H. Jardine, *Polyhedron* **1,** 569 (1982).
157. I. Hablot, Ph.D. Thesis, Institut National Polytechnique de Toulouse, Toulouse (1991).
158. J. M. Grosselin, C. Mercier, G. Allmang, and F. Grass, *Organometallics* **10,** 2126 (1991).
159. J. M. Grosselin, Fr. Patent 2,636,943 to Rhône-Poulenc Santé (09-26-1988).
160. C. Larpent and H. Patin, *J. Organomet. Chem.* **335,** C13 (1987).
161. M. M. Taqui Khan, S. A. Samad, and M. R. H. Siddiqui, *J. Mol. Catal.* **53,** 23 (1989).
162. M. M. Taqui Khan, S. A. Samad, Z. Shirin, and M. R. H. Siddiqui, *J. Mol. Catal.* **54,** 81 (1989).
163. F. Joó and A. Bényei, *J. Organomet. Chem.* **363,** C19 (1989).
164. A. Bényei and F. Joó, *J. Mol. Catal.* **58,** 151 (1990).
165. J. M. Grosselin, and C. Mercier, Fr. Patent 2,623, 799 to Rhône Poulenc Santé (12-01-1987).
166. J. M. Grosselin and C. Mercier, *J. Mol. Catal.* **63,** L25 (1990).
167. F. Joó, Z. Tóth, and M. T. Beck, *Inorg. Chim. Acta* **25,** L61 (1977).
168. E. Fache, F. Senocq, C. Santini, and J.-M. Basset, *J. Chem. Soc., Chem. Commun.*, 1776 (1990).

169. F. Joó, L. Somsak, and M. T. Beck, *J. Mol. Catal.* **24,** 71 (1984).
170. C. Larpent, R. Dabard, and H. Patin, *C. R. Acad. Sci. Ser. 2* **304,** 1055 (1987).
171. T. Hayashi, T. Mise, S. Mitachi, K. Yamamoto, and M. Kumada, *Tetrahedron Lett.* **14,** 1133 (1976).
172. Y. Amrani and D. Sinou, *J. Mol. Catal.* **24,** 231 (1984).
173. L. Lecomte, D. Sinou, J. Bakos, I. Tóth, and B. Heil, *J. Organomet. Chem.* **370,** 277 (1989).
174. I. Tóth, B. E. Hanson, and M. E. Davis, *Tetrahedron: Asymmetry* **1,** 913 (1990).
175. B. D. Vineyard, W. S. Knowles, M. J. Sabacky, G. L. Bachman, and D. J. Weinkauff, *J. Am. Chem. Soc.* **99,** 5946 (1977).
176. H. Brunner and W. Pieronczyk, *Angew. Chem., Int. Ed. Engl.* **18,** 620 (1979).
177. C. R. Landis and J. Halpern, *J. Am. Chem. Soc.* **109,** 1746 (1987), and references therein.
178. D. Sinou, M. Safi, C. Claver, and A. Masdeu, *J. Mol. Catal.* **68,** L9 (1991).
179. I. Tóth, B. E. Hanson, and M. E. Davis, *Catal. Lett.* **5,** 183 (1990).
180. M. Laghmari and D. Sinou, *J. Mol. Catal.* **66,** L15 (1991).
181. D. Chapman and P. J. Quinn, *Proc. Natl. Acad. Sci. U.S.A.* **73,** 3971 (1976).
182. C. Vigo, F. M. Goni, P. J. Quinn, and D. Chapman, *Biochim. Biophys. Acta* **508,** 1 (1978).
183. T. D. Madden, W. E. Peel, P. J. Quinn, and D. Chapman, *J. Biochem. Biophys. Methods* **2,** 19 (1980).
184. F. Farin, H. L. M. Van Gaal, S. L. Bonting, and F. J. M. Daemen, *Biochim. Biophys. Acta* **711,** 336 (1982).
185. L. Vigh, F. Joó, P. R. Van Hasselt, and P. J. C. Kuiper, *J. Mol. Catal.* **22,** 15 (1983).
186. P. J. Quinn and C. E. Taylor, *J. Mol. Catal.* **13,** 389 (1981).
187. D. Morel, Fr. Patent 2,486,525 to Rhône-Poulenc Industries (07-10-1980).
188. D. Morel, Fr. Patent 2,505,322 to Rhône-Poulenc Industries (05-11-1981).
189. D. Morel, "Progress in Terpen Chemistry." Proceedings of the Conference on Terpen Chemistry, Grasse, France, May, 1987.
190. D. Morel, Fr. Patent 2,541,675 to Rhône-Poulenc Santé (02-24-1983).
191. G. Mignani, and D. Morel, Fr. Patent 2,569,403 to Rhône-Poulenc Santé (08-23-1984).
192. E. Kuntz, Fr. patent 2,338,253 to Rhône-Poulenc Industries (01-13-1976).
193. M. M. Taqui Khan, S. B. Halligudi, and S. Shukla, *J. Mol. Catal.* **58,** 299 (1990).
194. J. Kiji, T. Okano, W. Nishiumi, and H. Konishi, *Chem. Lett.*, 957 (1988).
195. M. M. Taqui Khan, S. B. Halligudi, and S. H. R. Abdi, *J. Mol. Catal.* **44,** 179 (1988).
196. T. Okano, I. Uchida, T. Nakagaki, H. Konishi, and J. Kiji, *J. Mol. Catal.* **54,** 65 (1989).
197. T. Okano, Y. Moriyama, H. Konishi, and J. Kiji, *Chem. Lett.*, 1463 (1986).
198. P. Kalck and F. Monteil, unpublished results (1991).
199. M. M. Taqui Khan, S. B. Halligudi, and S. H. R. Abdi, *J. Mol. Catal.* **48,** 325 (1988).
200. M. M. Taqui Khan, S. B. Halligudi, S. H. R. Abdi, and S. Shukla, *J. Mol. Catal.* **48,** 25 (1988).
201. E. Kuntz, Fr. Patent 2,366,237 to Rhône-Poulenc Industries (07-27-1976).
202. F. J. Kristine and R. E. Shepherd, *J. Chem. Soc., Chem. Commun.*, 132 (1980).
203. M. M. Taqui Khan and G. Ramachandraiah, *Inorg. Chem.* **21,** 2109 (1982).
204. M. M. Taqui Khan, *Pure Appl. Chem.* **55,** 159 (1983).
205. M. M. Taqui Khan, A. Hussain, G. Ramachandraiah, and M. A. Moiz, *Inorg. Chem.* **25,** 3023 (1986).
206. M. M. Taqui Khan, A. Hussain, K. Venkatasubramanian, G. Ramachandraiah, and V. Oomen, *J. Mol. Catal.* **44,** 117 (1988).
207. M. M. Taqui Khan, M. R. H. Siddiqui, A. Hussain, and M. A. Moiz, *Inorg. Chem.* **25,** 2765 (1986).

208. M. M. Taqui Khan, S. A. Mirza, A. Prakash Rao, and C. Sreelatha, *J. Mol. Catal.* **44,** 107 (1988).
209. D. J. Cookson. T. D. Smith, J. F. Boas, P. R. Hicks, and J. R. Pilbrow, *J. Chem. Soc.*, 109 (1977).
210. D. Sinou, Eur. Patent 0,407,256 to Rhône-Poulenc Santé (06-22-1989).
211. M. Safi and D. Sinou, *Tetrahedron Lett.* **32,** 2025 (1991).
212. G. Mignani and D. Morel, Fr. Patent 2,561,641 to Rhône-Poulenc Santé (03-22-1984).
213. A. L. Casalnuovo and J. C. Calabrese, *J. Am. Chem. Soc.* **112,** 4324 (1990).
214. S. C. Tang, T. E. Paxson, and L. Kim, *J. Mol. Catal.* **9,** 313 (1980).
215. A. Skorobogaty and T. D. Smith, *J. Mol. Catal.* **16,** 131 (1982).
216. M. E. Ford and J. E. Premecz, *J. Mol. Catal.* **19,** 99 (1983).
217. F. Joó and M. T. Beck, *J. Mol. Catal.* **24,** 135 (1984).
218. I. Tóth, B. E. Hanson, and M. E. Davis, *J. Organomet. Chem.* **396,** 363 (1990).
219. I. Tóth, B. E. Hanson, and M. E. Davis, *J. Organomet.. Chem.* **397,** 109 (1990).
220. W. H. Quayle and T. J. Pinnavaia, *Inorg. Chem.* **18,** 2840 (1979).
221. J. P. Arhancet, M. E. Davis, J. S. Merola, and B. E. Hanson, *Nature* (*London*) **339,** 454 (1989).
222. I. T. Horváth, *Catal. Lett.* **6,** 43 (1990).
223. J. P. Arhancet, M. E. Davis, J. S. Merola, and B. E. Hanson, *J. Catal.* **121,** 327 (1990).
224. I. Guo, B. E. Hanson, I. Toth, and M. E. Davis, *J. Organomet. Chem.* **403,** 221 (1991).
225. Y. Degani and I. Willner, *J. Chem. Soc., Perkin Trans. 2*, 37 (1986).
226. I. Willner and R. Maidan, *J. Chem. Soc., Chem. Commun.*, 876 (1988).
227. A. J. Frank, Z. Goren, and I. Willner, *J. Chem. Soc., Chem. Commun.*, 1029 (1985).
228. M. M. Taqui Khan, R. C. Bhardwaj, and C. M. Jadhav, *J. Chem. Soc., Chem. Commun.*, 1690 (1985).
229. M. M. Taqui Khan, R. C. Bhardwaj, and C. Bhardwaj, *Angew. Chem., Int. Ed. Engl.* **27,** 923 (1988).
230. M. M. Taqui Khan and N. Nageswara Rao, *J. Mol. Catal.* **52,** L5 (1989).
231. M. M. Taqui Khan and N. Nageswara Rao, *J. Mol. Catal.* **58,** 323 (1990).

ADVANCES IN ORGANOMETALLIC CHEMISTRY, VOL. 34

Slowed Tripodal Rotation in Arene–Chromium Complexes: Steric and Electronic Barriers

MICHAEL J. McGLINCHEY

Department of Chemistry
McMaster University
Hamilton, Ontario L8S 4M1, Canada

I
INTRODUCTION

A. *Crystal Structures of Arene Chromium Complexes*

The syntheses of bis(arene)chromium and (arene)$Cr(CO)_3$ complexes (*1–3*) naturally prompted questions as to the mode of interaction of the metal with the aromatic ring system. An early X-ray crystallographic study on $(C_6H_6)_2Cr$ suggested an alternating pattern of carbon–carbon bond lengths, viz. 1.439 and 1.353 Å (*4*); however, subsequent determinations (*5,6*) did not support this view. Eventually, the crystal structure was determined at low temperature (100 K) by Keulen and Jellinek (*7*) and the molecule was unequivocally shown to possess D_{6h} symmetry. These latter reports have been reinforced by Haaland's electron diffraction measurements, which showed that the C–C distances differed by less than 0.02 Å (*8*).

X-Ray data on (arene)$Cr(CO)_3$ systems revealed a number of interesting features. The staggered isomer is found for (benzene)$Cr(CO)_3$ and it appears as though the octahedrally coordinated metal atom is attached to the midpoints of three C═C bonds, as in **1** (*9*). This observation was amplified considerably by a combined X-ray and neutron diffraction study at 78 K (*10*). These low-temperature measurements revealed that the C–C bonds in the ring alternate in length, the shortest bonds being trans to the carbonyls. The average difference between the "long" and "short" bonds was 0.017 Å. Moreover, the hydrogen atoms were found to be displaced by an average of 0.3 Å from the plane of the benzene in a direction toward the chromium atom.

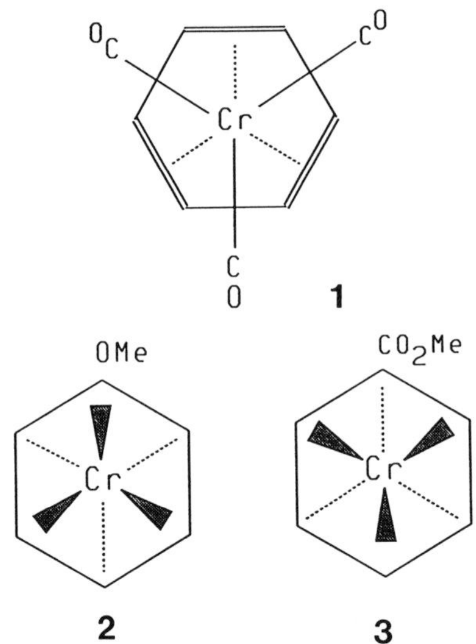

In contrast, in monosubstituted molecules $(C_6H_5R)Cr(CO)_3$ the carbonyls are found to eclipse three ring atoms. The favored conformations of these molecules in the solid state were enunciated by Carter *et al.* as the result of a series of X-ray crystallographic determinations. Specifically, incorporation of electron-donating or electron-withdrawing substituents yields the eclipsed rotamers **2** and **3**, respectively. The rationale offered to account for structure **2** is that the electron-donating group, *e.g.*, Me_2N or MeO, polarizes the charge distribution in the ring so as to render the *ortho* and *para* carbons relatively electron rich. This makes the C-2, C-4, and C-6 positions better donors to the metal atom and, maintaining the preferred octahedral geometry at chromium, the carbonyl ligands are found to eclipse the C-1, C-3, and C-5 sites (*11*). Conversely, complexes bearing electron-withdrawing substituents, as in $(C_6H_5CO_2Me)Cr(CO)_3$, adopt conformation **3** (*12*).

B. *NMR Measurements*

The initial nuclear magnetic resonance (NMR) data (*13,14*) on these half-sandwich molecules revealed that the proton chemical shifts in π-complexed rings are considerably shielded relative to their resonance positions in the free arenes. Subsequently, attention focused on whether the observed chemical shifts reflected the presence of the single rotamer found crystallographically in the solid state or whether the $M(CO)_3$ tripod was spinning freely on the NMR time scale (*15*). A particularly widely reported parameter was the *complexation shift*, that is, the chemical shift difference between corresponding protons in the free ligand and in the $M(CO)_3$ complex, where M = Cr, Mo, W. This in turn led to speculation concerning the ability of electron-donating or electron-withdrawing substituents to transmit their inductive and/or resonance effects through the coordinated ring (*16,17*).

In 1969 there appeared from Jackson's laboratory a particularly insightful report that had considerable significance for all future discussions of the tripodal rotation problem (*18*). It was noted that the resonance positions of the arene ring protons in *cis*- and *trans*-substituted (indane)Cr-$(CO)_3$ derivatives were quite different. In the monomethyl complex **4**

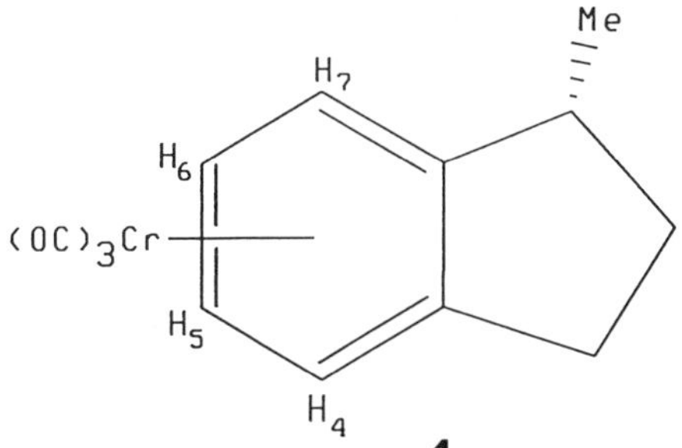

Me
H_7
H_6
$(OC)_3Cr$
H_5
H_4

4

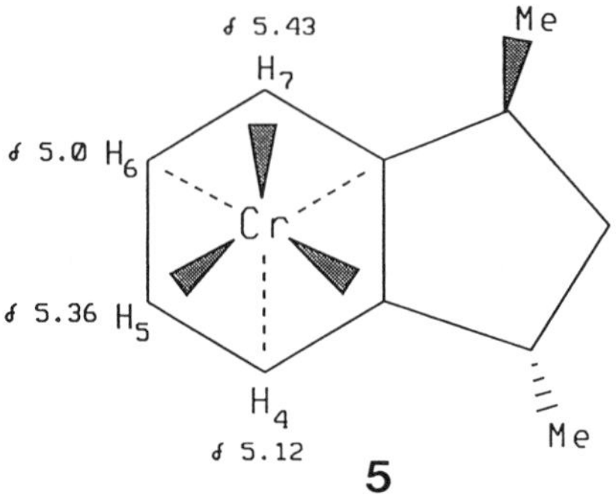

the arene protons H(4) through H(7) are clustered within a narrow range (0.18 ppm). In contrast, in (*trans*-1,3-dimethylindane)$Cr(CO)_3$, (**5**) the H-5 and H-7 protons are markedly deshielded relative to H-4 and H-6. This observation was rationalized in terms of a favored conformation that minimizes the adverse steric interactions between the tricarbonylchromium moiety and the *proximal* methyl group, as depicted in Fig. 1. Moreover, these authors went on to propose that "protons lying under superimposed metal carbonyl bonds are deshielded relative to the exposed protons."

This hypothesis gained rapid acceptance and is still frequently invoked more than two decades later (*19*). Among the recent examples that strongly reinforce Jackson's early observations are those involving the α- and β-$Cr(CO)_3$ complexes of methyl O-methylpodocarpate (**6** and **7**, respectively). These diastereomers have been characterized both by X-ray crystallography (*20*) and by two-dimensional NMR techniques at 500 MHz (*21*). In each case, the arene protons eclipsed by carbonyl ligands resonate

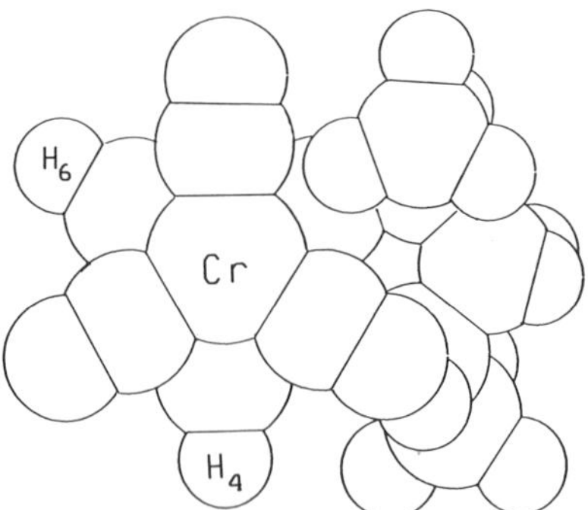

FIG. 1. Space-filling model of (*trans*-1,3-dimethylindane)$Cr(CO)_3$ (**5**).

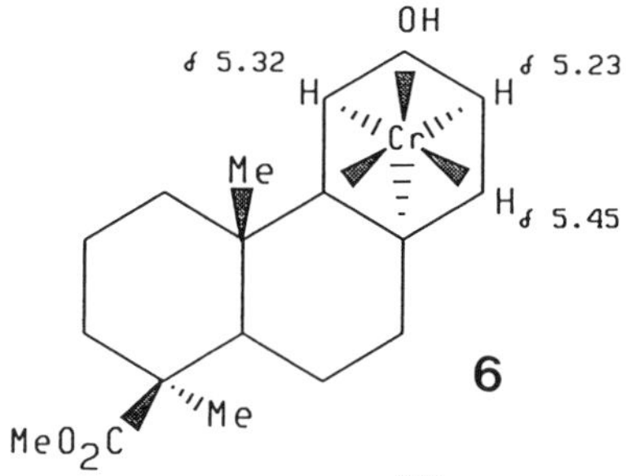

at a higher frequency than do their noneclipsed partners. Furthermore, the diamagnetic anisotropy of the $Cr(CO)_3$ group (*22*) has been used to distinguish the α- and β-isomers of (estradiol)$Cr(CO)_3$ (*23*).

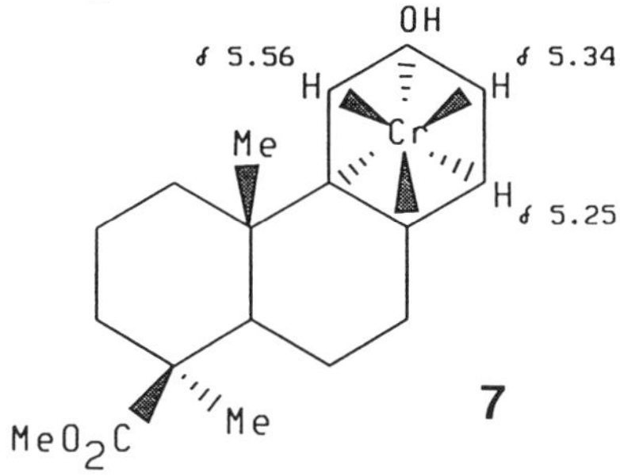

II

KINETIC AND THERMODYNAMICALLY RESTRICTED ROTATION

The interconversion of two symmetry-equivalent rotamers, such as **8** and **9** in Fig. 2, will leave the arene nuclei, be they ^1H, ^{13}C, ^{19}F, *etc.*, unchanged because the effective molecular point group is C_s. Moreover, if the activation energy for tripodal rotation is low, the ^{13}C or ^{17}O resonances of the three carbonyl ligands will be averaged to singlets. However, if the barrier to rotation about the ring centroid–metal axis is raised by steric or electronic factors that disfavor the transition state, then the $Cr(CO)_3$ moiety can give rise to a 2:1 ^{13}CO peak pattern.

In cases where the rotamers **10** and **11** coexist in solution and are separated by a small barrier, as in Fig. 3, not only will there be rapid exchange of the carbonyl environments but also the corresponding proton or carbon-13 resonances in the two isomers will be averaged. Now if **10** is

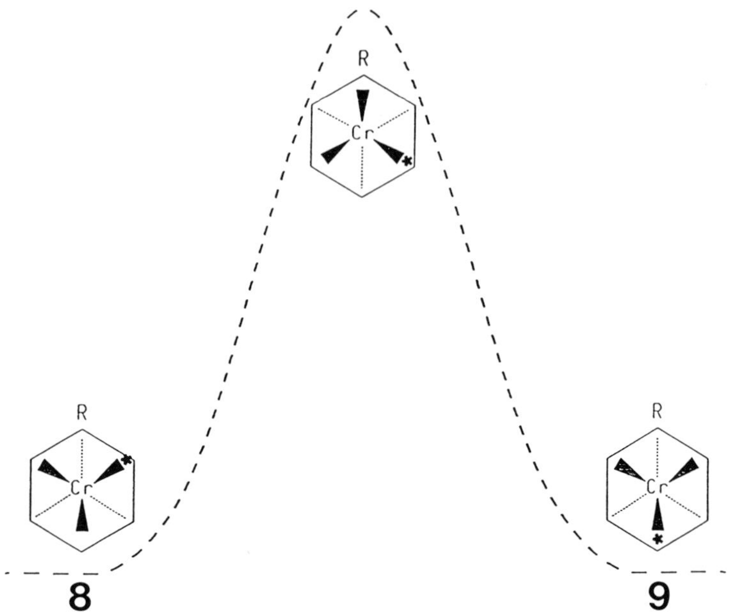

FIG. 2. Tripodal rotation that interconverts symmetry-equivalent isomers **8** and **9**.

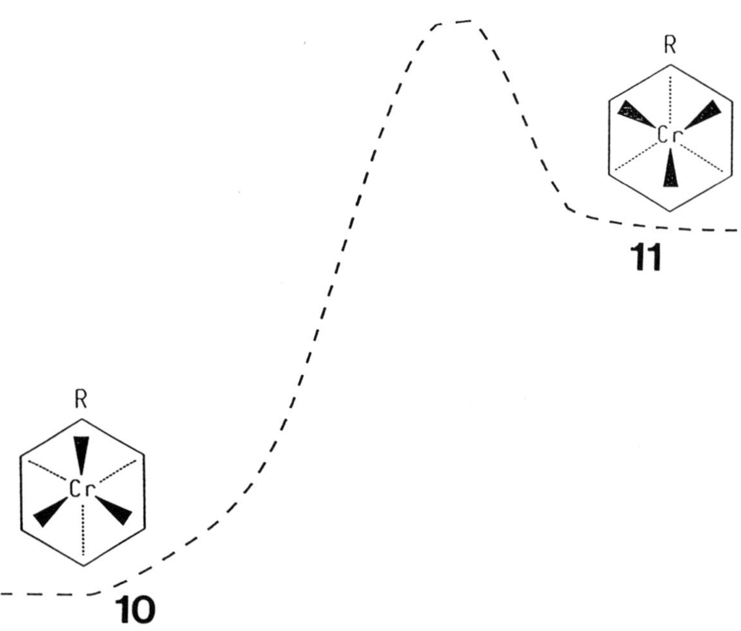

FIG. 3. Tripodal rotation that interconverts different eclipsed isomers **10** and **11**.

preferred for electronic reasons, but steric factors favor **11**, then increasing the temperature will shift the equilibrium toward **11**. [This occurs because the entropy factor ($T\Delta S$) will gradually overcome the lower enthalpy term (ΔH) of the electronically favored isomer. Consequently, at higher temperature the free energy, ΔG, favors the sterically less hindered molecule **11**.] These changing equilibria are reflected in temperature-dependent chemical shifts. To be specific, lowering the temperature increases the fraction of conformer **10** in which the CO ligands eclipse the C-3 and C-5 positions; thus the observed chemical shift of the *meta* protons moves gradually downfield (*i.e.*, to higher frequency) as the temperature is decreased.

These concepts, first clearly delineated by Jackson and co-workers (*18,24*), were soon applied to a wide variety of (arene)$Cr(CO)_3$ systems but led to much controversy; a series of claims, counterclaims, and rebuttals appeared in the literature (*25–30*), but some of these contributions may have resulted from a misunderstanding of the situation. Thus, the failure to observe temperature-dependent ^1H NMR chemical shift changes in (C_6H_5-*t*-Bu)$Cr(CO)_3$ does not necessarily imply that tripodal rotation has slowed on the NMR time scale. If there is a single predominant rotamer, *e.g.*, **11,** and the tripod executes jumps of 120°, as in Fig. 2, then we would not expect to see changes in the proton spectrum at different temperatures. Jackson *et al.* tried to clarify the situation in terms of "kinetically restricted rotation" and "thermodynamically restricted rotation" (*30*). It is the former that is invoked for cases of slowed tripodal rotation; this "kinetic restriction" to conformational exchange involves a barrier to interchange between two conformers and is quantitatively expressed as a free energy of activation for that conformational interchange (*31,32*). "Thermodynamic restriction" to conformational exchange, as defined by Mislow and Raban, refers to the situation when the equilibrium favors a single rotamer; even if the conformational change is rapid enough on the NMR time scale to provide signal averaging, the conformational change is still restricted in the sense that the "conformer populations" are not equal (*31,32*). Under ideal circumstances, kinetically restricted rotation will manifest itself by a splitting of the degeneracy of one or more of the NMR resonances; thus a 2:1 ^{13}CO pattern for a (C_6H_5X)$Cr(CO)_3$ complex would be indicative of slowed tripodal rotation on the NMR time scale. As we shall see presently, the experimental realization of these predictions had to await the advent of very-high-field spectrometers with their enormous spectral dispersion.

In an attempt to construct a system in which a single conformer predominates in solution, van Meurs *et al.* (*33*) prepared the tricarbonylchromium complex of (t-Bu)$_2$CH–Ph, **12**. This was assumed to adopt structure

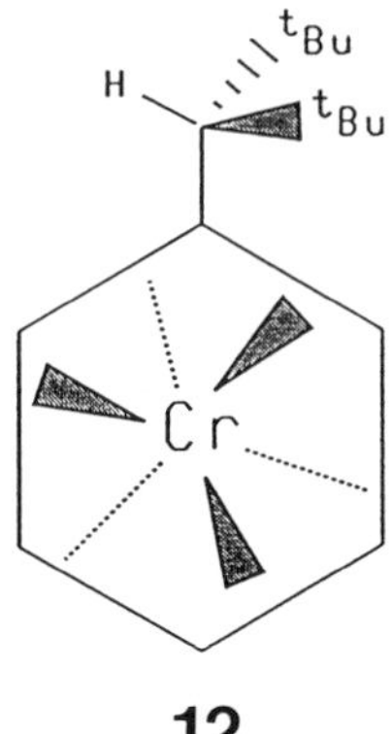

12

11 exclusively and was taken as the standard to which other conformational equilibria could be compared. However, a subsequent X-ray crystallographic determination on **12** revealed that, at least in the solid state, the tripod did not lie directly below positions C-2, C-4, and C-6 but was in fact twisted by 16° away from the perfectly eclipsed orientation (*34*).

The first unequivocal example of slowed tripodal rotation relative to a metal-to-arene ring axis was reported by Pomeroy and Harrison (*35*), who observed a splitting of the degeneracy of the ring carbons in [1,4-bis(*t*-butyl)benzene]Ru(CO)$(SiCl_3)_2$ (**13**) at low temperature. This work has

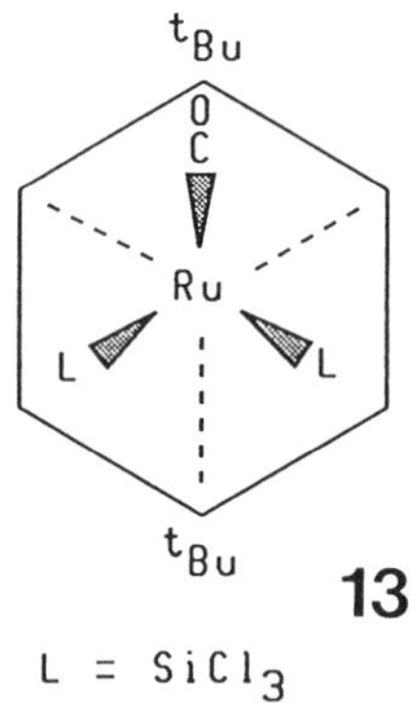

13

L = $SiCl_3$

since been extended to other related molecules (*36*). Nevertheless, even taking advantage of the much larger ^{13}C chemical scale, similar splittings were not detectable at −60°C in the case of [1,3-bis(*t*-butyl)benzene]-$Cr(CO)_3$ (**14**) (*37*), nor even with [1,3,5-tris(*t*-butyl)benzene]$Cr(CO)_2$-PPh_3 (**15**) (*38*). Clearly, either the barriers in these systems are too low or the chemical shift differences in the chromium complexes are rather small

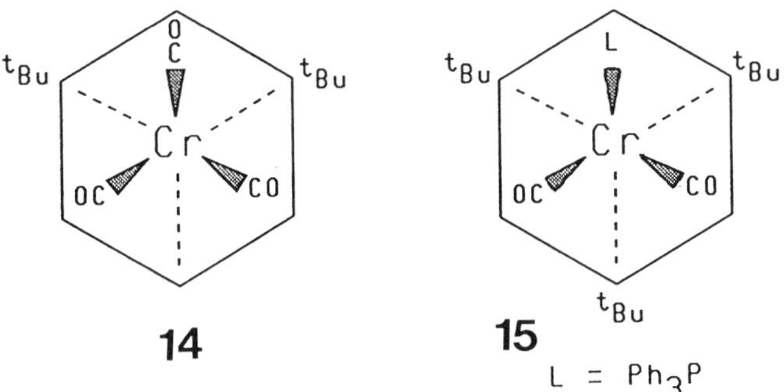

(*31*). Apparently, these experiments need to be carried out at very low temperatures at the highest magnet strength available.

III

CHEMICAL REACTIVITY AND TRIPODAL ORIENTATION

A. *Electrophilic Attack on (Arene)Cr(CO)$_3$ Complexes*

The studies on conformational equilibria involving the preferred orientations of tripodal moieties raised the question as to whether these factors affected the chemistry of the π-complexed arene. These went beyond the obvious points that the electron-withdrawing character of the $Cr(CO)_3$ group increased the acidity of (benzoic acid)$Cr(CO)_3$ relative to C_6H_5-CO_2H, decreased the basicity of (aniline)$Cr(CO)_3$ with respect to C_6H_5-NH_2, and enhanced the susceptibility of (chlorobenzene)$Cr(CO)_3$ toward nucleophilic displacement of halide (*19*). Interest soon focused on more subtle effects such as isomer distributions on acetylation of (alkylbenzene)-$Cr(CO)_3$ complexes. It was shown (*39*) that the percentage of *ortho* and *meta* substitution was very much higher than that observed in the free ligands. Moreover, increasing the bulk of the alkyl group in the arene ring (from methyl, through ethyl and isopropyl to *tert*-butyl) enhanced the degree of *meta* substitution at the expense of the *para* isomers. The rationale proposed was that the transition state for *para* attack (**16**) was disfavored when the tripod adopted an orientation that avoided adverse steric interactions with the bulky alkyl substituent. As shown in Fig. 4, the phenonium ion **17** arising from acetylation at the *meta* position does not suffer from such problems and so should be the favored intermediate for electrophilic attack.

Ac
H···AlCl3
O
C
Cr
OC
CO
tBu
Cl3Al···H
Ac
O
C
Cr
OC
CO
tBu

16 **17**

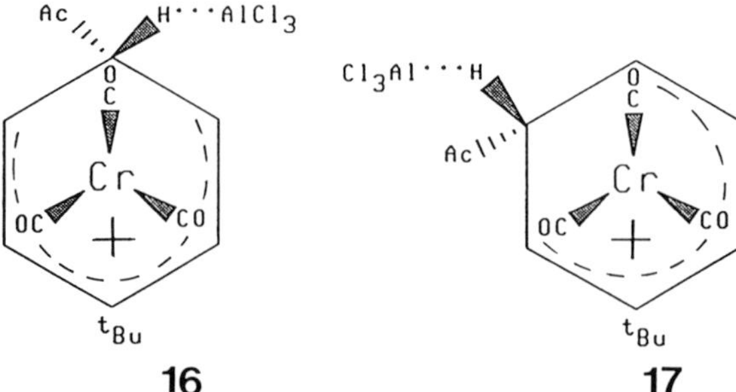

FIG. 4. Phenonium ions formed during acetylation of $(t\text{-Bu-}C_6H_5)Cr(CO)_3$.

B. *Nucleophilic Attack on (Arene)Cr(CO)$_3$ Complexes*

These concepts have been extended to rationalize the favored sites for nucleophilic attack in (arene)$Cr(CO)_3$ molecules. It has been suggested that the regioselectivity of attack of a nucleophile is controlled not only by the nature of the arene substituents but also by the conformation of the tripod (*40–43*). There is now ample experimental evidence to support this assertion (*43*). Typically, Rose and co-workers observed 79% attack at the eclipsed position of the spiro compound **18**. The orientation of the $Cr(CO)_3$ tripod in **18** was established X-ray crystallographically, and an analysis of the NMR spectrum in solution leads one to believe that the same conformer is favored as was found in the solid state (*44*). Nevertheless, the situation may not be quite as straightforward as it appears at first sight. Brocard and Lebibi (*45*) have noted that when strongly electron-donating substituents (MeO, Me_2N, *etc.*) or strongly electron-withdrawing groups (CO_2R) are present, then the site of attack appears to be controlled more by the electronic effects of the substituents rather than by the orientation of the tripod. Clearly, this aspect of (arene)$Cr(CO)_3$ chemistry merits further study.

3%
79%
12%
Cr
O
1. LiCHMeCN
2. oxidize
R
O
R = CH(Me)CN

18

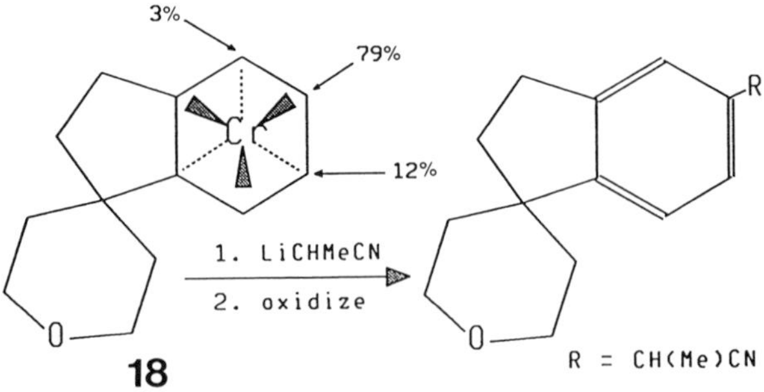

IV

THEORETICAL CALCULATIONS

Hoffmann and co-workers (*46, 47*) used the extended Hückel molecular orbital (EHMO) approach to delineate some of the factors that influence the barriers to tripodal rotation in (arene)$Cr(CO)_3$ systems. These seminal contributions not only led to a rationalization of many of the previously reported observations but also prompted synthetic chemists to devise experimental tests for the predictions. It was shown initially that in (benzene)$Cr(CO)_3$ the tripod should behave as an almost unhindered rotor; this is in accord with the vapor-phase electron diffraction data of Schäfer and co-workers, who described the structure as "a mixture of several conformations differing only in the rotational arrangement of the six-membered ring with respect to the carbonyl groups" (*48*). Moreover, Hoffmann and co-workers predicted that slowed $Cr(CO)_3$ rotation might

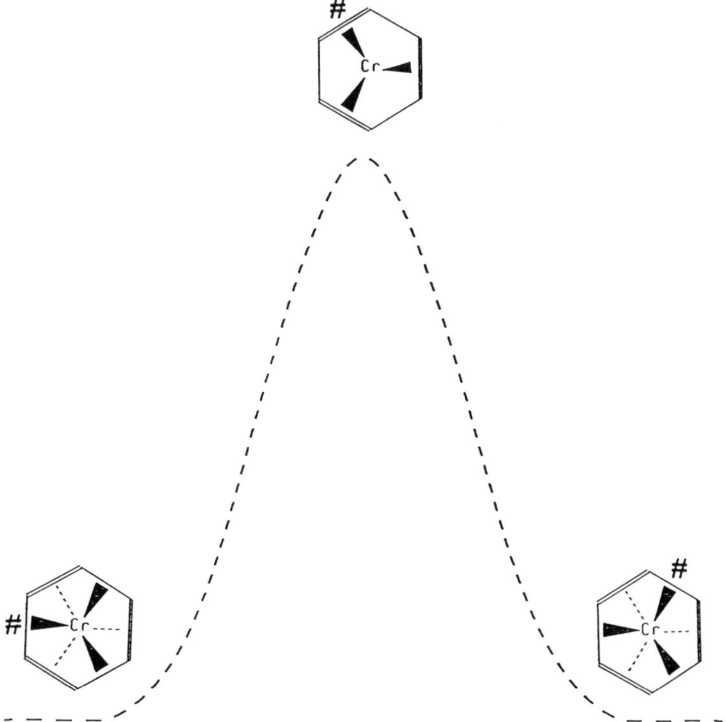

FIG. 5. Tripodal rotation in (1,3,5-cyclohexatriene)$Cr(CO)_3$.

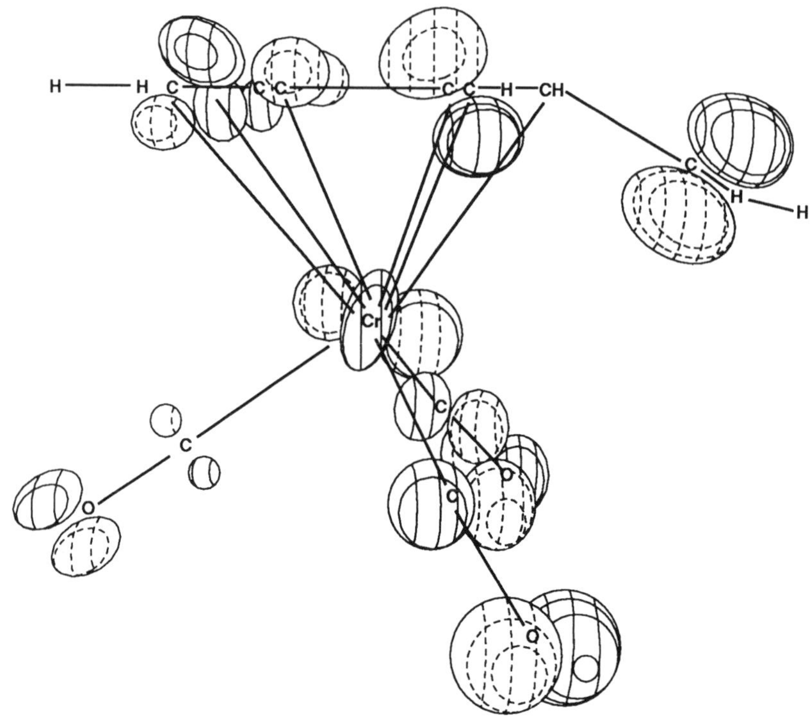

FIG. 6. $[(\text{Benzyl})Cr(CO)_3]^+$ in which the methylene group is bent down so as to optimize the orbital overlap between p_z on the benzylic carbon and $d_{x^2-y^2}$ on chromium.

be observable if certain electronic criteria could be satisfied. For example, if one could bring about bond fixation in an arene ring such that one Kekulé structure was strongly preferred, as in Fig. 5, then the barrier to rotation of the $Cr(CO)_3$ moiety could be as high as 19 kcal mol^{-1}.

Furthermore, it was proposed that molecules possessing a benzylic-type carbon with a vacant p_z orbital should interact with a filled orbital of the $Cr(CO)_3$ fragment and so stabilize the nonplanar structure shown in Fig. 6. These calculations suggested not only that the benzylic carbon would bend down toward the metal through an angle of approximately 12° but also that conformer **19a** should be markedly favored over **19b**.

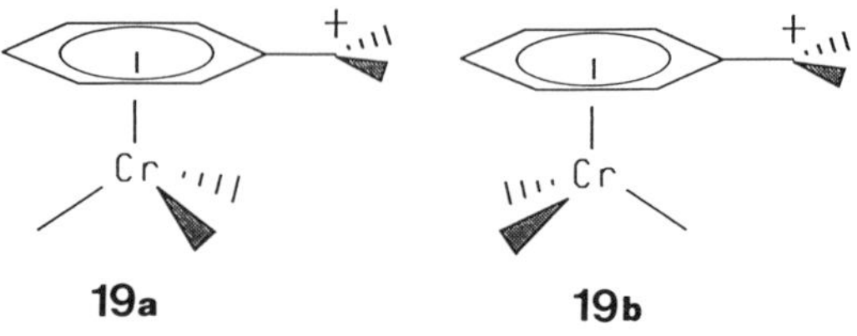

The predictions of Hoffmann and co-workers have since been experimentally verified and are discussed in detail in Section VI.

V

STERIC BARRIERS TO TRIPODAL ROTATION

A. ^{13}C NMR Spectra of (Hexaethylbenzene)$Cr(CO)_3$

As mentioned in Section I, the relative abilities of free and of π-complexed arene rings to transmit the inductive and resonance effects of substituents have been the subject of numerous kinetic measurements and spectroscopic studies (*19,49–51*). Such reports have not been restricted to ^1H NMR spectra but have also included ^{13}C and ^{19}F investigations (*52–54*). Among the factors that have been invoked to account for the observed data are (1) the magnetic anisotropy of the metal–arene bond, (2) σ participation in the transmission of substituent effects, and (3) a reduction in the arene ring current.

The evaluation of ring current contributions to chemical shifts of protons in the vicinity of arenes requires an understanding of the local anisotropic contributions of the ring carbons (*55,56*); these data can be obtained only by solid-state ^{13}C mesurements (*57,58*). To this end, the cross-polarization magic angle spinning (CPMAS) ^{13}C solid-state NMR spectra of a series of (arene)$Cr(CO)_3$ complexes, where the arene was C_6H_6, C_6Me_6, and C_6Et_6, were measured by Maricq *et al.* (*59*). A surprising result emerged for the hexaethylbenzene complex, and this serendipitous observation ultimately led to an unequivocal resolution of the tripodal rotation problem.

Before one can extract the principal elements of the ^{13}C shielding tensor for the ring carbons of the above-mentioned (arene)$Cr(CO)_3$ complexes, one must obtain the isotropic chemical shifts of these carbons from the magic angle spinning experiment. In the complex (hexaethylbenzene)-$Cr(CO)_3$ (**20**), the spectrum exhibited two equally populated environments for the methyl carbons, for the methylene carbons, and also for the aromatic carbon nuclei. The conclusion drawn was that the sixfold

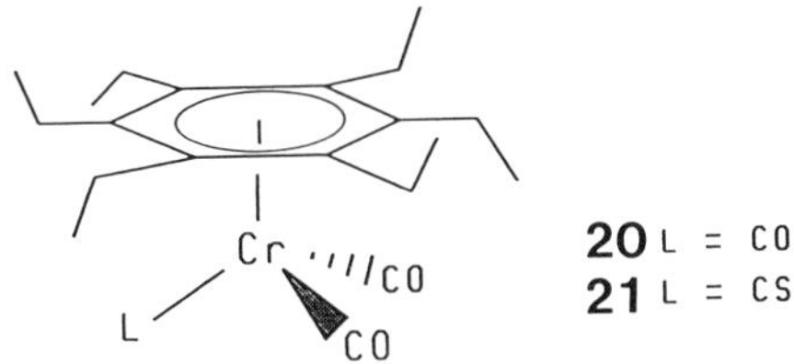

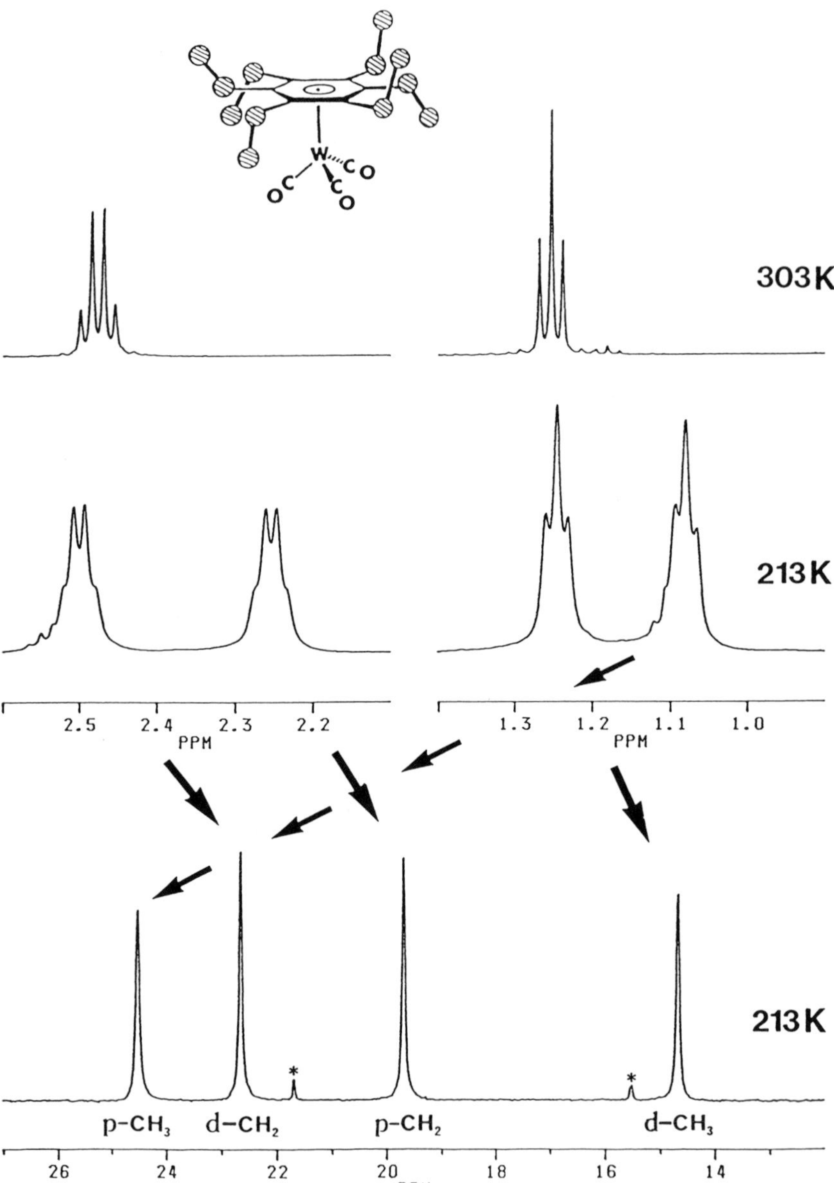

FIG. 7. Variable-temperature ^{1}H and ^{13}C NMR spectra of $(HEB)W(CO)_3$ indicating the *proximal* and *distal* ^{1}H and ^{13}C environments.

symmetry of the ring had been broken to C_3, thus making three alternating ring positions inequivalent to the intervening ones (*59*). Because a preliminary X-ray structure (*60*) of noncomplexed hexaethylbenzene (HEB) indicated that the ethyl groups projected alternately above and below the plane of the arene ring (giving a molecular point group of D_{3d}), it was proposed that the arene adopted the same conformation in (HEB)Cr$(CO)_3$ (**20**). Thus the methyl carbons positioned *proximal* and *distal* to the Cr$(CO)_3$ group are rendered magnetically nonequivalent. Similarly, the methylene and arene ring carbons should be split in accordance with the overall C_{3v} symmetry of the complex.

Subsequently, these same effects were observed in the ^{13}C NMR spectra of (HEB)Cr$(CO)_3$ (**20**) at low temperature in solution (*61*); again the methyl, the methylene, and the arene ring carbons are each split into equally intense resonances. The analogous tungsten complex, (HEB)-W$(CO)_3$, behaves similarly and its ^{1}H and ^{13}C variable-temperature NMR spectra are depicted in Fig. 7; once again the C_{3v} structure of the molecule is evident.

B. *(HEB)Cr(CO)$_2$L Complexes, L = CO, CS, PR$_3$*

In 1980, a very significant contribution was made by Hunter *et al.*, who established that (HEB)Cr$(CO)_3$ in the solid state adopts the alternating *proximal–distal* arrangement of ethyl groups (*62*). Moreover, they also showed that the organometallic tripod is oriented such that the metal carbonyl ligands lie directly beneath the *distal* ethyls so as to minimize any steric interactions with the *proximal* alkyl chains. Furthermore, the authors of this important paper discussed the relative energies of the numerous conformers of HEB (see Fig. 8, conformers **a**, **c**, **e**, and **h**) and showed the 1,3,5-*distal*–2,4,6-*proximal* isomer (**a**) to be favored over those structures that place adjacent ethyls on the same face of the arene (*63*). The highest energy conformation (**h**) is that in which all six ethyls are oriented the same way and the molecule has C_{6v} symmetry; of course, in a metal complex of type **h**, these six ethyls must all be *distal* with respect to the organometallic moiety. (The labeling of the conformers of HEB follows the notation introduced in Ref. 63.)

One fluxional process in **20** that was unequivocally established was the barrier to ethyl rotation, *i.e.* the interconversion of *proximal* and *distal* ethyl groups. This mechanism, for which $\Delta G^{\ddagger} = 11.5$ kcal mol^{-1}, equilibrates not only the two arene ring carbon environments but also the two methylene and the two methyl peaks, as shown in Fig. 7.

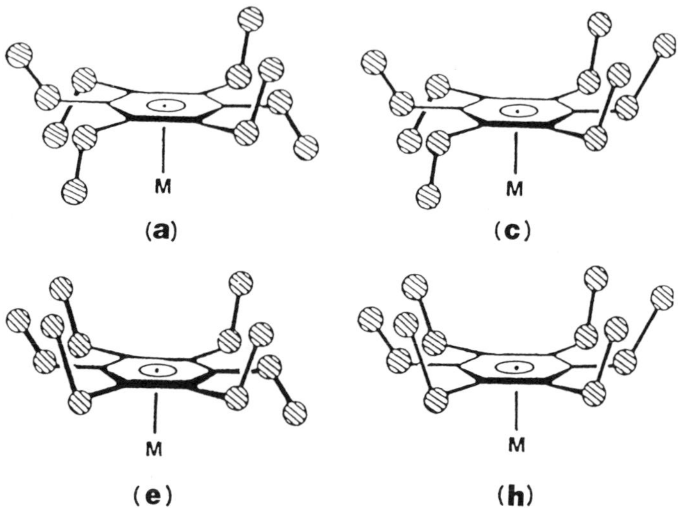

FIG. 8. $(C_6Et_6)M$ conformers with differing arrangements of *proximal* and *distal* ethyl substituents.

The large magnitude (8 ppm) of the chemical shift difference between the two arene ring carbon environments in $(HEB)Cr(CO)_3$, both in solution and in the solid state, is suggestive of very different chromium-to-ring interactions. This might be interpreted as circumstantial evidence in favor of slowed tripodal rotation on the NMR time scale because three ring carbons would be directly bonded to the metal center. Nevertheless, such a claim remains open to the rebuttal that, once the ethyl rotation has slowed, the molecule has effective C_{3v} symmetry whether the tripod spins or not!

To resolve this problem, McGlinchey *et al.* chose to break the symmetry of the tripod by incorporating a thiocarbonyl ligand (*64*). The X-ray crystal structure of $(HEB)Cr(CO)_2CS$ (**21**) closely parallels that of $(HEB)Cr(CO)_3$ in that the tripodal ligands eclipse the three *distal* ethyl groups; the molecular point group in **21** is C_s—a single mirror plane. This lowered symmetry is reflected in the ^{13}C NMR spectrum of the hexaethylbenzene ligand that exhibits 2:1:2:1 patterns for the methyls, the methylene carbons, and the aromatic ring nuclei, as shown in Fig. 9. These data, together with the observation that the ^{13}C NMR spectra for **21** were very similar in CD_2Cl_2 solution and in the solid state, led McGlinchey *et al.* to propose that tripodal rotation had indeed been slowed on the NMR time scale (*64*). Such a scenario would impose C_s symmetry on the molecule and break the three fold degeneracy of the *proximal* and *distal* ethyl groups.

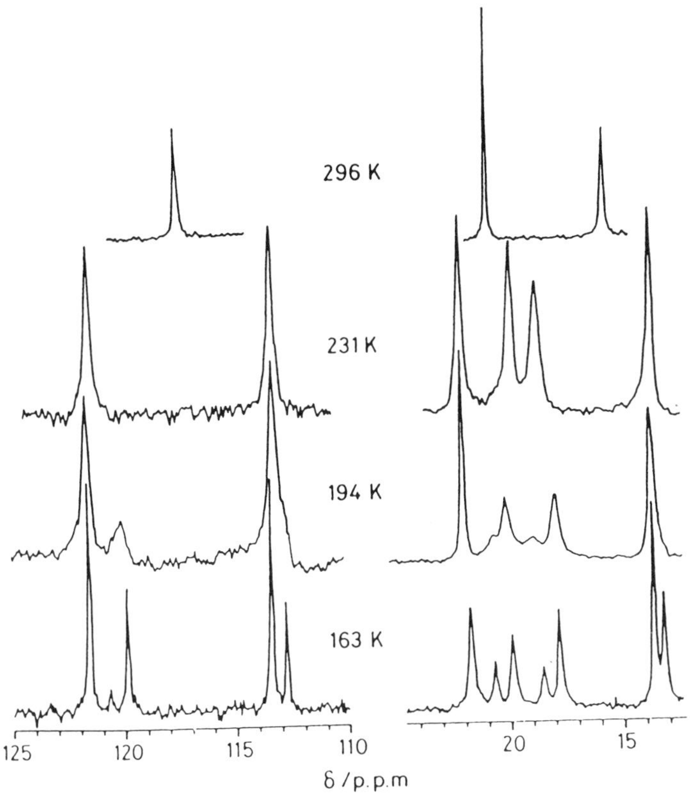

FIG. 9. Variable-temperature ^{13}C NMR spectra of $(HEB)Cr(CO)_2CS$ (**21**).

However, this view was challenged by Hunter and Mislow (*65*), who pointed out that another explanation had to be considered. They raised the possibility that the hexaethylbenzene ligand in **21** might adopt a conformation in solution different from that found X-ray crystallographically in the solid state. In particular, they noted that in conformers such as **c** or **e** in Fig. 8 the HEB ligand possesses C_s symmetry and thus it is not necessary to invoke slowed rotation of the $Cr(CO)_2CS$ tripod. This objection could not be dismissed lightly because by this time several metal complexes of hexaethylbenzene were known in which a variety of ethyl orientations could be detected both in solution and in the solid state.

Astruc *et al.* found that $[(HEB)Fe(C_5H_5)]^+$ PF_6^- (**22**) adopts conformation **c** in the crystalline state (*66*), but Zaworotko and co-workers demonstrated that merely changing the counterion, as in $[(HEB)Fe(C_5H_5]^+$

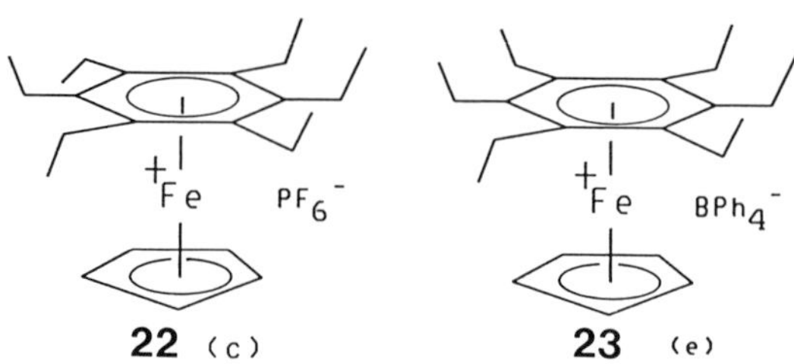

BPh_4^- (**23**), yielded structure **e** (*67*). Moreover, it was shown by variable-temperature NMR measurements that these rotamers coexisted with conformer **a** in solution (*68*). Mislow and Hunter found that replacement of one carbonyl ligand in (HEB)Cr$(CO)_3$ by a series of phosphines of increasing steric bulk yielded the following structures: (HEB)Cr$(CO)_2PMe_3$ (**24; c**) (*68*); (HEB)Cr$(CO)_2PEt_3$ (**25; e** and **h**) (*69*); and (HEB)-Cr$(CO)_2PPh_3$ (**26; h**) (*63*). Similarly, Hunter *et al.* have established the structure of $(C_6Pr_6)Cr(CO)_2PPh_3$ in which all six *n*-propyl substituents are *distal* (*70*). More recently, Herrmann *et al.* have reported that in $[(HEB)Mo(CO)_3Cl]^+MoCl_6^-$ (**27**), the chlorine and the three carbonyl ligands occupy the vertices of a square-based pyramid capped by the molybdenum atom (*71*). In this complex, the hexaethylbenzene exists as yet another stereoisomer, *viz.*, the 1,2,4,5-*distal*–3,6-*proximal* conformer

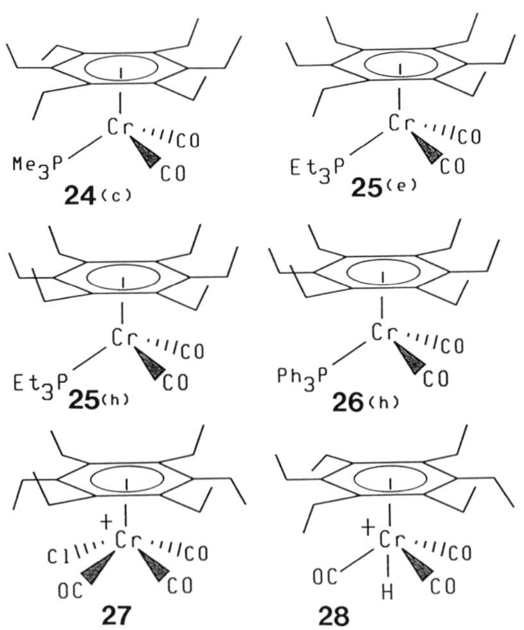

(**b**), which presumably minimizes the interactions between the ethyl substituents and the ligands on the metal without having to resort to the high-energy all-*distal* arrangement. No variable-temperature NMR data have yet been reported on this fascinating molecule. However, in the closely analogous protonated species $[(HEB)M(CO)_3H]^+$ (**28**), where M = Cr, Mo, and W, the cation has C_{3v} symmetry because the hexaethylbenzene ligand retains conformation **a** but the proton attacks along the threefold axis (*72*).

The choice was thus clearly delineated: if $(HEB)Cr(CO)_2CS$ (**21**) could be unequivocally shown to adopt conformation **a** in solution, then the assertion of restricted rotation would be on firm ground. For **a** it is only by freezing out the $Cr(CO)_2CS$ rotation that one can generate a molecule possessing a single mirror plane. On the other hand, if it could be shown that the predominant isomers of **21** present in solution were **c** and/or **e**, then the cessation of tripodal rotation would remain an open question. Subsequent discussion centered on whether one could unequivocally assign particular methyl, methylene, and ring carbon ^{13}C NMR resonances to *proximal* or to *distal* ethyl groups (*73, 74*). Such attributions are obviously crucial to an unambiguous determination of the particular HEB conformer in any given complex.

C. *(Pentaethylacetophenone)Cr(CO)$_3$ and Related Complexes*

The original strategy for resolving the problem of arene–metal rotation was to split the threefold symmetry of the tripod and then detect this lowered symmetry by monitoring the chemical shifts of the methyl, methylene, and ring carbons of the HEB ligand. A second approach involved the use of $(C_6Et_5COMe)Cr(CO)_3$ (**29**), in which the highest possible symmetry is C_s; the crystal structure of **29** reveals that this molecule adopts a conformation that closely mimics that of $(C_6Et_6)Cr(CO)_3$ (**20**) (*75*).

As shown in Fig. 10, the tricarbonylchromium moiety in **29** is nestled in an environment made up of three *proximal* methyl groups whereas the

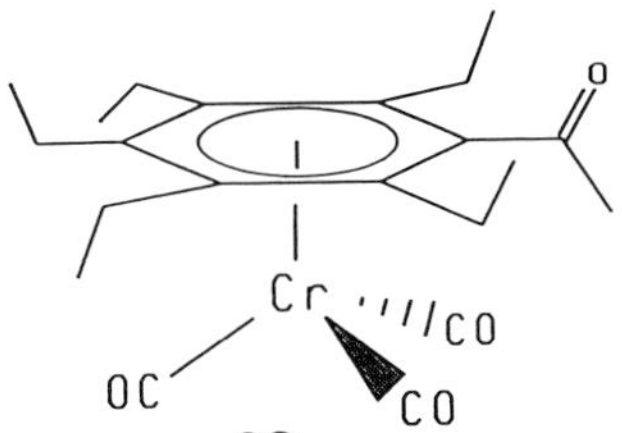

O
Cr
CO
OC
CO

29

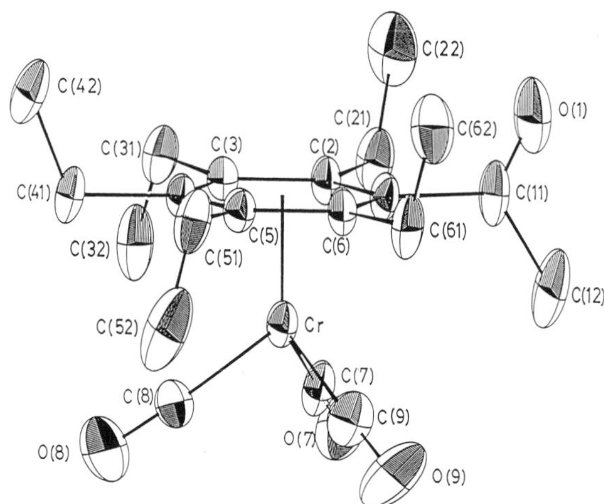

FIG. 10. X-Ray crystal structure of $(C_6Et_5COCH_3)Cr(CO)_3$ (**29**). Reproduced from Ref. 75, with permission of the American Chemical Society.

intervening ethyls are disposed in a *distal* manner. In this case, the 500-MHz ^1H and 125-MHz ^{13}C NMR resonances of the pentaethylacetophenone (PEAP) ligand are unaffected by the spinning rate of the tripod; however, restricted rotation of the $Cr(CO)_3$ fragment in a molecule possessing a single mirror plane must split the threefold degeneracy of the carbonyl ligands. The spectra shown in Fig. 11 provide an experimental realization of the hypothetical situation previously alluded to in Section II. The 2:1 splitting of the $Cr(^{13}CO)_3$ signal in the solid-state spectrum or in solution at $-100°C$ unequivocally demonstrated the cessation of tripodal rotation on the NMR time scale (*74*); the activation energy for this tripodal rotation process is $\approx$9 kcal mol^{-1}.

Moreover, the striking correspondence between the ^{13}C NMR data in both phases left no doubt that the arene substituents adopted the alternating *proximal–distal* arrangement, as for conformer **a** in HEB. We note here the chemical shifts of the *distal* 4-ethyl substituent in **29**. These assignments are rendered completely unambiguous in this case because these resonances are only half as intense as those of the other ethyl groups. These methyl, methylene, and ring carbons are found at δ 14.68, 22.96, and 118.62, respectively, and establish the *distal* resonances in (HEB)-$Cr(CO)_3$ at δ 14.2, 22.8, and 117.2. The corresponding *proximal* peaks in **29** (**20**) appear at δ 20.8 (20.1), 19.3 (19.4), and 109.1 (108.8). The assignment of given ethyl substituents to *proximal* or *distal* sites in an (HEB)-ML_3 complex can thus be made with some confidence, and data for all the known systems have been collected (*76*).

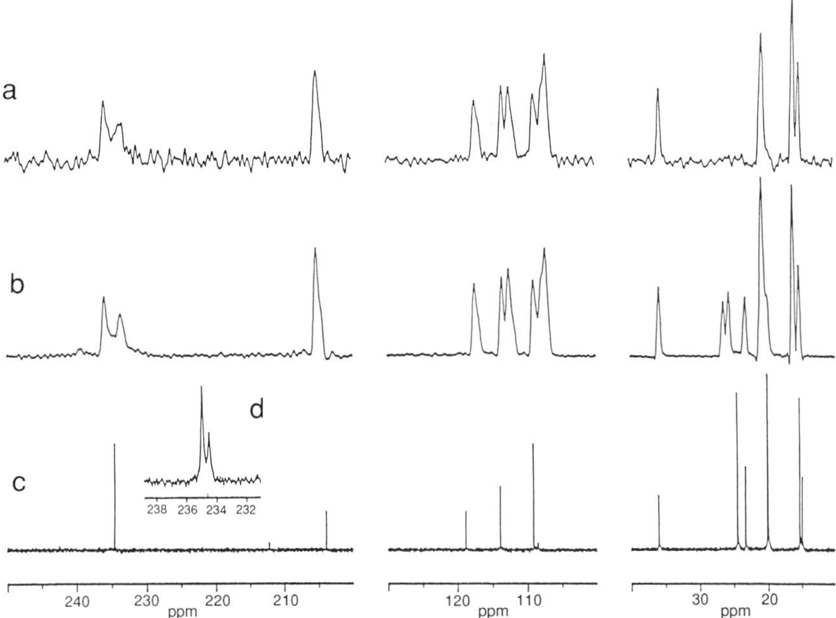

FIG. 11. ^{13}C NMR spectra of $(C_6Et_5COCH_3)Cr(CO)_3$ (**29**). (a) Dipolar dephased 25.1-MHz CPMAS solid-state spectrum in which the methylene resonances have been suppressed. (b) Normal 25.1-MHz CPMAS solid-state spectrum, recorded at +30°C. (c) 125.7-MHz high-resolution spectrum in CD_2Cl_2 at +30°C. (d) Metal carbonyl resonances at −100°C. Reproduced from Ref. 75, with permission of the American Chemical Society.

A particularly elegant extension of this concept has been reported by Kilway and Siegel (77), who prepared 1,4-bis(4,4-dimethyl-3-oxopentyl)-2,3,5,6-tetraethylbenzene (**30**; Fig. 12) and its tricarbonyl complex (**31**). The free ligand has C_{2h} symmetry, which renders inequivalent the methylene protons of each ethyl group. However, as depicted in Fig. 12, rotation of the alkyl groups in **30** interconverts these proton environments. That is, methylene protons of the ethyl groups are **diastereotopic** at the static limit but are **enantiotopic** under dynamic exchange; they serve as a probe of the ethyl group rotation in the free arene. In contrast, the complex **31** has only C_s symmetry at the static limit but has effective C_{2v} symmetry when alkyl and tripodal rotations are rapid on the NMR time scale. Thus one can distinguish the two ketonic environments in **31** when alkyl rotation is slow (at −6°C). Moreover, the $Cr(CO)_3$ tripod yields a 2:1 ^{13}CO pattern at −90°C. This ingenious experiment allowed the independent measurement of the alkyl rotation process (11.3 kcal mol^{-1} in the free ligand; 11.5 kcal mol^{-1} in the complex) and the barrier to tripodal

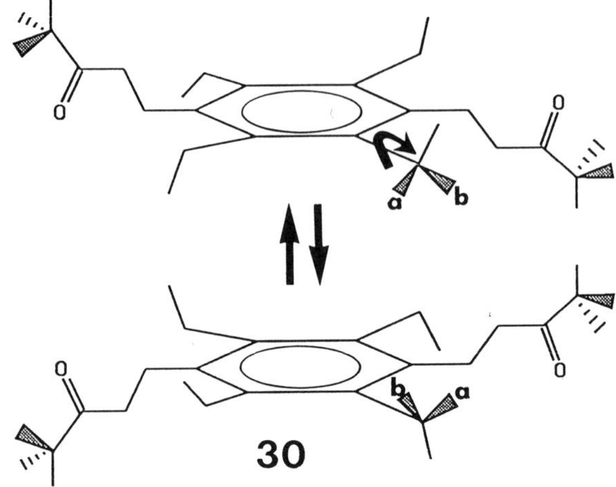

FIG. 12. Alkyl rotation in **30** equilibrates the "inner" and "outer" methylene protons of the ethyl groups.

rotation (9.5 kcal mol^{-1}). Although there is now a consensus that tripodal rotation can be slowed in these systems, the question of steric complementarity between the arene and the metal tripod continues to attract attention (*78*).

D. *(HEB)Cr(CO)L′L″ Complexes, L = NO^+, CS*

Having established that one could readily assign *proximal* and *distal* ethyl groups in these systems, Mailvaganam and co-workers returned to the (HEB)ML_3 problem (*75*). Clearly, it is necessary to incorporate sterically nondemanding ligands because bulky phosphines complicate the situation by generating mixtures of stereoisomers (*74*). To this end, they chose to synthesize other molecules of the type (HEB)Cr(CO)L′L″ in which the new ligands were small cylindrical ones possessing essentially the same steric properties as the carbonyl group. The new systems examined were $[(HEB)Cr(CO)_2NO]^+BF_4^-$ (**32**) and $[(HEB)Cr(CO)(CS)NO]^+BF_4^-$ (**33**). This latter molecule is a tribute to our French colleagues, whose elegant use of chirality to elucidate mechanisms derives not merely from the discoveries of Louis Pasteur but also from the topology of their country, which is known colloquially as "L'Hexagone." The enantiomers of $(FRANCE)Cr(CO)_3$ appear as Fig. 13.

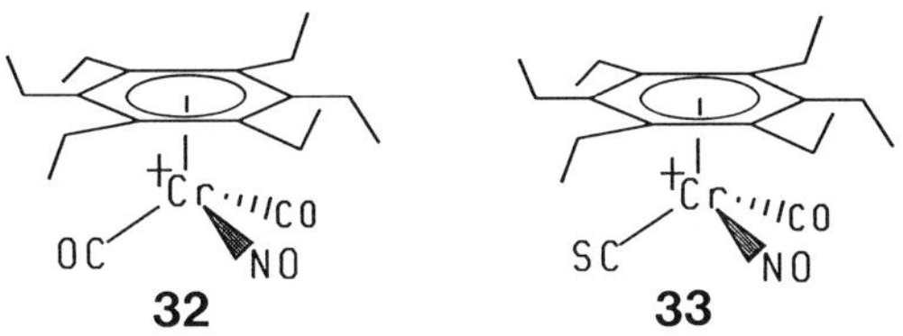

The ions **32** and **33** were both characterized by X-ray crystallography (see Fig. 14) and were shown to adopt the 1,3,5-*distal*–2,4,6-*proximal* conformation (**a**; Fig. 8) entirely analogous to the previously characterized complexes $(HEB)Cr(CO)_3$ (**20**) and $(HEB)Cr(CO)_2CS$ (**21**). The nitrosyl complex, **32**, exhibited a low-temperature ^{13}C NMR spectrum with 2:1:2:1 patterns for the methyl, methylene and ring carbon environments, in complete accord with the behavior of the thiocarbonyl system **21**. Again, these are consistent only with a molecule of C_s symmetry showing restricted rotation about the ring–tripod axis.

The *pièce de résistance* is the low-temperature 125-MHz ^{13}C NMR spectrum of $[(HEB)Cr(CO)(CS)NO]^+$ (**33**), in which the chiral tripod renders different all 18 carbons in the hexaethylbenzene ligand. In this spectacular

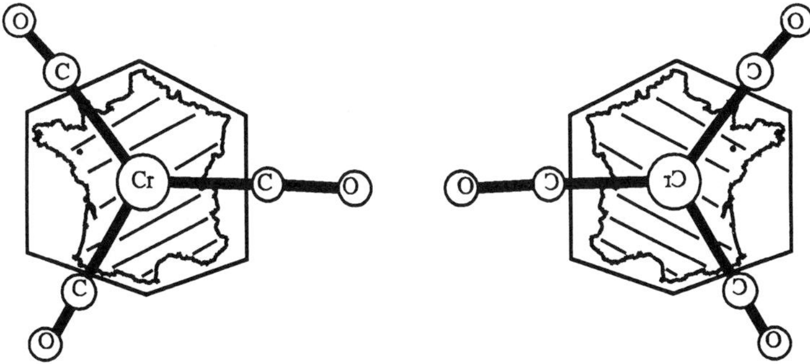

FIG. 13. A map of France superimposed on $(benzene)Cr(CO)_3$ renders the molecule chiral. Pure enantiomers of chiral arene–chromium complexes were first characterized by G. Simonneaux, A. Meyer, and G. Jaouen [*Tetrahedron* **31,** 1891 (1975)] at the University of Rennes, which is marked by a dot on the map.

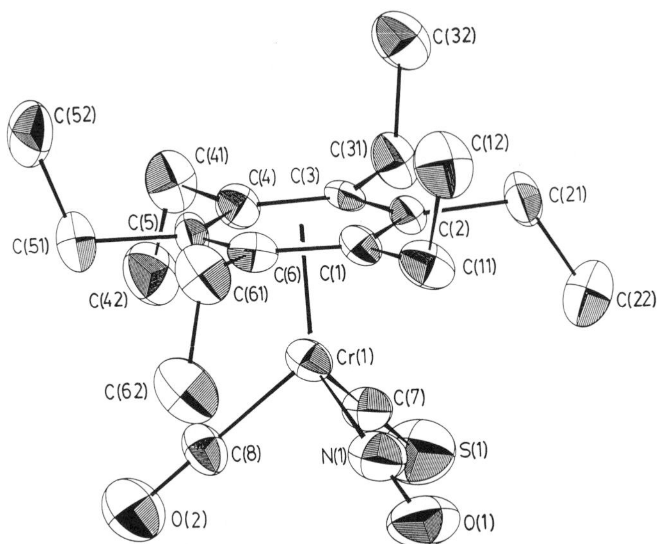

FIG. 14. X-Ray crystal structure of the cation $[(HEB)Cr(CO)(CS)NO]^+$ (**32**). Reproduced from Ref. 76, with permission of the American Chemical Society.

spectrum, shown as Fig. 15, not only does one see six peaks for the methyl carbons, the methylene carbons, and the ring carbons, but also these 18 resonances are found at chemical shifts appropriate for three *proximal* and three *distal* ethyl groups. This latter point is crucial because the mere observation of six different ethyl resonances is insufficient to show unequivocally that tripodal rotation has slowed. In conformer **c** (Fig. 8), the 1,3-*distal* and 2,4-*proximal* groups would be diastereotopic even when the chiral tripod is rotating. The crucial requirement is that the 1,3,5-*distal*–2,4,6-*proximal* peak pattern be retained. Simulation of the variable-temperature spectra of **21**, **32**, and **33** reveals that they are best fit by invoking two rotational barriers. In each case, the tripodal rotation has an activation energy of approximately 9.5 kcal mol^{-1} whereas rotating ethyl groups must surmount a barrier of ≈11.5 kcal mol^{-1} (*76*). The large difference in activation energies for these two processes tells us that they are not correlated. One might have envisioned a cogwheel process (*79,80*) by which a correlated interconversion of *proximal* and *distal* ethyl substituents allows the tripod to spin through 60°. Such a mechanism would require that the coalescence behavior in the chiral complex **33** show exchange *between proximal* and *distal* sites (*i.e.*, ethyl rotation) with the same activation energy as the exchange *within* these *proximal* and *distal* environments; the latter process is brought about by tripodal rotation. Such a scenario is not in

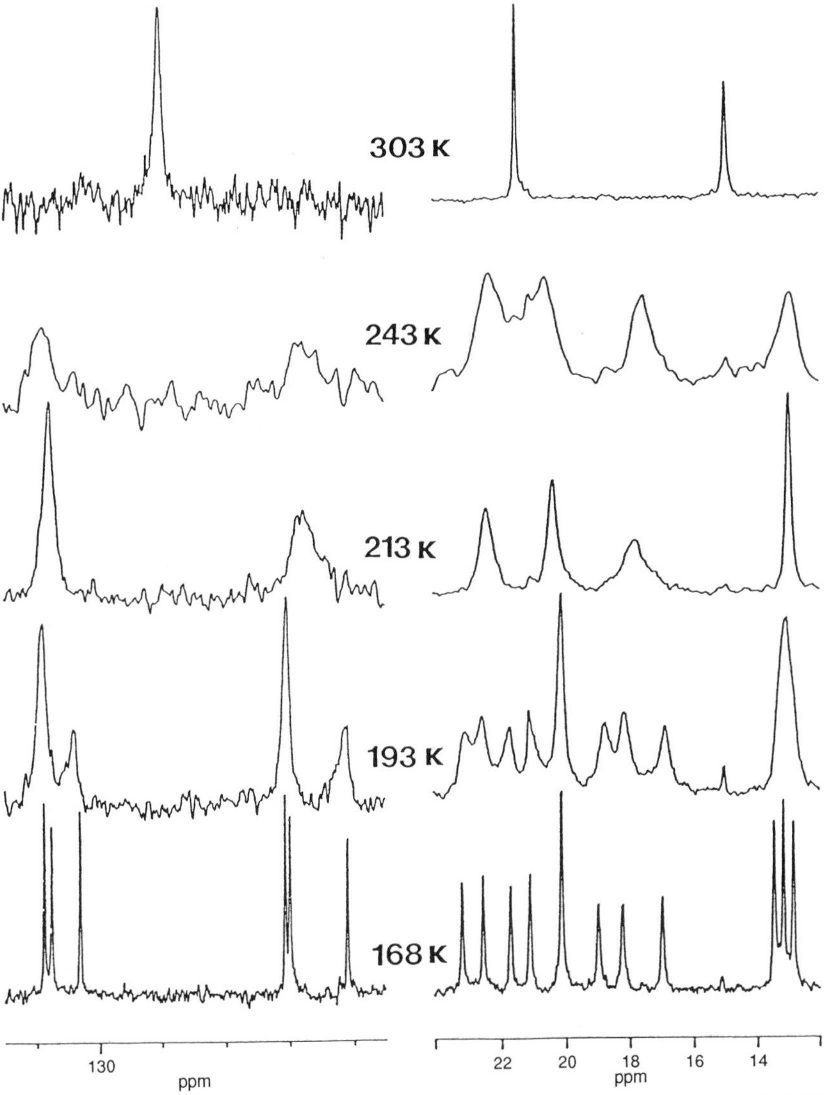

FIG. 15. Variable-temperature 125.8-MHz ^{13}C NMR spectra of $[(HEB)Cr(CO)(CS)NO]^+$ (**32**). Reproduced from Ref. 76, with permission of the American Chemical Society.

accord with the experimental observations (*76*). One is therefore led to the conclusion that the tripod executes 120° jumps over a barrier of $\approx$9.5 kcal mol^{-1} so as to maintain its orientation below the *distal* ethyls. Subsequently, as the temperature is raised, ethyl rotation becomes fast on the NMR time scale and the combination of these two fluxional processes equilibrates all six methyl, all six methylene, and all six ring carbon sites. The first experimental evidence indicating that *proximal–distal* exchange proceeds *via* uncorrelated ethyl rotations came as a result of simulations by Mislow and co-workers (*63*) and this claim has been vindicated by all subsequent reports.

E. *Other Sterically Hindered Arenes*

Two other examples have appeared in which tripodal rotation of an $M(CO)_2L$ moiety can be slowed because of steric hindrance. Hunter and co-workers prepared and crystallographically characterized the very hindered complex dicarbonyltriphenylphosphine(η^6-1,3,5-triethyl-2,4,6-tris-(trimethylsilylmethyl)benzene)molybdenum(0) (**34**) (*81*). In this system,

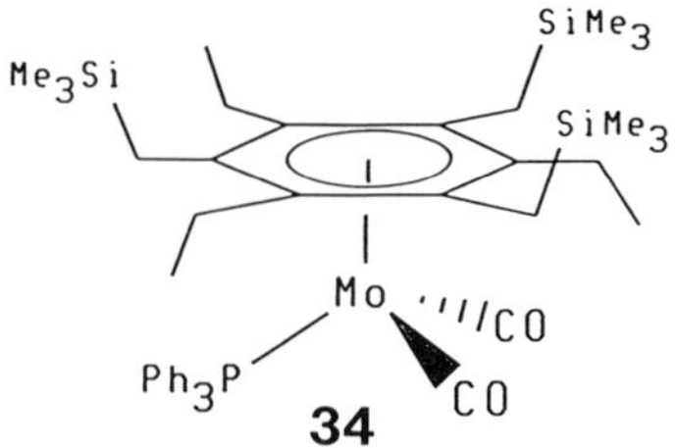

they demonstrated that three fluxional processes can become slow on the NMR time scale at low temperatures: slowed *proximal–distal* ethyl exchange (*63, 64, 75–77*) and restricted rotation about a metal–phosphorus axis (82–85) have been observed elsewhere, but the splitting of the carbonyls to give two resonances indicates that rotation of the $Mo(CO)_2PPh_3$ tripod about the ring centroid-to-molybdenum axis must also have a barrier detectable in the NMR regime. Because of the low intensities of the decoalesced resonances, the authors were unable to report a precise value of the activation energy for tripodal rotation, but an estimate of $\approx$8.5 kcal mol^{-1} was offered (*81*).

Another beautiful example reported by Hunter and co-workers involves the chromium tricarbonyl complexes of the *all-syn* and the *syn–anti-syn* isomers of trimeric bicyclo[2.2.1]hept-2-yne, **35** and **36**, respectively (*86*). Both of these molecules have been characterized X-ray crystallographically

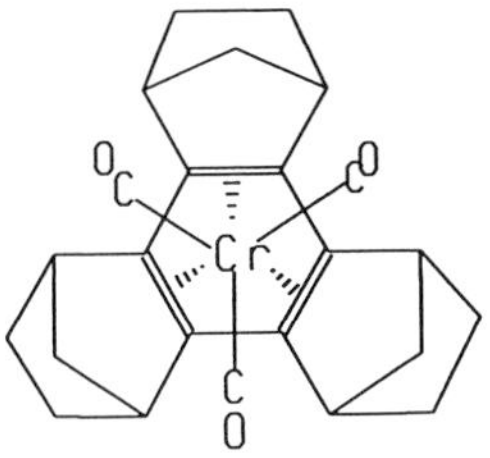

syn-syn-syn (C_{3v})
35

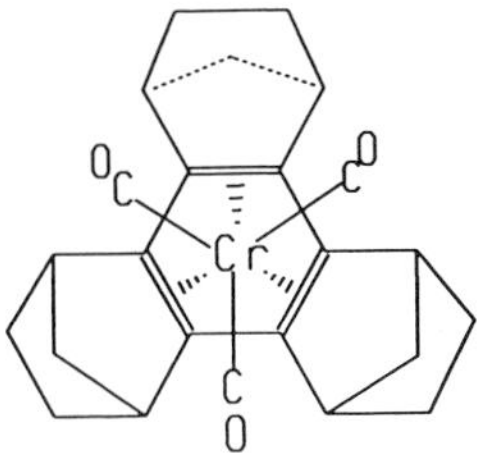

syn-anti-syn (C_s)
36

and also by ^{13}C NMR spectroscopy at 75.5 MHz. In the *all-syn* isomer (**35**), the three methylene bridges are *proximal* with respect to the metal; the steric hindrance is minimal and the ^{13}CO spectrum remains a singlet even at −130°C. In contrast, the $Cr(CO)_3$ tripod in the *syn–anti-syn* complex **36** encounters the much more bulky two-carbon bridge. The carbonyl resonance in **36** broadens very markedly below −120°C and in the limit appears to be approaching the 1:2 pattern expected for the C_s symmetry found in the solid state. The estimated barrier of ≈6.5 kcal mol^{-1} is close to the lower limit attainable experimentally by using NMR line-broadening techniques.

VI

ELECTRONIC BARRIERS TO TRIPODAL ROTATION

A. *Localized Double Bonds in Arene Complexes*

The calculatory approach of Hoffmann and co-workers (*46,47*), which was mentioned in Section IV, advanced two scenarios by which electronic influences might be brought to bear on the tripodal rotation barrier. The first possibility raised was that bond fixation in an arene ring would impose a substantial barrier to tripodal rotation. This concept, brilliantly exploited by Nambu and Siegel (*87*), saw its experimental realization in the terphenylene complex **37**. The crucial factors here were (1) the partial bond fixation caused by the presence of the benzocyclobutadiene moieties and (2) the lowering of the symmetry from C_{3v} to C_s, which resulted in a 2:1 splitting of the $Cr(^{13}CO)_3$ resonance at low temperature. The tripodal rotation barrier in **37** is approximately half of that predicted by Hoffmann and co-workers for a fully localized system. Thus, one is tempted to use this measured barrier to $Cr(CO)_3$ spinning as a measure of the extent of

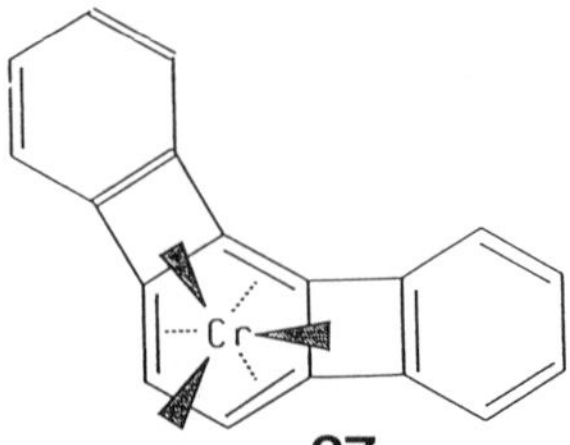

Cr

37

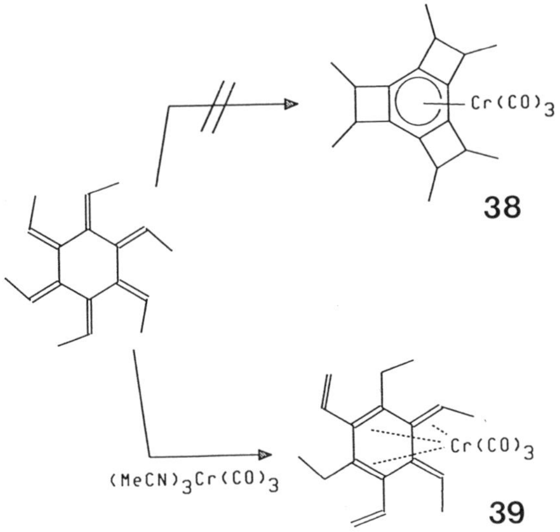

$Cr(CO)_3$

38

$Cr(CO)_3$

$(MeCN)_3Cr(CO)_3$

39

bond fixation in the terphenylene molecule. An earlier attempt to prepare the $Cr(CO)_3$ complex of benzotricyclobutene (**38**) by cyclization of a radialene led instead to the polyene **39** (*88*).

B. *Benzyl chromium Tricarbonyl Cations*

Hoffmann and co-workers also proposed that $[(benzyl)Cr(CO)_3]^+$ would preferentially adopt structure **19a**, analogous to **3**, which would be approximately 6.8 kcal mol^{-1} more stable than rotamer **19b**. Moreover, it was suggested that the exocyclic methylene group should bend down by 11° toward the metal atom so as to optimize the overlap between the vacant p_z orbital on carbon and a suitable filled *d* orbital on chromium (*47*). [X-Ray crystallographic data on the closely analogous ferrocenyl cation **40** revealed a methylene bend angle of 20° (*89, 90*); moreover NMR data on the tricobalt cluster **41** have confirmed that the $C{=}CR_2$ moiety leans toward a $Co(CO)_3$ vertex (*91*).]

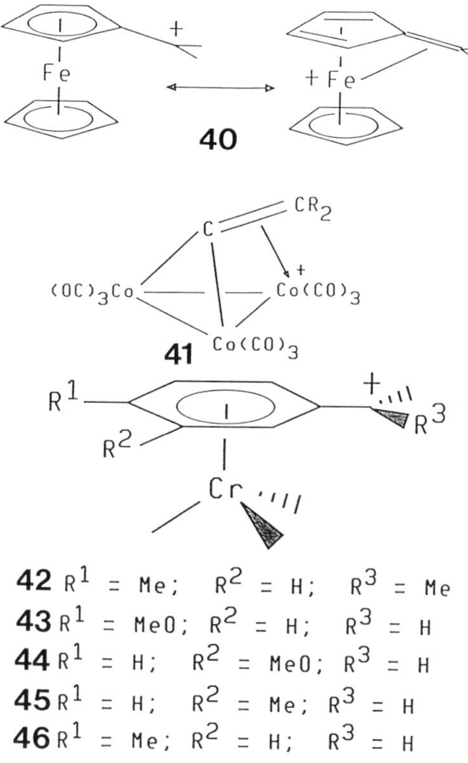

Preliminary evidence supporting the proposal by Hoffmann and co-workers was provided by Ceccon and co-workers, who reported that the 22.6-MHz ^{13}CO spectrum of the cation **42**, derived by protonating (methyl-*p*-tolyl carbinol)$Cr(CO)_3$ with FSO_3H, shows three carbonyl resonances at low temperature (*92*). However, no spectra were shown nor was an activation energy barrier calculated. Subsequently, this experiment was repeated and the NMR spectra were recorded using a 500-MHz spectrometer (*93*). Figure 16 depicts the 125.7-MHz ^{13}C spectrum of **42** over the range −60 to −25°C (at which point the coordinated cation shows clear signs of decomposition) and fully confirms the original claim by Ceccon and co-workers. Simulation of the spectrum gives a $\Delta G^{\ddagger}_{248}$ value of ca. 11.7 $kcal^{-1}$ for the tripodal rotation process.

More recent EHMO calculations found a minimum energy structure for **19a** when the chromium is displaced by 0.12 Å toward the benzylic carbon and the methylene bend angle is 22°. As shown already in Fig. 6, the principal stabilizing interaction allows the delocalization of the positive charge onto the $Cr(CO)_3$ unit *via* overlap with the filled $d_{x^2-y^2}$ orbital on

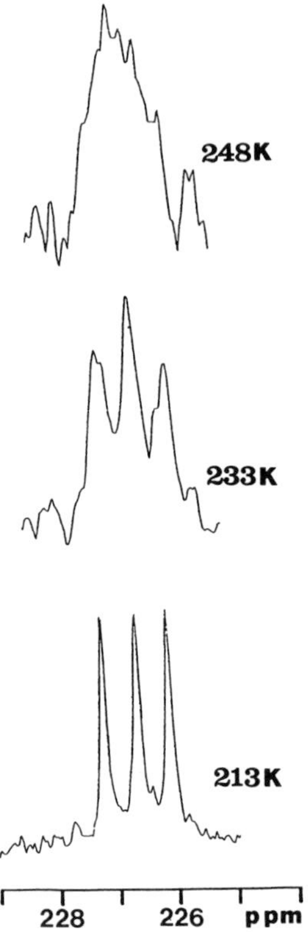

FIG. 16. Variable-temperature 125.8-MHz ^{13}C NMR spectra of $[(4\text{-Me-}C_6H_4\text{–CHMe})\text{-}Cr(CO)_3]^+$ (**42**) in the metal carbonyl region.

chromium. At the optimum position the gain in stability resulting from the bending of the methylene carbon is $\approx$1.7 kcal mol^{-1}. To the extent that one chooses to believe the charge distribution given by the EHMO calculations, the benzylic carbon in the planar cation bears a charge of +0.34 but this is reduced to +0.23 as a result of methylene bending and interaction with the chromium center. The experimentally determined activation energy of 11.7 kcal mol^{-1} presumably includes not only the barrier to rotation of the tricarbonylchromium moiety through 120°, but also the bending of the methylene fragment. The EHMO derived barrier for this fluxional process in $[(benzyl)Cr(CO)_3]^+$ was calculated to be 7.4 kcal mol^{-1} (*93*).

As pointed out by Nambu and Siegel (*87*), the barrier to tripodal rotation in **37** can be related to the the degree of bond fixation in the central aromatic ring. Likewise, the extent of charge localization in $[(benzyl)Cr(CO)_3]^+$ should control the barrier to $Cr(CO)_3$ rotation. The incorporation of electron-donating or -withdrawing substituents in the π-complexed phenyl ring should modify the electron density distribution at the ring carbons and so affect the barrier to tripodal rotation. Thus, placement of a methyl or methoxy group *para* to the benzylic cation should help to localize the relatively electron-rich sites at the C-1, C-3 and C-5 positions. In contrast, incorporation of a methyl or methoxy substituent *meta* to the benzyl cation would be expected to counteract the charge localization and lower the barrier to rotation of the $Cr(CO)_3$ fragment because no single orientation is greatly favored over the others. This picture gains support from EHMO calculations that yield rotational barriers of 7.5 and 5.1 kcal mol^{-1}, respectively, for the *p*- and *m*-(methoxybenzyl)$Cr(CO)_3$ cations. Indeed, tripodal rotation in the *p*-methoxybenzyl complex **43** is slowed at −70°C, but the onset of decoalescence appears only at −100°C for the analogous *m*-methoxy system **44**.

Interestingly, an EHMO calculation on the tertiary cation $[(C_6H_5–CMe_2)Cr(CO)_3]^+$ shows that bending of the α-carbon toward the metal is energetically slightly disfavored. There are no major steric problems and one might speculate that the tertiary cation has much less need for anchimeric assistance from the chromium than does a primary cation. Such a proposal has been advanced previously with respect to cluster-stabilized cations such as $[Co_3(CO)_9C\text{-}CR_2]^+$ (*94,95*). The ^{13}C NMR spectra of the primary cations [(3-methylbenzyl)$Cr(CO)_3]^+$ (**45**) and [(4-methylbenzyl)-$Cr(CO)_3]^+$ (**46**), which are stable below −50°C, have been recorded. The slightly increased rotational barrier in **45** relative to that in **42** perhaps indicates a stronger interaction of the primary cationic center with the electron-rich metal atom than is the case for the secondary cation **42**.

Olah and Yu have reported a marked deshielding of the benzylic carbon when going from (cumyl alcohol)$Cr(CO)_3$, δ 71.42, to $[(C_6H_5\text{-}CMe_2)Cr(CO)_3]^+$, δ 170.92; however, the complexed cumyl cation is still shielded by 83 ppm relative to the noncoordinated system, viz. $[C_6H_5CMe_2]^+$ (*96*). It was suggested that this downfield shift of only 100 ppm could be interpreted in terms of decreased localization of the positive charge on the benzylic site because the electron deficiency could be alleviated by donation from the tricarbonylchromium moiety. Seyferth *et al.* (*97*) and Ceccon and co-workers (*92*) have noted that transformation of secondary alcohol complexes into secondary carbocations, e.g., the mono- or bis-$Cr(CO)_3$ complexes of $[(MeC_6H_4)_2CH]^+$, brings about a deshielding of the benzylic carbon nucleus of ≈65 ppm. However, protonation of primary alcohols to yield the corresponding primary carbocation complexes, such as **43–46**, results in a deshielding of only ≈25 ppm (*93*). This is a clear indication that the delocalization of positive charge onto the $Cr(CO)_3$ fragment is very considerable in the primary carbocationic complexes.

It appears, therefore, that the ^{13}C NMR chemical shifts of the benzylic carbons together with the activation energy barriers toward tripodal rotation are useful probes of the localized nature of the arene-to-chromium interactions.

C. *Fulvene Complexes*

As noted above, the ferrocenylmethyl cation **40** adopts a bent geometry (*89,90*) such that one extreme canonical form would represent the cation as a $(C_5H_5)Fe^+$ fragment bonded to a nonplanar fulvene ligand. This concept is buttressed by the X-ray crystallographically determined structure of (6,6-diphenylfulvene)$Cr(CO)_3$ in which the exocyclic carbon is bent down through 30° (*98*). Hoffmann and co-workers have calculated that for (fulvene)$Cr(CO)_3$ the staggered rotamer **47a** should be 7.3 kcal mol^{-1} more stable than the eclipsed rotamer **47b**; this difference is augmented to 9.3 kcal mol^{-1} when the exocyclic methylene is bent down by 30° toward the metal (*46*). Preliminary results by Kreiter (*99*) and by Behrens and co-workers (*100*) suggest that tripodal rotation may be slowed on the NMR time scale at low temperature.

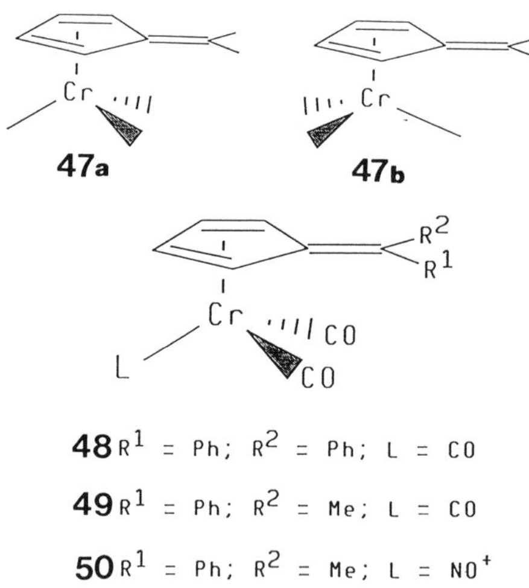

The ^{13}CO NMR spectra of (6,6-diphenylfulvene)$Cr(CO)_3$ (**48**) and of (6-methyl-6-phenylfulvene)$Cr(CO)_3$ (**49**) each show decoalescence behavior at low temperature. The former yields a 1:2 peak pattern whereas the latter, as befits a chiral molecule, exhibits three carbonyl resonances;

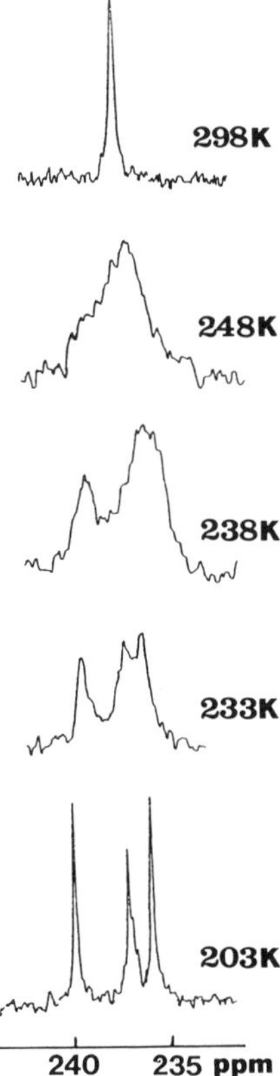

FIG. 17. Variable-temperature 125.8-MHz ^{13}C NMR spectra of (6-methyl-6-phenylfulvene)$Cr(CO)_3$ (**49**) in the metal carbonyl region.

the experimental spectra for **49** are shown in Fig. 17. The Arrhenius plots in each case yield energy barriers of ca. 11 kcal mol^{-1}. These data on tripodal rotation barriers support the concept of parallelism between benzyl and fulvene ligands (*93*).

Finally, it is noteworthy that in their pioneering studies on half-sandwich molecules Rausch *et al.* reported the ^{1}H and ^{13}C NMR spectra of a series of $[(C_5H_4{-}CR'R'')Cr(CO)_2NO]^+$ complexes (**50**) (*101*). It was noted that in some cases the carbonyl ligands gave rise to two resonances even at ambient temperature; this was interpreted in terms of restricted rotation of the ML_3 tripod. In view of the low barriers to $Cr(CO)_3$ rotation observed in the benzyl and fulvene complexes discussed here, the CO splitting in the nitrosyl systems requires an explanation. In fact, we note that the only cations **50** that yield two ^{13}CO peaks are those in which $R' \neq R''$; such molecules are chiral and the CO ligands are rendered diastereotopic. Thus, these carbonyls are always different and one cannot use this observation as a valid argument for restricted rotation *of the tripod.* It does, however, support the assertion by Rausch *et al.* that there is restricted rotation about the C-5–C-6 bond. Likewise, as shown by Ceccon and co-workers, (*92*) rotation about the aryl–CR_2 bond in benzyl systems has a very high barrier. Indeed, the retention of stereochemical integrity in such cations has been exploited synthetically (*102*).

EHMO calculations on **50**, where R′ = R″ = H, show that the symmetric isomer **51a** is favored by 2.7 kcal mol^{-1} over the unsymmetrical rotamer **51b** and also that the barrier to rotation is only 7.3 kcal mol^{-1}. This is consistent with observations that rotation of $Cr(CO)_3$, $[Cr(CO)_2NO]^+$, $Cr(CO)_2CS$, and $[Cr(CO)(CS)NO]^+$ tripods attached to hindered arenes have very similar barriers (*76*).

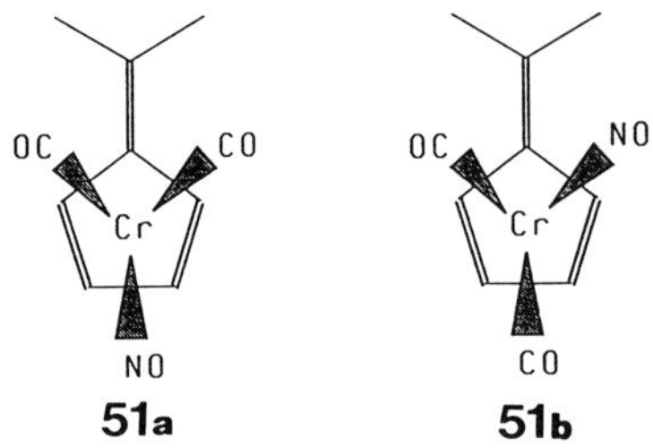

It is relevant to recall that, some years ago, Graves and Lagowski prepared a series of bis(arene)chromium complexes and recorded their NMR spectra. In some molecules, such as bis(*o*-fluorotoluene)chromium, the methyl resonances are split; in each case, this splitting is not exhibited by the corresponding *para* isomers (*103*). The authors were careful to point out that such observations do not require restricted rotation about the arene–metal bond. The existence of *meso* and *d,l* sandwich compounds in the *ortho* or *meta* series can lead to the appearance of extra resonances. Analogously, 1,1′-di-*t*-butyl-3,3′-bis(2,2-dimethylpropionyl)ferrocene shows a multitude of ^1H and ^{13}C peaks. These were initially explained in

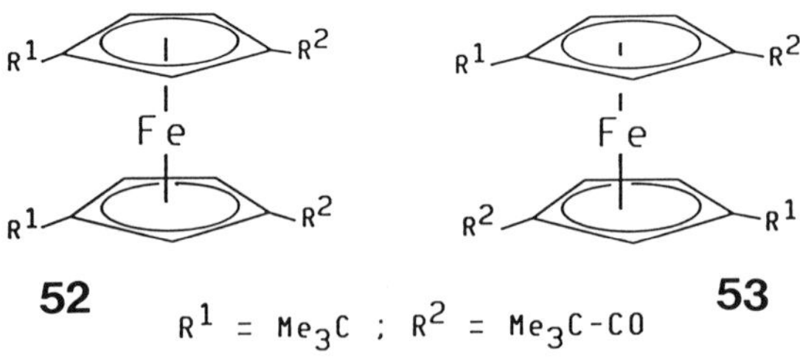

terms of restricted rotation of the cyclopentadienyl rings (*104*), but the spectrum was subsequently rationalized on the basis of a mixture of *meso* and *d,l* forms, **52** and **53**, respectively (*105*). Clearly, one must always take careful account of molecular symmetry when considering the possibility of restricted rotation.

VII

RELAXATION TIME STUDIES AND SOLID-STATE NMR MEASUREMENTS

In $(C_6H_6)Cr(CO)_3$ and in $(t\text{-Bu-}C_6H_5)Cr(CO)_3$ the barriers to tripodal rotation in solution have been calculated from ^{13}C spin-lattice relaxation data and are $\approx$1 kcal mol^{-1} (*106*). These values compare with estimates of $\approx$4 kcal mol^{-1} for the corresponding process for $(C_6H_6)Cr(CO)_3$ in the crystalline state (*107*). Wagner and Hansen have reported values of 15 and 17 kcal mol^{-1}, respectively, for the ring rotation in (toluene)$Cr(CO)_3$ and the molybdenum analog (*108*). However, these latter results have been reevaluated by Braga and Grepioni, who prefer to interpret the data in terms of small-amplitude vibrations rather than full reorientational motion of the $M(CO)_3$ moiety relative to the arene ring (*109,110*).

VIII

STERICALLY CROWDED CYCLOPENTADIENYL COMPLEXES

The comprehensive studies of Erker *et al.* on diene complexes of zirconium have yielded numerous molecules of considerable interest to NMR spectroscopists. In particular, it has been shown that bis(η^5-alkylcyclopentadienyl)(η^4-butadiene)zirconium (**54**) exhibits restricted rotation of the cyclopentadienyl rings when bulky alkyl substituents are used (*111*). This phenomenon was first noted in the solid-state NMR spectrum but is also detectable in solution; the barrier is $\approx$10.1 kcal mol^{-1}. In related work, Winter *et al.* have measured the barriers to ring rotation in the molecules $[C_5(SiMe_3)_nH_{5-n}]MCl_2$ (55), where M is Zr or Hf. Cyclopentadienyl ring rotation is much more restricted in the tris- rather than in the bis-trimethylsilyl complexes; the authors attribute this result to the "gear-meshing" of these bulky groups (*112*).

Ferrocenes bearing bulky substituents might be expected to exhibit restricted rotation of the cyclopentadienyl rings. Preliminary studies on

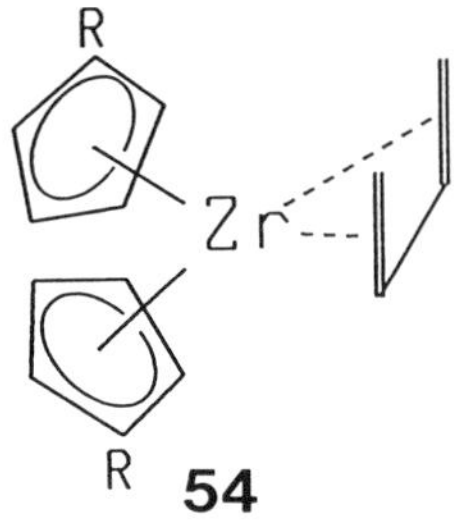

54 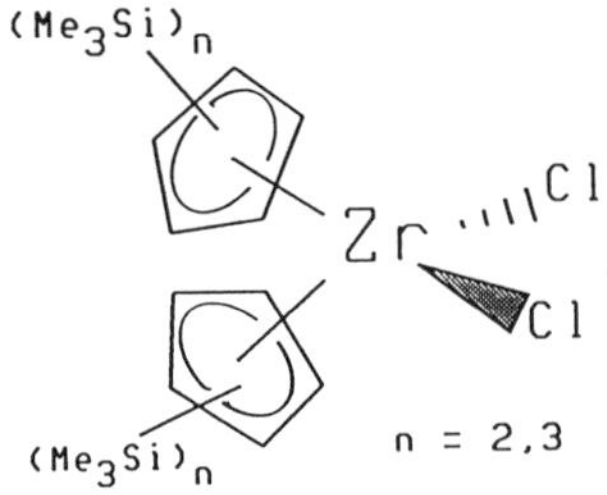 **55**

$n = 2,3$

1,1′,3,3′-tetrakis(trimethylsilyl)ferrocene, the corresponding cobalticinium system, and on tetraphenyl and octaphenyl ferrocenes have been reported (*113–120*). Very recently, such systems have been subjected to the more rigorous approach of total line-shape analysis, which yielded accurate values for the rotational barriers in ferrocenes and their ruthenium analog (*121*). Depending on the bulk of the substituents, these barriers range from 10 to 20 kcal mol^{-1}; in the ruthenium systems, the greater inter-ring separations lead to activation energies of approximately 10 to 14 kcal mol^{-1}.

We finish by describing the multiple fluxional processes available to the chiral complex $(C_5Ph_5)Fe(CO)(CHO)(PMe_2Ph)$ (**56**) (*122*). The metallocenic chirality arising from the clockwise or anticlockwise canting arrangement of alkyl substituents in peralkylated planar cyclic ligands has already been exploited; a particularly fine example is $[(i\text{-}Pr_5C_5)Co(C_5H_5)]^+$ (**57**), in which interconversion of enantiomers occurs *via* rotation of the five isopropyl substituents (*123*). In **56**, the combination of the canting of the phenyls, together with the inherent chirality associated with the iron center, yields a diastereomeric mixture. At low temperature on a 500-MHz instrument, one can see not only restricted rotation of the phenyls but also of the tripodal moiety. The former process ($\Delta G^\ddagger =$ 11.8 kcal mol^{-1}) is not correlated with the tripodal fluxionality, which has

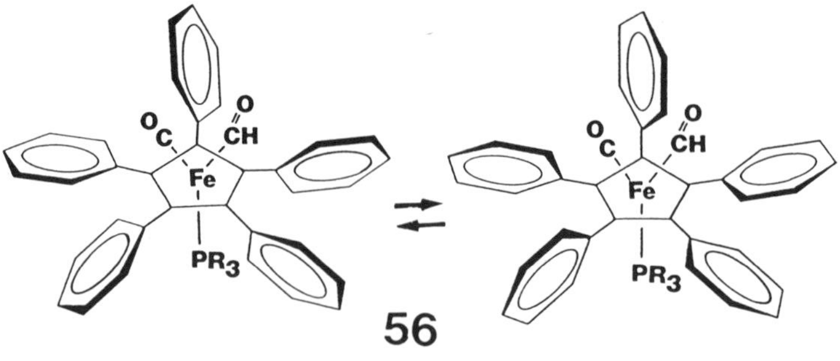

56

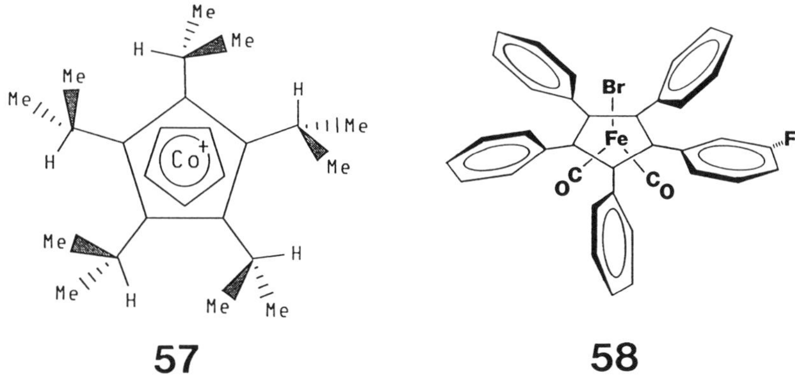

57

58

a barrier almost 3 kcal mol^{-1} lower than that required to spin the phenyls (*124*). The rotational behavior of the phenyl rings in (C_5Ar_5)–metal complexes can also be conveniently followed by ^{19}F NMR spectroscopy; this has been successfully accomplished for $[C_5(C_6H_5)_4(m\text{-}C_6H_4F)]Fe(CO)_2Br$ (**58**) (*125*).

Acknowledgments

The author's work in this area has been supported by the Natural Sciences and Engineering Research Council of Canada, and by the Petroleum Research Fund, administered by the American Chemical Society. It is a pleasure to acknowledge numerous discussions, lively correspondence, and/or exchange of data prior to publication with Professors Geoffrey Hunter (Dundee, Scotland), Kurt Mislow (Princeton University), Ian Rae (Monash University, Australia), Roland Pomeroy (Simon Fraser University, Vancouver), Claude Lapinte and Jean-René Hamon (Rennes, France), and Jay Siegel (San Diego).

References

1. E. O. Fischer and W. Hafner, *Z. Naturforsch. B: Anorg. Chem. Org. Chem. Biochem. Biophys. Biol.* **10b,** 665 (1955).
2. E. O. Fischer and K. Öfele, *Chem. Ber.* **90,** 2532 (1957).
3. B. Nicholls and M. C. Whiting, *J. Chem. Soc.*, 551 (1959).
4. F. Jellinek, *Nature* (*London*) **187,** 871 (1960); F. Jellinek, *J. Organomet. Chem.* **1,** 43 (1963).
5. F. A. Cotton, W. A. Dollase, and J. S. Wood, *J. Am. Chem. Soc.* **85,** 1543 (1963).
6. J. A. Ibers, *J. Chem. Phys.* **40,** 3129 (1964).
7. E. Keulen and F. Jellinek, *J. Organomet. Chem.* **5,** 490 (1966).
8. A. Haaland, *Acta Chem. Scand.* **19,** 41 (1965).
9. B. Rees and P. Coppens, *Acta Crystallogr., Sect. B: Struct. Crystallogr. Cryst. Chem.* **B29,** 2516 (1973).
10. M. F. Bailey and L. F. Dahl, *Inorg. Chem.* **4,** 1314 (1965).
11. O. L. Carter, A. T. McPhail, and G. A. Sim, *J. Chem. Soc. A*, 822 (1966).
12. O. L. Carter, A. T. McPhail, and G. A. Sim, *J. Chem. Soc. A*, 1619 (1967).
13. W. Strohmeier and H. Hellmann, *Chem. Ber.* **97,** 1877 (1964).

14. W. McFarlane and S. O. Grim, *J. Organomet. Chem.* **5,** 147 (1966).
15. D. E. F. Gracey, W. R. Jackson, W. B. Jennings, S. C. Rennison, and R. Spratt, *Chem. Commun.*, 231 (1966).
16. J. T. Price and T. S. Sorensen, *Can. J. Chem.* **46,** 515 (1968).
17. R. V. Emanuel and E. W. Randall, *J. Chem. Soc. A*, 3002 (1969).
18. D. E. F. Gracey, W. R. Jackson, W. B. Jennings, S. C. Rennison, and R. Spratt, *J. Chem. Soc. B.*, 1210 (1969).
19. A. Solladié-Cavallo, *Polyhedron* **4,** 910 (1985), and references therein.
20. R. C. Cambie, G. R. Clark, A. C. Gourdie, P. S. Routledge, and P. D. Woodgate, *J. Organomet. Chem.* **297,** 177 (1985).
21. B. Mailvaganam, R. E. Perrier, B. G. Sayer, B. E. McCarry, R. A. Bell, and M. J. McGlinchey, *J. Organomet. Chem.* **354,** 325 (1988).
22. M. J. McGlinchey, R. C. Burns, R. Hofer, S. Top, and G. Jaouen, *Organometallics* **5,** 104 (1986).
23. S. Top, G. Jaouen, A. Vessières, J.-P. Abjean, D. Davoust, C. A. Rodger, B. G. Sayer, and M. J. McGlinchey, *Organometallics* **4,** 2143 (1985).
24. W. R. Jackson, W. B. Jennings, S. C. Rennison, and R. Spratt, *J. Chem. Soc. B.*, 1214 (1969).
25. T. F. Jula and D. Seyferth, *Inorg. Chem.* **7,** 1245 (1968).
26. G. Barbieri and F. Taddei, *J. Chem. Soc. D*, 312 (1970).
27. C. Segard, B. P. Roques, C. Pommier, and G. Guiochon, *Anal. Chem.* **43,** 1146 (1971).
28. F. van Meurs and H. van Bekkum, *J. Organomet. Chem.* **133,** 321 (1977).
29. B. P. Roques, C. Segard, S. Combrisson, and F. Wehrli, *J. Organomet. Chem.* **73,** 327 (1974).
30. W. R. Jackson, W. B. Jennings, and R. Spratt, *J. Chem. Soc. D*, 593 (1970).
31. K. Mislow and M. Raban, *in* "Topics in Stereochemistry" (N. Allinger and E. L. Eliel, eds.), Vol. 1, p. 1. Wiley (Interscience), New York, 1967.
32. Y. Shvo, E. C. Taylor, K. Mislow, and M. Raban, *J. Am. Chem. Soc.* **89,** 4910 (1967).
33. F. van Meurs, J. M. van der Toorn, and H. van Bekkum, *J. Organomet. Chem.* **113,** 341 (1976).
34. F. van Meurs and H. van Koningsveld, *J. Organomet. Chem.* **118,** 295 (1976).
35. R. K. Pomeroy and D. J. Harrison, *J. Chem. Soc., Chem. Commun.*, 661 (1980).
36. X. Hu, J. Duchowski, and R. K. Pomeroy, *J. Chem. Soc., Chem. Commun.*, 362 (1988).
37. W. R. Jackson, C. F. Pincombe, I. D. Rae, and S. Thapebinkarn, *Aust. J. Chem.* **28,** 1535 (1975).
38. W. R. Jackson, C. F. Pincombe, I. D. Rae, D. Rash, and B. Wilkinson, *Aust. J. Chem.* **29,** 2431 (1976).
39. W. R. Jackson and W. B. Jennings, *J. Chem. Soc. B.*, 1221 (1969).
40. M. F. Semmelhack, G. Clark, R. Farina, and M. Saeman, *J. Am. Chem. Soc.* **101,** 217 (1979).
41. T. A. Albright and B. K. Carpenter, *Inorg. Chem.* **19,** 3092 (1980).
42. A. Solladié-Cavallo and G. Wipff, *Tetrahedron Lett.*, 3047 (1980).
43. F. Rose-Munch, K. Aniss, E. Rose, and J. Vaisserman, *J. Organomet. Chem.* **415,** 223 (1991), and references therein.
44. J. C. Boutonnet, L. Mordenti, E. Rose, O. Le Martret, and G. Precigoux, *J. Organomet. Chem.* **221,** 147 (1981).
45. J. Brocard and J. Lebibi, *J. Organomet. Chem.* **320,** 295 (1987).
46. T. A. Albright, P. Hofmann, and R. Hoffmann, *J. Am. Chem. Soc.* **99,** 7546 (1977).
47. T. A. Albright, R. Hoffmann, and P. Hofmann, *Chem. Ber.* **111,** 1591 (1978).

48. N.-S. Chiu, L. Schäfer, and R. Seip, *J. Organomet. Chem.* **101,** 331 (1975).
49. A. Wu, E. R. Biehl, and P. C. Reeves, *J. Organomet. Chem.* **33,** 53 (1971).
50. D. A. Brown and J. R. Raju, *J. Chem. Soc.* (*A*), 1617 (1966).
51. G. Klopman and K. Noack, *Inorg. Chem.* **7,** 579 (1968).
52. G. M. Bodner and L. J. Todd, *Inorg. Chem.* **13,** 360 (1974).
53. J. L. Fletcher and M. J. McGlinchey, *Can. J. Chem.* **53,** 1525 (1975).
54. J. L. Adcock and G. L. Aldous, *J. Organomet. Chem.* **201,** 411 (1980).
55. M. Barfield, D. M. Grant, and D. Ikenberry, *J. Am. Chem. Soc.* **97,** 6956 (1975).
56. A. Agarwal, J. A. Barnes, J. L. Fletcher, M. J. McGlinchey, and B. G. Sayer, *Can. J. Chem.* **55,** 2575 (1977).
57. A. Pines, M. G. Gibby, and J. S. Waugh, *Chem. Phys. Lett.* **59,** 373 (1972).
58. S. Pausak, J. Tegenfeldt, and J. S. Waugh, *J. Chem. Phys.* **61,** 1338 (1974).
59. M. M. Maricq, J. S. Waugh, J. L. Fletcher, and M. J. McGlinchey, *J. Am. Chem. Soc.* **100,** 6902 (1978).
60. H. K. Pal and A. C. Guha, *Z. Kristallogr. Kristallgeom. Kristallphys. Kristallchem.* **92,** 392 (1935).
61. J. L. Fletcher, Ph.D. Thesis, McMaster University, Hamilton, Ontario (1980).
62. G. Hunter, D. J. Iverson, K. Mislow, and J. F. Blount, *J. Am. Chem. Soc.* **102,** 5942 (1980).
63. D. J. Iverson, G. Hunter, J. F. Blount, J. R. Damewood, Jr., and K. Mislow, *J. Am. Chem. Soc.* **103,** 6073 (1981)
64. M. J. McGlinchey, J. L. Fletcher, B. G. Sayer, P. Bougeard, R. Faggiani, C. J. L. Lock, A. D. Bain, C. Rodger, E. P. Kündig, D. Astruc, J.-R. Hamon, P. LeMaux, S. Top, and G. Jaouen, *J. Chem. Soc., Chem. Commun.*, 634 (1983).
65. G. Hunter and K. Mislow, *J. Chem. Soc., Chem. Commun.*, 172 (1984).
66. J.-R. Hamon, J.-Y. Saillard, A. LeBeuze, M. J. McGlinchey, and D. Astruc, *J. Am. Chem. Soc.* **104,** 7549 (1982).
67. R. H. Dubois, M. J. Zaworotko, and P. S. White, *J. Organomet. Chem.* **362,** 155 (1989).
68. J. F. Blount, G. Hunter, and K. Mislow, *J. Chem. Soc., Chem. Commun.*, 170 (1984).
69. G. Hunter, J. F. Blount, J. R. Damewood, Jr., D. J. Iverson, and K. Mislow, *Organometallics* **1,** 448 (1981).
70. G. Hunter, T. J. R. Weakley, and W. Weissensteiner, *J. Chem. Soc., Dalton Trans.*, 1545 (1987).
71. W. A. Herrmann, W. R. Thiel, and E. Herdtweck, *Polyhedron* **7,** 2027 (1988).
72. B. Mailvaganam, B. G. Sayer, and M. J. McGlinchey, *Organometallics* **9,** 1089 (1990).
73. M. J. McGlinchey, P. Bougeard, B. G. Sayer, R. Hofer, and C. J. L. Lock, *J. Chem. Soc., Chem. Commun.*, 789 (1984).
74. G. Hunter, J. R. Weakley, K. Mislow, and M. G. Wong, *J. Chem. Soc., Dalton Trans.*, 577 (1986).
75. P. A. Downton, B. Mailvaganam, C. S. Frampton, B. G. Sayer, and M. J. McGlinchey, *J. Am. Chem. Soc.* **112,** 27 (1990).
76. B. Mailvaganam, C. S. Frampton, S. Top, B. G. Sayer, and M. J. McGlinchey, *J. Am. Chem. Soc.* **113,** 1177 (1991).
77. K. V. Kilway and J. S. Siegel, *J. Am. Chem. Soc.* **113,** 2332 (1991).
78. K. V. Kilway and J. S. Siegel, *J. Am. Chem. Soc.* **114,** 255 (1992).
79. H. Iwamura and K. Mislow, *Acc. Chem. Res.* **21,** 175 (1988).
80. K. Mislow, *Chemtracts—Org. Chem.* **2,** 151 (1989).
81. J. A. Chudek, G. Hunter, R. L. MacKay, G. Fäber, and W. Weissensteiner, *J. Organomet. Chem.* **377,** C69 (1989).

82. J. A. Chudek, G. Hunter, R. L. MacKay, P. Kremminger, K. Schlögl, and W. Weissensteiner, *J. Chem. Soc., Dalton Trans.*, 2001 (1990).
83. J. A. S. Howell, M. G. Palin, M.-C. Tirvengadun, D. Cunningham, P. McArdle, Z. Goldschmidt, and H. E. Gottlieb, *J. Organomet. Chem.* **413,** 269 (1991).
84. S. G. Davies, A. E. Derome, and J. P. McNally, *J. Am. Chem. Soc.* **113,** 2854 (1991).
85. L. Li, M. F. D'Agostino, B. G. Sayer, and M. J. McGlinchey, *Organometallics* **11,** 477 (1992).
86. P. Kremminger, W. Weissensteiner, C. Kratky, G. Hunter, and R. L. MacKay, *Monatsh. Chem.* **120,** 1175 (1989).
87. M. Nambu and J. S. Siegel, *J. Am. Chem. Soc.* **110,** 3675 (1988).
88. M. Yalpani, R. Benn, R. Goddard, and G. Wilke, *J. Organomet. Chem.* **240,** 49 (1982).
89. S. Lupon, M. Kapon, M. Cais, and F. H. Herbstein, *Angew. Chem., Int. Ed. Engl.* **11,** 1025 (1971).
90. U. Behrens, *J. Organomet. Chem.* **182,** 89 (1979).
91. R. T. Edidin, J. R. Norton, and K. Mislow, *Organometallics* **1,** 561 (1981).
92. A. Acampora, A. Ceccon, M. Dal Farra, G. Giacometti, and G. Rigatti, *J. Chem. Soc., Perkin Trans. 2*, 483 (1977).
93. P. A. Downton, B. G. Sayer, and M. J. McGlinchey, *Organometallics*, in press (1992).
94. S. L. Schreiber, M. T. Klimas, and T. Sammakia, *J. Am. Chem. Soc.* **109,** 5749 (1987).
95. M. F. D'Agostino, C. S. Frampton, and M. J. McGlinchey, *J. Organomet. Chem.* **394,** 145 (1990).
96. G. A. Olah and S. H. Yu, *J. Org. Chem.* **41,** 1694 (1976).
97. D. Seyferth, J. S. Merola, and C. S. Eschbach, *J. Am. Chem. Soc.* **100,** 4124 (1978).
98. V. G. Andrianov and Yu. T. Struchkov, *J. Struct. Chem.* **18,** 2551 (1977).
99. C. G. Kreiter, personal communication, quoted in Ref. *47*.
100. O. Koch, F. Edelmann, and U. Behrens, *Chem. Ber.* **115,** 1313 (1982).
101. M. D. Rausch, D. J. Kowalski, and E. A. Mintz, *J. Organomet. Chem.* **342,** 201 (1988).
102. S. Top, G. Jaouen, and M. J. McGlinchey, *J. Chem. Soc., Chem. Commun.*, 1110 (1980).
103. V. Graves and J. J. Lagowski, *J. Organomet. Chem.* **120,** 397 (1976).
104. W. Bell and C. Glidewell, *J. Chem. Res.* **5,** 44 (1991).
105. W. Bell and C. Glidewell, *J. Organomet. Chem.* **411,** 251 (1991).
106. A. Gryff-Keller, P. Szczeciński, and H. Koziel, *Magn. Reson. Chem.* **28,** 25 (1990).
107. P. Delise, G. Allegra, E. R. Mognaschi, and A. Chierico, *J. Chem. Soc., Faraday Trans. 2*, 207 (1975).
108. G. W. Wagner and B. E. Hansen, *Inorg. Chem.* **26,** 2019 (1987).
109. D. Braga and F. Grepioni, *J. Chem. Soc., Dalton Trans.*, 3143 (1990).
110. S. Aime, D. Braga, R. Gobetto, F. Grepioni, and A. Orlandi, *Inorg. Chem.* **30,** 951 (1991).
111. G. Erker, R. Nolte, C. Krüger, R. Schlund, R. Benn, H. Grondey, and R. Mynott, *J. Organomet. Chem.* **364,** 119 (1989).
112. C. H. Winter, D. A. Dobbs, and X.-X. Zhou, *J. Organomet. Chem.* **403,** 145 (1991).
113. J. Okuda, *J. Organomet. Chem.* **356,** C43 (1988).
114. J. Okuda, *J. Organomet. Chem.* **367,** C1 (1989).
115. J. Okuda and E. Herdtweck, *J. Organomet. Chem.* **373,** 99 (1989).
116. W. D. Luke and A. Streitweiser, Jr., *J. Am. Chem. Soc.* **103,** 3241 (1981).
117. J. Okuda and E. Herdtweck, *Chem. Ber.* **121,** 1899 (1988).
118. J. Okuda, *Chem. Ber.* **122,** 1075 (1989).
119. H. Sitzmann, *J. Organomet. Chem.* **354,** 203 (1988).

120. M. P. Castellani, J. M. Wright, S. J. Geib, A. L. Rheingold, and W. C. Trogler, *Organometallics* **5,** 1116 (1986).
121. E. W. Abel, N. C. Long, K. G. Orrell, A. G. Osborne, and V. Sik, *J. Organomet. Chem.* **403,** 195 (1991).
122. P. Bregaint, J.-R. Hamon, and C. Lapinte, *Organometallics* **11,** 1417 (1992).
123. B. Gloaguen and D. Astruc, *J. Am. Chem. Soc.* **112,** 4607 (1990).
124. L. Li, A. Decken, M. J. McGlinchey, P. Bregaint, J.-R. Hamon, and C. Lapinte, unpublished results (1992).
125. A. Decken and M. J. McGlinchey, unpublished results (1992).

Index

D

O

P

V

W

Cumulative List of Contributors

ISBN 0-12-031134-8

90051

9 780120 311347